Stem Cells and CNS Development

Contemporary Neuroscience

Stem Cells
and
CNS Development

Edited by

Mahendra S. Rao

Department of Neurobiology and Anatomy
University of Utah Medical School
Salt Lake City, UT

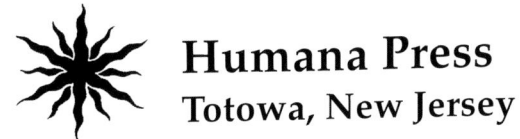

Humana Press
Totowa, New Jersey

Cover design by Patricia F. Cleary.

Cover illustration taken from Fig. 6 in Chapter 9. It is a scanning EM showing a solitary oligodendrocyte in a mixed culture or ES cell-derived neural cells.

For additional copies, pricing for bulk purchases, and/or information about other Humana titles, contact Humana at the above address or at any of the following numbers: Tel.: 973-256-1699; Fax: 973-256-8314; E-mail: humana@humanapr.com

Printed in the United States of America. 10 9 8 7 6 5 4 3 2 1

Library of Congress Cataloging in Publication Data

Main entry under title:

Stem cells and CNS development / edited by Mahendra S. Rao.
      p.   cm. — (Contemporary neuroscience)
      Includes bibliographical references and index.
      ISBN 0-89603-886-6 (alk. paper)
      1. Developmental neurobiology. 2. Stem cells. I. Rao, Mahendra S. II. Series.

  QP363.5 .S75 2001
  612.8—dc21

                                         00-058081

Preface

Neural differentiation is an early embryonic event that occurs soon after germ layer specification from the blastula. The early formed ectoderm undergoes further patterning to separate into two identifiable components, the presumptive neural ectoderm and the presumptive epidermis. Neural tissue segregates as a clearly demarcated epithelium termed the neuroepithelium (or neuroectoderm). This neuroepithelium generates the central nervous system (CNS), whereas cells at the margins of the neuroepithelium will generate the peripheral nervous system (PNS). A variety of evidence has been accumulated to show that the process of neural differentiation involves a sequential restriction in differentiation potential.

A fundamental breakthrough in our understanding of nervous system development was the identification of multipotent neural stem cells (neurospheres) about 10 years ago. Dr. Samuel Weiss and colleagues showed that EGF (epidermal growth factor)-dependent stem cells could be harvested from different brain regions at different developmental stages and that these could be maintained over multiple passages in vitro. The original finding that EGF-dependent neural stem cells exist has been replicated and extended by many investigators, and there has been a veritable explosion of research on stem cells, their role in normal development, and their potential therapeutic uses. Different classes of neural stem cells have been identified, new markers described, cell lines generated, and factors that regulate the differentiation process characterized. Other investigators have shown that these pluripotent stem cells likely generate CNS and PNS derivatives via the generation of intermediate lineage restricted precursors that differ from each other and from multipotent stem cells. The therapeutic implications of accessing a virtually unlimited population of homogenous progenitor cells to treat CNS disorders or for gene and drug discovery has not escaped investigators, and several companies have been formed to exploit stem cell technology and several research institutions have initiated transplant studies. This rapid transition from a basic discovery to clinical trials is both surprising and unprecedented.

In *Stem Cells and CNS Development,* I have invited some of the leading authorities in the field of neural stem cell biology to summarize their findings and describe how these results may lead to novel therapies. The first part of the book surveys the various kinds of stem cells, progenitor cells, and precursors that have been described, while the second half describes how these cells are beginning to be used for therapeutic purposes. It is my hope that this book will serve as a valuable compendium of practical information on the current state of the field for all those engaged in this research.

Mahendra Rao

Contents

Contributors

ARTURO ALVAREZ-BUYLLA • *The Rockefeller University, New York, NY*

LUCA BONFANTI • *Department of Veterinary Morphophysiology, University of Turin, Turin, Italy*

KATE A. BOWER • *Harvard Medical School and The Children's Hospital, Boston, MA*

MARIANNE BRONNER-FRASER • *Division of Biology, California Institute of Technology, Pasadena, CA*

SOPHIA COLAMARINO • *Laboratory of Genetics, The Salk Institute for Biological Studies, La Jolla, CA*

FRED H. GAGE • *Laboratory of Genetics, The Salk Institute for Biological Studies, La Jolla, CA*

ROSSELLA GALLI • *Institute for Stem Cell Research, S. Raffaele Hospital, Milan, Italy*

STEVEN A. GOLDMAN • *Department of Neurology and Neuroscience, Cornell University Medical Center, New York, NY*

ANGELA GRITTI • *Institute for Stem Cell Research, S. Raffaele Hospital, Milan, Italy*

ALEXANDER KAMB • *Arcaris Inc., Salt Lake City, UT*

JOHN A. KESSLER • *Davee Department of Neurology, Northwestern University Medical School, Chicago, IL*

DANIEL A. LIM • *The Rockefeller University, New York, NY*

MARLA B. LUSKIN • *Department of Cell Biology, Emory University School of Medicine, Atlanta, GA*

PETER C. MABIE • *Albert Einstein College of Medicine, Bronx, NY*

MARGOT MAYER-PRÖSCHEL • *Center for Cancer Biology, University of Rochester Medical Center, Rochester, NY*

JOHN W. MCDONALD • *Center for the Study of Nervous System Injury & Department of Neurology, Washington University School of Medicine, St. Louis, MO*

MARK F. MEHLER • *Albert Einstein College of Medicine, Bronx, NY*

TANYA A. MORENO • *Division of Biology, California Institute of Technology, Pasadena, CA*

MARK NOBLE • *Center for Cancer Biology, University of Rochester Medical Center, Rochester, NY*

THEO D. PALMER • *Department of Neurosurgery, Stanford University, Palo Alto, CA*

MAHENDRA S. RAO • *Department of Neurobiology and Anatomy, University of Utah Medical School, Salt Lake City, UT*

EVAN Y. SNYDER • *Department of Neurology, The Children's Hospital, Boston, MA*

BARBARA A. TATE • *Harvard Medical School and The Children's Hospital, Boston, MA*

SALLY TEMPLE • *Center for Neuroscience and Neuropharmacology, Albany Medical College, Albany, NY*

GIRI VENKATRAMAN • *Department of Cell Biology, Emory University School of Medicine, Atlanta, GA*

ANGELO L. VESCOVI • *Institute for Stem Cell Research, S. Raffaele Hospital, Milan, Italy*

1

Defining Neural Stem Cells and Their Role in Normal Development of the Nervous System

Sally Temple

INTRODUCTION

Stem cells are key players in the development and maintenance of specific tissues in most animals. The discovery of stem cells in the central and peripheral nervous systems (CNS and PNS) is a relatively recent event. First, continued neurogenesis (neuron generation) in the adult pointed to a long-lived progenitor cell *(1)*. Isolation of stem-like cells from the embryonic CNS, including basal forebrain *(2,3)*, cerebral cortex *(4)*, hippocampus *(5)*, spinal cord *(6)*, and the PNS *(7)* as well as evidence for multipotent, stem-like progenitors in vivo *(8–10)*, have indicated that they are important components of the developing nervous system (Fig. 1). Much excitement surrounded the isolation of adult stem cells from known neurogenic (neuron-generating) zones (the subventricular zone and hippocampal dentate gyrus) in rat, primate, and human (reviewed in ref. *11*). More recent evidence for continued neurogenesis in areas not previously considered to be neurogenic, such as the spinal cord *(12,13)* and neocortex *(14,15)*, suggests that stem cells may be a more widespread feature of the adult nervous system than previously imagined (Fig. 1).

Current research is focused on identifying the characteristics and functions of neural stem cells, in both developing and adult systems, to reveal their place in CNS biology and to facilitate the harnessing of these remarkable cells for repairing damaged nervous systems. To help us understand more about neural stem cells, we can explore a wealth of knowledge concerning stem cells in other systems and organisms, looking for common themes that might explain the essential stem cell state, as well as differences that might reveal the uniqueness of neural stem cells.

From: *Stem Cells and CNS Development*
Edited by: M. S. Rao © Humana Press Inc., Totowa, NJ

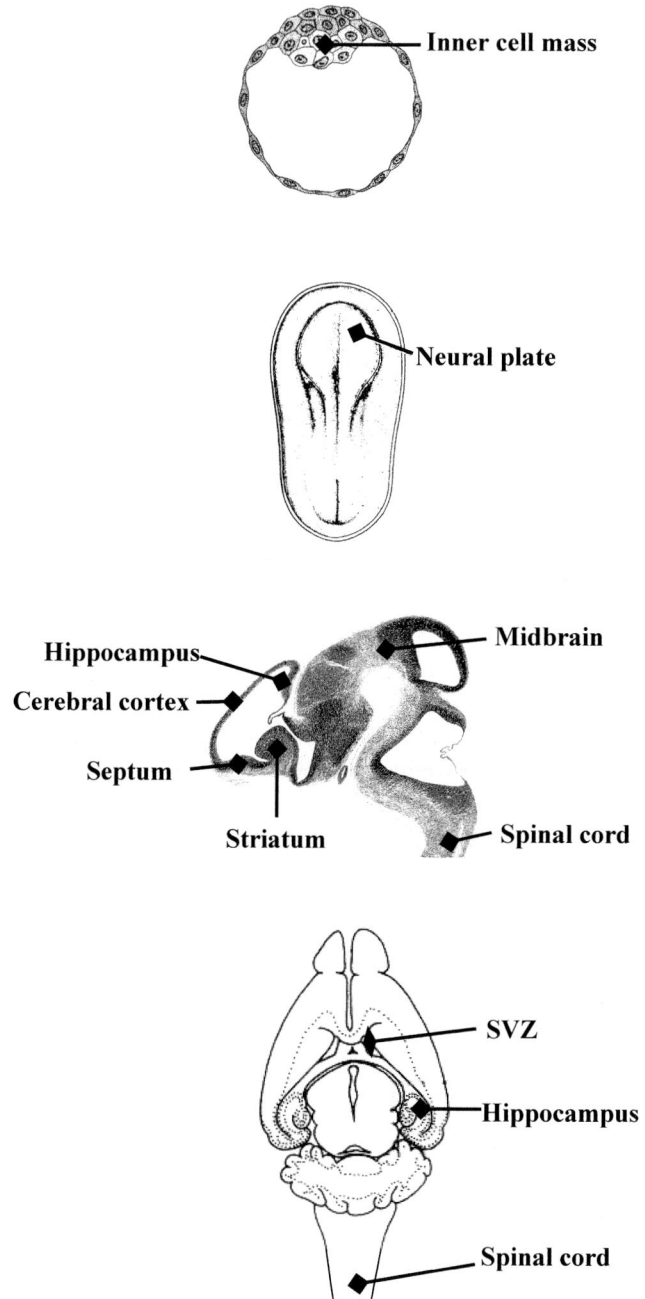

Fig. 1. Multipotent stem cells generating neurons and glia have been isolated from the early embryo through to adult.

The term *stem cell* has a number of different meanings to different people — depending on the system being analyzed and the perspective of the researcher using the term. A general definition is "a cell that is capable of both self-renewal and differentiation." Most researchers in the stem cell field would agree with this baseline definition. In this chapter, progenitor is used as a blanket term to describe any dividing cell that can generate differentiated progeny, whether or not it can self-renew, and the term precursor is used to describe a cell that is committed to a specific fate.

Besides the two fundamental features of self-renewal and differentiation, other specific characteristics have been attributed to stem cells, some of which apply only to stem cells in particular systems, rather than being a general feature of all stem cells, as illustrated in a recent review *(16)*. This chapter discusses some of the general terminology used to describe and define stem cells, focusing on the terms that might apply to neural stem cells, especially during normal neural development.

DOES THE NERVOUS SYSTEM ARISE FROM A SINGLE STEM CELL TYPE?

The *potency* of a progenitor cell represents the range of cell types it can generate. In a general model, stem cells are the basic cell type from which all others emanate through restriction of potency. The most primitive cells are considered *totipotent* — able to generate the entire organism or the entire tissue that is being considered. Subsequent restriction of potency can occur within a stem lineage, so that stem cells may be *pluripotent* or *multipotent* (making many different cell types, but perhaps not all), *oligopotent* (having a few choices), or *unipotent* (making one type of progeny). The epidermal stem cell in the adult skin is thought of as unipotent, as it appears to generate solely keratinocytes. Stem cells may also release multipotent or oligopotent or unipotent restricted progenitor cells, which serve as transit populations to expand the stem cell progeny prior to terminal differentiation. Formation of blood is believed to follow this general model, in which a primitive hemopoietic stem cell proceeds via restriction of potency within its stem and progenitor progeny toward final hemopoietic cell fates *(17)*.

It is not clear at this point whether there is a stem cell type that can give rise to the entire nervous system with its wealth of neuronal and glial cell types. Embryonic stem cells (ES), which are considered totipotent themselves, can generate neural stem-like cells *(18)*. In tissue culture, or on transplantation into the embryo, the neural derivatives of ES cells can integrate and generate a wide variety of cell types *(19–22)*. Whether they can generate the entire range of neural cell types remains to be resolved, but these cells may emerge as the most plastic neural stem cells. A stem cell that can generate both PNS and CNS derivatives exists in the early embryo, suggesting it has a broad range of potency *(23)*. Multipotent stem cells have been isolated from various regions of the developing and adult CNS (reviewed in refs. *11* and *24–29*) and from the neural crest, which gives rise to the PNS *(30–32)*. Dissociated embryonic neural progenitor populations that include some stem cells can be transplanted from one region of the developing CNS into other regions and show remarkable properties of integration *(33)*, producing cells that resemble those endogenous to the transplant site.

The contribution of stem cells versus other progenitor cells to the differentiated cells in the new site has not yet been determined, but these data suggest that the stem cell may be able to respond to regional information and differentiate accordingly. As with the ES cells, it is not yet clear whether nervous system-derived stem cells can generate the entire range of neural cell types. The results of some studies indicate that their potency may be limited. After transplantation, differentiating cells may acquire the morphology typical of their new location but do not always express its characteristic molecular markers, suggesting a lack of complete integration. For example, telencephalic cells grafted into the embryonic diencephalon or mesencephalon continue to express telencephalic markers, even into adulthood *(34)*. Similarly, hippocampus-derived stem cells on transplantation into the adult retina expressed appropriate neuronal and glial morphologies but not end-stage markers of retinal differentiation *(35)*. Furthermore, cells from one CNS region may incorporate more successfully in some regions than others. In one study, mouse progenitor cells from the lateral ganglionic eminence (LGE) or from the ventral mesencephalon (VM) were dissociated and injected into the lateral ventricles of embryonic rats at a similar stage of development. The LGE cells preferentially incorporated into the striatum, whereas the VM cells preferentially incorporated into the hypothalamus and midbrain. None of these cells, derived from basal CNS regions, incorporated efficiently into dorsal structures such as the cerebral cortex and hippocampus *(36)*.

These data may indicate specific regional restrictions in potency. Progenitor cells from embryonic ferret cerebral cortex transplanted into an older cortex can produce age-appropriate cells, but cells from an older cortex, upon transplantation into a younger cortex, are unable to make age-appropriate cell types *(37)*. Similarly, mid-hindbrain progenitors show a wider degree of regional incorporation at embryonic day (E)10.5 than at E13.5 *(38)*. These data indicate a temporal restriction in potency; this has particular significance when we consider the range of potency of adult neural stem cells. They can generate neurons, astrocytes, and oligodendrocytes, but the types of neurons and glia generated may be limited. In vivo, adult neural stem cells are primed to generate interneurons, and this appears to be their behavior after transplantation to adult neurogenic zones *(39,40)*. When placed in developing nervous system areas, adult stem cells can generate more cell types than they can after transplantation into the adult (for example, as in the retina; *35*). It is important, however, to establish whether they are capable of generating the major projection neurons in the CNS, most of which arise early in embryonic development. Furthermore, adult neural stem cells cultured for long periods may become increasingly biased toward production of glial cells, in some cases eventually losing neurogenic potential *(41)*.

Besides generating a wide variety of neural cells, stem cells derived from the nervous system may also be capable of producing cells of other tissues. In one remarkable study, it has been shown that a stem cell derived from the adult CNS could generate blood cells after transplantation into the bone marrow of an irradiated host *(42)*. Although we marvel at the plasticity of the stem cell involved, this experiment does not speak to its potency in generating *neural* cell types. Also, it was not clear what the characteristics of the starting cell were. Was it a neural stem cell that acquired the features of a hemopoietic stem cell through transdifferentiation or dedifferentiation? Or could there be a small population of totipotent stem cells in the brain, perhaps even derivatives of migratory germ cells that did not reach the germinal ridges *(43)*, that was responsible for blood cell production in this experiment? If the cell was indeed a neural stem cell, is its remarkable plasticity a reflection of its normal biology, or could it be the result of growing for long periods in tissue culture prior to transplantation? Undoubtedly answers to these questions will be found soon and will help us understand the types of stem cells present in the adult brain, their normal potency, and how long-term cell culture might alter them.

Although the presence of a neural stem cell capable of generating all neural tissue has not yet been documented, the idea that normal development might proceed through gradual restriction of potency, as occurs in the blood system, is supported by studies of developing nervous system stem cells. In the CNS and PNS, multipotent progenitors generate restricted progenitors for neurons and glial cells *(28)*. The heterochronic cortical transplantation studies mentioned previously also support this model. How might restriction of potency occur within the stem cell? It has been suggested for stem cells in a number of systems that the more primitive stem cells express a wide variety of transcripts at a low level, perhaps maintaining genes in an "open" chromatin configuration that is poised for transcription. Restriction of potency would proceed by turning off some genes and enhancing expression of others *(44,45)*. For the nervous system, this might explain why fetal glutamatergic and γ-aminobutyric acid (GABA)ergic cortical neurons both express glutamic acid decarboxylase (GAD) transcripts *(46)*, or why neural progenitor cells in the spinal cord express genes characteristic of both interneurons and motorneurons before selecting one or the other phenotype *(47,48)*. Restriction of potency might involve a hierarchy of transcription factors that drive the cell toward a particular fate. In *Drosophila*, proneural genes, e.g., achaete, scute, and atonal, confer competence for neural differentiation, via a chain of transcription factor activation *(49)*.

Homologs of these genes may operate similarly in vertebrates. For example, Mash-1, a mouse homolog of *Drosophila* achaete/scute, stimulates expression of the transcription factor Phox2a, which in turn stimulates expression of panneuronal properties and of the receptor c-RET, specifying subtypes of autonomic lineage cells *(30,50,51)*. Mash-1 initiates a cascade with different components in the olfactory system *(52)*. It also appears important for generating neurons in the ventral embryonic forebrain, perhaps via influence on Notch signaling *(53,54)*. Given the prevalence of achaete/scute and atonal homologs as well as other members of the bHLH transcription factor family, in the developing vertebrate nervous system, there is undoubtedly much to be learned about how these factors might interact within stem cell lineages to generate diverse neural cell fates.

STEM CELL POTENCY AND REGIONAL IDENTITY WITHIN THE EMERGING NERVOUS SYSTEM

It is important to note that potency is empirically determined — a cell is challenged with specific environmental signals, and we examine what types of cells it can generate. Hence, a cell that develops in a normal context may not show its full range of potency. A cell present in the cerebellum may be capable of making motor neurons if transplanted into the spinal cord — this reveals its plasticity, but not its normal developmental biology. Furthermore, it cannot be concluded from this experiment that early neural stem cells are undifferentiated and do not possess regional information, only that the information that they might have can be changed. In fact, it is likely that stem cells normally acquire regional information very early. If progenitor cells are removed from different regions of the early embryonic nervous system and placed in tissue culture, they develop into cell types characteristic of the region from which they were derived. Thus embryonic retina progenitors give retinal cells, embryonic cerebellar progenitors produce cerebellar cells, and embryonic neural crest progenitors generate typical PNS derivatives.

We know that in normal neural development positional information that presages regionalization of the nervous system is imparted very early, probably concomitantly with the neural induction process in the gastrula *(55)*. Given their behavior after isolation in tissue culture, one can hypothesize that positional information is embodied in neural stem cells. Thus, an important role of stem cells in normal development might be to interpret positional information and to read it out by generating cells appropriate to their location. The plasticity that we see exhibited in transplantation experiments may be important in normal development, for example, in the initial interpretation of positional signals, in designating the fate of progenitor cells at the borders between neural regions, or in regulative events that coordinate development throughout the embryo. It may also help the embryo compensate for disease or damage that in the natural environment are normal developmental events. In fact, the evolution of developmental mechanisms may be closely linked with the evolution of repair processes, as suggested by the similarity of some signaling pathways operating in disease and development in *Drosophila (56)*.

We can ask further whether within each region of the developing nervous system, there is one type of progenitor cell — one type of cerebellar progenitor cell, one type of cortical progenitor, etc. — or a number of types with different specificities. In the early cerebral cortex, for example, the cells cycle with apparently uniform dynamics, and there are no overt features that suggest diversity within the population. However, even at early stages, clonal analysis reveals that only about 10% of the cells behave like stem cells in culture *(57,58)*. Perhaps this is a shortcoming of the clonal culture system, which might not allow the stem cell phenotype to be fully expressed. Alternatively, this could indicate that stem cells are in fact a subpopulation, even in the primitive neuroepithelium. Similar to the early cerebral cortex, the neural crest contains a mixture of different types of progenitor cells *(32)*. Are these crest progenitors related by a more primitive common precursor, or did they arise from the dorsal neuroepithelium as distinct entities?

To date few lineage studies (for example, using retroviral labeling) have been conducted early enough to establish the possible heterogeneity of the primitive neuroepithelium. Although conclusions are tentative, several current studies suggest that heterogeneity does exist. In this case, the vertebrate neuroepithelium might be more like that of *Drosophila*. In the fly, each neuroblast has a discrete identity — based on which segment it is in, and where in the segment it arises — and generates appropriate types and numbers of progeny accordingly *(59)*. Furthermore, in *Drosophila*, laser ablation experiments have shown that there is a degree of plasticity within the neuroepithelium. Progenitor cells within a so-called proneural cluster or equivalence group can replace a neuroblast if ablated, but the group is limited in size, and hence reprogramming is restricted, as might well be the case in vertebrates *(59–61)*.

It is clear that multipotent stem cells are definitely present in the developing CNS, and although they are plastic, there is no conclusive evidence as yet for a stem cell that is capable of making all neural cell types. It is certainly true that restriction in potency of multipotent stem cells is involved in normal neural development, but this may not be the only mechanism for cell generation. There may be, from the earliest stage (perhaps designated by positional information), distinct types of neural stem cells, each of which generates more restricted progeny. There may also be, from the earliest stage, restricted progenitor cell types that produce certain classes of neural cells that eventually interweave with stem cell products. Rather than being a sheet of equivalent, uncommitted cells, perhaps the early neuroepithelium is a mosaic of progenitor types with defined roles influenced by positional information and with limited developmental plasticity that is necessary to generate a complete functioning organism.

SELF-RENEWAL IN RELATION TO STEM CELLS IN THE DEVELOPING AND ADULT NERVOUS SYSTEMS

The central defining feature ascribed to stem cells is the ability to self-renew, sometimes called self-maintenance *(16,62,63)*. This is the essence of the stem cell state — maintaining the ability to generate more stem cells for future generations of progeny. Self-renewal may be a feature of each individual stem cell. Alternatively, it might be an emergent property of a population of stem cells in which, for example, there is a certain probability of dividing or differentiating, so that the maintenance of the stem cell state may be stochastically determined by the dynamics of the population. Demonstration of self-renewal is the litmus test — the functional definition — of stem cells. In the nervous system, self-renewal has been demonstrated in vitro by allowing a stem cell to develop, and then subcloning its progeny to show that it made at least some progeny that behave as stem cells. This has been done in adhesion-based culture systems, by showing that the subcloned cells make secondary clones, and in non-adhesion-based culture systems by showing that the subcloned cells make secondary neurospheres — the large floating spheres of cells that are believed to represent stem cell products *(12)*.

Two ideas are implied by the term *self-renewal*, first that the stem cell maintains its developmental potency (i.e., the range of types of progeny it is capable of generating) and second that it maintains its proliferative capacity. These features, which are interrelated, perhaps describe an ideal stem cell, but they are not represented by normal stem cells. Regarding potency, as discussed earlier, production of diverse blood cell types appears to involve successive restriction of stem cell potency from a totipotent cell; this also appears true of the developing neural stem cell. Thus, rather than the exact maintenance of potency, changes in this property may actually be central to stem cell function during tissue production. Each stage of stem cell development, from the most primitive multipotent to unipotent stem cells, may have some capacity to maintain potency through cell division. In the adult, maintenance of the potency of the most primitive cells present is important to ensure homeostasis. However, as seen in the next section, stem cell division is not without limit.

The proliferative capacity of stem cells is not perfectly maintained during self-renewal. Blood stem cells can be transplanted into, and repopulate, a new host, but this repopulation can only be accomplished a certain number of times, indicating that the stem cell's impressive proliferative capacity is finite *(17)*. In fact, different types of blood stem cells defined by surface markers have specific characteristic proliferative capacities, implying that it is both a finite and an intrinsically determined characteristic *(17)*. Blood stem cells present in the embryo have a larger division potential than those of the adult *(64)*. Both embryonic and adult neural stem cells can be maintained for long periods in tissue culture, but the limits of this maintenance, and comparison of the two stages, have not been fully explored. A recent study indicates that most fetal spinal cord-derived stem cells divide for just three to six passages, and the few cells that divide for longer periods become biased toward generating nonneuronal progeny *(41)*.

Changes in potency and/or proliferative capacity in stem cell systems appear to be the norm in many tissues. Besides the examples given, there are age-related changes in stem cells, observed, for example, in the adult intestinal crypt *(65)* and in neurogenic cells in the adult hippocampus *(66)*, that further dispute the concept of perfect self-renewal.

Employment of a strict definition of self-renewal has challenged the inclusion of certain types of cells in the stem cell class. It has been suggested by some researchers that *Drosophila* neuroblasts are not really stem cells because they change over time and because they only undergo a few asymmetric divisions that generate neuroblasts *(67)*. Others call them stem cells because they are multipotent, undergo asymmetric division, and are the primary source of CNS tissue. Instead of using the austere definition, one might think of self-renewal as the finite capacity of a stem cell to maintain the stem cell *state*, rather than the stem cell *per se*. This allows for changes in potency that might be critical for normal development and repair, as well as age-related changes that might be inevitable. We can think of self-renewal as a modifiable property of stem cells that is tailored to the job that the stem cell has to accomplish — to make appropriate progeny according to the demands of the developing or the adult system.

The extent of self-renewal might be linked to telomerase activity. In most dividing somatic cells successive divisions involve progressive shortening of the telomeres at the ends of chromosomes, and telomere erosion correlates with cessation of cell division. In contrast, telomere shortening progresses much more slowly in certain types of stem cells, such as those in the germline, owing to the activity of a specific telomerase enzyme. The telomerase holoenzyme consists of an RNA template and protein components, including a cellular reverse transcriptase *(68)*. Its activity is high in certain proliferative cells and in the vast majority of neoplasms, including neural tumors, again providing a correlation with extended proliferative capacity *(69–71)*. Transvection of telomerase has in some cases conferred immortalization and allowed the establishment of cell lines, for example from skin *(72)*. Expression of telomerase has been reported in the developing nervous system, although expression significantly downregulates after birth, and activity may be undetectable in the adult *(71,73)*. However, it is possible that rare stem cells in the adult nervous system may retain a low level of expression, as do blood stem cells *(74)*. There are indications that telomerase activity is required for maintenance of normal nervous system development. Mice that lack telomerase RNA show progressively worse symptoms with generations, largely associated with defects in highly proliferative tissues *(75)*. After around six generations, the embryos die very early and show defects in neural tube closure *(76)*.

THE SIGNIFICANCE OF QUIESCENCE AND PROLIFERATION RATE TO THE STEM CELL STATE

It has been suggested that embryonic neural stem cells are not stem cells because they divide too rapidly, whereas stem cells are slowly dividing or quiescent, but this concept is erroneous. In fact, proliferation rates among stem cells vary widely. Intestinal crypt stem cells divide about once a day; other stem cells, such as hemopoietic and epidermal cells, divide much more slowly; others, such as the muscle satellite cell, may be genuinely quiescent *(77,78)*. Furthermore, the idea that when actively dividing stem cells are lost they are replaced by a quiescent population of dormant reserve stem cells might be an overgeneralization. While in blood, plant meristem, and muscle this may be the case *(79–81)*, reserve stem cells in the intestinal crypt are actually rapidly dividing progeny of stem cells that can dedifferentiate and revert to the stem cell state.

Given the large proliferative capacity of stem cells, it is clear that the rate of stem cell division must normally be highly regulated. In the intestinal crypt, there are a small number of stem cells, perhaps between five and seven. This is tightly controlled: one cell too many or one cell too few is detected and fixed by apoptosis or cell division *(82)*. Exactly how the changes in stem cell number are detected is unclear, but environmental factors must be key. This is true in the blood system, where quiescent blood stem cells can be rapidly stimulated to divide by cytokines *(83)*.

During normal development of the nervous system, the rates of proliferation of cells in germinal zones change with region and with time. Division rates within the neural germinal zones may be as rapid as every 7–10 h *(83a)* but may be as infrequent as 18 h by late gestation *(84)*. Adult stem cells may have a cycle time that is on the order of many days *(85–87)*. It seems likely then that the proliferation rate of neural stem cells changes during normal neural development, in different regions of the embryo and into adulthood. How these regulative events are accomplished is not clear. They most likely involve regional and age-related changes in environmental factors, such as stem cell mitogens. Fibroblast growth factor-2 (FGF-2) and epidermal growth factor (EGF) or the related factor transforming growth factor (TGF-α) are present in the CNS throughout life and profoundly stimulate neural stem cell proliferation in vitro and in vivo *(88)*. There is likely to be a large number of as yet undiscovered regulatory molecules that stimulate or inhibit neural stem cell division and hold the promise for expanding stem cells in vivo or in vitro; perhaps they also inhibit division of neural tumors.

THE ROLE OF ASYMMETRIC AND SYMMETRIC CELL DIVISIONS DURING NORMAL NEURAL DEVELOPMENT

Mitotic cell divisions produce two daughter cells that acquire identical genetic material but not necessarily identical epigenetic components. These components may include *cytoplasmic determinants* — molecules that can direct cell fate. Hence, by altering the way these molecules are distributed during the cell division process, it is possible to generate diverse cell fates. When a progenitor cell divides to generate two daughters with essentially the same fate, the process is called *symmetric cell division*; when it divides to generate two daughters with different fates, it is called *asymmetric cell division*. Sometimes a cell division generating two equivalent daughters that subsequently differentiate differently because of environmental influences has been called asymmetric. However, this may be an overextension of the definition, the crux of which is to show that the division process itself is actively involved in producing two distinct daughter cells.

The ability to divide asymmetrically is often described as a fundamental feature of a stem cell. If stem cells are to undergo both self-renewal and the generation of differentiated progeny, one way to do this is to divide asymmetrically. However, there are other ways to achieve this end. For example, self-renewal may be a stochastically determined, intrinsic property of a population of stem cells in which each cell has a given probability to make more stem cells or generate differentiated daughters. In this case, the two functions result from a population, rather than a single stem cell lineage, so it is not necessary to invoke asymmetric divisions to achieve them. Another way is for the stem cell to generate equivalent daughters that move into different environments, some promoting self-renewal and others promoting differentiation.

Nevertheless, asymmetric cell division may be utilized by stem cells, and there is direct evidence for asymmetric cell division within a few stem cell populations. In some species the lineages of progenitor cells have been reconstructed, providing direct evidence for asymmetric cell divisions. For example, in the *Drosophila* CNS, asymmetric cell divisions of the stem-like neuroblast result in the production of a smaller daughter called the ganglion mother cell (GMC) that goes on to generate two neurons or glial cells and another neuroblast. Repeated divisions of the neuroblast result in a chain of GMCs, and thence from each a chain of pairs of differentiated daughter cells *(59)*.

Well-characterized asymmetric cell lineages in *Drosophila* and *Caenorhabditis elegans* nervous systems render them ideal model systems for understanding how asymmetric divisions are achieved. Studies of mutations in both systems have revealed genes that are involved in this process (Fig. 2). In *Drosophila* neural development, Prospero and Numb proteins are cell fate determinants that directly influence neural fate decisions at asymmetric cell divisions. Prospero is a transcription factor with a homeodomain, and Numb is an adapter protein with a phosphotyrosine binding (PTB) domain. Both proteins become asymmetrically localized in the basal cortex of the stem cell neuroblast at metaphase and then preferentially segregate into the GMC *(89,90)*. In addition, *prospero* mRNA is localized in a basal crescent at mitosis and segregated into the GMC. Once in the GMC, Prospero is released from the cortex and translocates into the nucleus, where it controls transcription of certain neuroblast- and GMC-specific genes *(91)*. Prospero also prevents cell division and stimulates differentiation so that the GMC only divides once, generating two neurons or glia *(92)*. Numb's function in the GMC is not clear; however, at the following division,

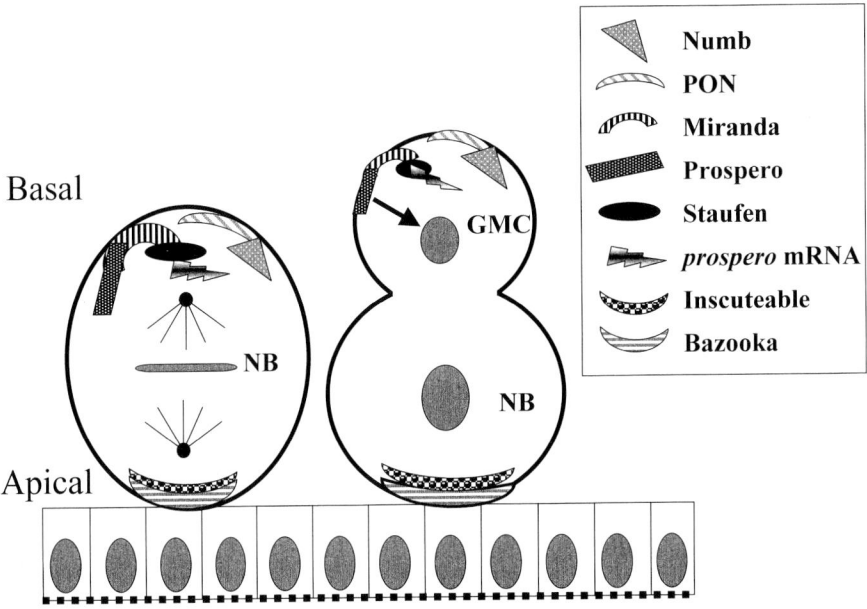

Fig. 2. Summary of components involved in asymmetric cell divisions in *Drosophila* neuroblasts.

Numb segregates asymmetrically into the two daughter cells, where it can create two different neuronal fates by inhibiting Notch function in one but not in the other cell *(93,94).*

Complexes of cytoplasmic components bring about the asymmetric localization of Prospero and Numb. Bazooka and Inscuteable are required for correct mitotic spindle orientation in the neuroblast and maintain apical-basal polarity from epithelial cells to neuroblasts *(95–97).* They are themselves localized asymmetrically at the apical side of the neuroblast before mitosis and provide the positional information necessary for other components (Miranda, Staufen, Prospero, Partner of Numb [PON], and Numb) to be localized basally at mitosis and to be further preferentially segregated into the GMC (Fig. 2) *(98).*

The asymmetric distribution of the Prospero complex is cell- cycle dependent. At interphase, Inscuteable forms a prominent crescent along the apical cell cortex. Miranda, a membrane-associated, multidomain adapter protein, inter-acts with Inscuteable and Prospero and tethers Staufen, which in turn binds *prospero* mRNA *(99).* When the neuroblast enters metaphase, Miranda, Staufen, Prospero, and *prospero* mRNA move as a group to the basal side of the dividing neuroblast. After mitosis, this complex of protein and mRNA is segregated into the GMC *(100–102)*. It is not clear how the movement of this complex occurs. Similarly, apical localization of Inscuteable provides a positional guide for the Numb complex, which concentrates at the basal side of the neuroblast during mitosis. Although Miranda can interact with Numb protein in vitro, it is not necessary for asymmetric numb localization. PON colocalizes with Numb at the basal cortex of the mitotic neuroblast and loss of *pon* function causes defects in asymmetric Numb localization *(103)*.

Although a number of the apical components that are retained in the neuroblast rather than moving into the GMC have been isolated, none of them have been shown to be cytoplasmic determinants of the neuroblast fate. It would be very interesting to find such molecules, if they exist, because they might tell us which gene functions are essential to maintain the stem cell state. These molecules will hopefully be uncovered in the near future, as we build on knowledge of the basal neuroblast complexes.

Studies of the dynamics of cell proliferation within vertebrate CNS proliferative zones suggest that early cell divisions are largely symmetric, perhaps allowing expansion of the stem cell population; as neurogenesis gets under way, cell divisions appear to become largely asymmetric. An interesting correlation has been noted between this purported change in the type of cell division and the direction of the plane of division of mitosing cells at the ventricular surface *(104)*. Early divisions usually have the division plane oriented perpendicular to the ventricular surface, whereas later divisions have the plane oriented preferentially in the horizontal direction. Perhaps the perpendicular divisions are symmetric and the horizontal divisions asymmetric? In slices of ferret cortex it has been shown that at least at early stages the products of perpendicular divisions appear to behave similarly, migrating at similar rates within the ventricular zone, whereas the products of horizontal divisions do not *(104)*. It will be important to follow these cells for longer periods to establish whether the change in division plane correlates with final symmetric or asymmetric cell fates. There are situations in which it does not; for example, in plant meristem, the division plane alters when progenitor cells generate different types of products, yet the cell divisions involved are asymmetric *(105)*.

In the vertebrate, especially in mammalian systems that develop *in utero*, it is currently impossible to follow the lineage trees of progenitor cells in vivo. We can identify the components of a clone that developed in vivo, for example, by labeling individual cells within the ventricular zone of the CNS using retroviral markers, waiting for a period, and then revealing the clonal contents by a histochemical technique. In labeling experiments conducted in the developing cerebral cortex, clone distribution is sometimes spread between cortical layers, suggesting a stem-like lineage tree with repeated asymmetric divisions, and sometimes confined to a single layer, suggesting a symmetric, proliferative type of lineage tree *(106–109)*. However, we cannot currently determine with these methods exactly how symmetric and asymmetric divisions contribute to clonal development, and how these clone members are generated over time in the animal.

In tissue culture, it has been possible to follow the development of individual isolated embryonic mouse ventricular zone cells for long periods. Continuous recording of the divisions of these cells, combined with immunostaining of the progeny, provides lineage trees for mammalian cells developing in vitro. These data show directly that mouse cortical progenitor cells undergo largely asymmetric cell divisions when generating neurons *(110)*. The major neuroblast divides to give a minor neuroblast that produces a small "packet" of about 10 neurons and another stem-like progenitor that divides asymmetrically again, producing another "packet" of neurons, and so on (Fig. 3). Occasionally the minor neuroblast clones appear to have a symmetric lineage, generating their progeny at the same time. (Perhaps these are equivalent to the retrovirally labeled clones that reside within a cortical layer.) Symmetric lineages are associated with the expansion of glioblast clones. However, in the vast majority of cases the neuroblast clones are generated by asymmetric cell divisions. Interestingly, the lineage trees of these cortical neuroblasts are similar to those described for neural progenitors in *C. elegans* and *Drosophila*, suggesting an evolutionary conservation of the mechanisms underlying neural cell generation *(111)*.

It will be important to find out whether the division patterns actually play a role in defining types of cortical cells, as they do in invertebrates. It will also be important to establish the mechanisms for generating asymmetric cell divisions in vertebrates. Homologs of Numb have been described and appear to segregate unevenly within dividing cells in the cortical ventricular zone, suggesting a similar function *(112)*. Interestingly, the fact that the asymmetric mouse cortical lineage trees are seen in a culture setting devoid of normal environmental cues suggests that the presence of other cells may not be necessary for the asymmetry in CNS progenitor cell divisions to occur. The same has been shown for *Drosophila:* cultured neuroblasts retain an asymmetric lineage pattern in vitro *(113,114)*. It is surprising that in a highly regulative embryo such as the mouse, in which cell–cell signaling allows for plasticity in the molding of the final organism, invertebrate-type lineage trees, which are thought of as invariant and characteristic of mosaic development, are employed in making the CNS. One has to remember, however, that although cell division-based mechanisms might operate to

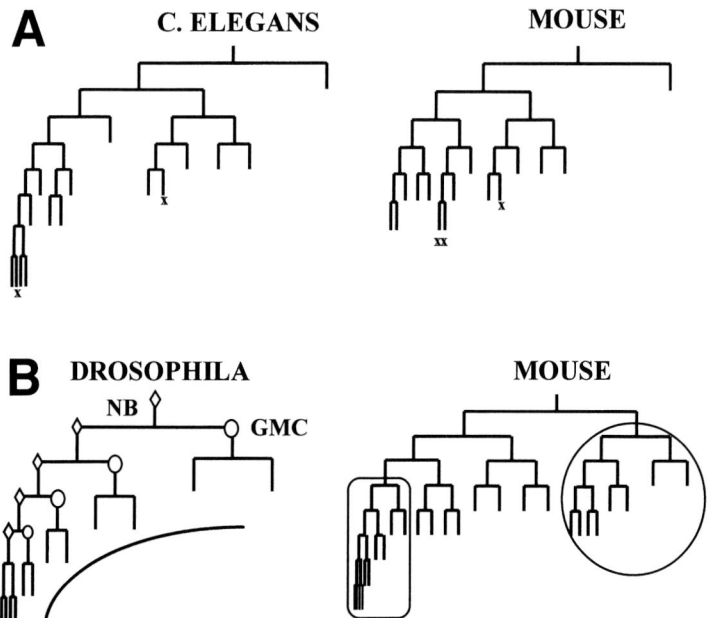

Fig. 3. (A, B) Mouse embryonic cortical progenitor cells undergo asymmetric division patterns, resulting in lineage trees that resemble those of invertebrate neural lineages.

generate neural cell types, these processes may still be environmentally responsive and capable of change.

STEM CELL NICHE

From extensive studies of blood stem cells and other stem cell types, it has become clear that the immediate environment of a stem cell — its niche — is critical for determining its behavior. It is in this primary environment that the stem cell acquires information about whether or not to divide and what types of progeny to generate. Blood stem cells need to be kept in a bone marrow stromal niche to maintain their self-renewal. The bone marrow niche is a complex environment of extracellular matrix components, e.g. tenascin, and growth factors including stem cell factor, granulocyte/macrophage colony stimulating factor, and FGFs. Specific cytokines release blood stem cells from their adhesion in the niche, mobilizing them for circulation and proliferation *(115)*. In other systems, stem cells also reside in complex niches that regulate their behavior *(116)*. When stem cells derived from the nervous system incorporate into bone marrow, they can make blood cells *(117)*.

Similarly, muscle stem cells can make blood, and blood stem cells can produce muscle and bone if they arrive in the right niche *(118,119)*.

During normal development of the nervous system, neural stem cells reside in the germinal zones — notably the ventricular zone and the subventricular zone. In the adult, neurogenesis is limited primarily to a few specific areas of the CNS, around the lateral ventricles and in the dentate gyrus (Fig. 1). As reviewed elsewhere *(120–124)*, many extracellular matrix molecules (including tenascin) and growth factors (such as EGF and FGFs) have been described in these areas. Clearly, there are also region-specific niche molecules — as shown by the ability of cells in a particular region to dictate new differentiation programs in transplanted cells. Removal of surface components from embryonic progenitor cells prior to transplantation can alter their ability to recognize regional signals *(125)*.

In the adult nervous system, although there are defined areas of neurogenesis, stem cells can be isolated from nonneurogenic areas, e.g., the spinal cord, and grown in culture to generate both neurons and glial progeny. However, when put back into the spinal cord, these stem cells only make glia. When placed in a neurogenic area such as the subventricular zone, these spinal cord stem cells can make neurons as well as glia *(126)*. This implies that key environmental molecules within different adult CNS niches regulate the ability of stem cells to generate neuronal or glial products. The search for factors that define the niche (and allow the specification of stem cells, their self-renewal and regulation) is a key area for study: the niche may ultimately define the stem cell.

IS THERE A MOLECULAR DEFINITION OF THE NEURAL STEM CELL?

Molecular characterization of stem cells allows the identification of cell components that are critical for stem cell regulation and that help us understand stem cell biology. In addition, employment of these markers for cell selection allows researchers to study a purer population of stem cells and to provide a uniform population for therapeutic use. Hence considerable efforts have been made to find markers for stem cells, especially surface markers that allow live cell isolation. Hemopoietic stem cells are probably the best characterized at this point, and they can be isolated (based on the expression of particular surface antigens) for study and transplantation *(127)*. Epidermal stem cells have high levels of β1-integrin, which also allows their selection *(128)*.

Characterization of stem cells from a number of systems reveals some common features that help define the general class. For example, the Notch signaling system, initially identified in *Drosophila*, appears to play a role in regulating a variety of stem cells. Notch is expressed on blood stem cells, and Jagged is expressed in the bone marrow; activation of the receptor might keep the blood stem cell in a quiescent state *(129–133)*. Notch is expressed in muscle cell precursors and also regulates their differentiation *(134)*. In the nervous system of *Drosophila,* Notch is involved in switching epidermal progenitor cells between two alternative fates, nonneural versus the neuroblast fate, or between two types of neuron *(135,136)*. By feedback regulation (called lateral inhibition) between Notch and its ligand Delta, the number and placement of neuroblasts in the neuroepithelial sheet are determined. In the vertebrate nervous system, Notch1 and its ligands Delta and Jagged are widely expressed in germinal zones during development, and constitutive expression of Notch in an embryonic cell line suggests that its activation might make progenitor cells differentiate into glial cells rather than neurons *(137,138)*. In the adult CNS, Notch is present in neurogenic zones, in both ependymal and subependymal areas *(139–142)*. Notch is also expressed on long-lived adult oligodendrocyte O-2A progenitors that have stem cell-like properties, and Jagged is expressed by mature oligodendrocytes and neurons. Activation of Notch on the adult O-2A may maintain them in an immature state *(143)*.

There are other examples of common features shared by stem cells from a variety of sources. For example, FGF-1 and -2, closely related members of the FGF family, EGF, or its relative TGF-α appear mitogenic for a wide variety of stem cells, including epidermal, bone, blood, gut, and neural stem cells as well as primordial germ cells. Similarly, members of the bone morphogenetic family (BMPs) influence cell division in these stem cell systems, often negatively *(123,144–151)*. Components of these signaling pathways may thus help identify a number of classes of stem cells.

Other markers appear to define particular classes of stem cells, even across the plant and animal kingdoms. For example, the gene *piwi* is specifically expressed in *Drosophila* germinal stem cells. In a remarkable show of evolutionary conservation, *piwi* is structurally similar to a plant gene called *zwille*, which is seen in stem cells in the plant meristem, the equivalent of germinal tissue that generates flowers *(152,153)*. Hence, this family of related transcription factors, whose function remains to be elucidated, might help define the germ stem cell class and reveal evolutionarily conserved mechanisms of germ cell maintenance.

Neural stem cells, besides expressing Notch and various components of the EGF/FGF/BMP signaling pathways, have a number of cell-intrinsic markers. They possess the RNA binding protein Musashi, the mouse homolog of *Drosophila* Musashi. In the fly, Musashi is involved in neuron development *(154)*, and its presence in mouse neural stem cells indicates that it may also play an important role in these cells *(155,156)*. There is persuasive evidence that adult stem cells in the subventricular zone express the intermediate filament proteins Nestin *(156,157)* and glial fibrillary acidic protein (GFAP) *(158)*. As yet, none of these markers appear to be exclusive to neural stem cells. For example, in the adult, Notch is also expressed on subpopulations of postmitotic neurons *(139,140)*, and Musashi and GFAP are expressed by astrocytes. Nevertheless, it is possible that particular combinations of these markers may define neural stem cells at the molecular level. More information is needed regarding surface markers on neural stem cells, as these will help in their live isolation and purification. Neural crest stem cells carry the low-affinity p75 neurotrophin receptor, which has allowed their retrieval, e.g., from the sciatic nerve, suggesting that they may be available for longer periods than previously thought, for both for potential repair and potential tumor formation *(159)*.

CONCLUSION

Stem cells are critically involved in the normal development and maintenance of a great variety of tissues, from plants to animals. It is interesting to consider why stem cells are used so prevalently to make tissues. Perhaps they provide a compact solution to disease prevention or to wear and tear: the information to generate a wide variety of cell types can be held in a succinct package, waiting for activating signals. Perhaps the embodiment of a variety of developmental possibilities within a single cell type allows effective integration of developmental and homeostatic signals to occur. Consistent with this idea, there is evidence that multipotent progenitors appear to integrate the input of combinations of factors that can act on them singly *(160,161)*. Thus information from different sources can play on a single cell, which reads the inputs and responds appropriately.

At this point, the best definition of neural stem cells is still probably the sparsest — self-renewing (and this in its broader meaning) and differentiating. However, as research continues apace, we may soon be able to refine this definition, to subdivide it as distinct types of neural stem cells are revealed, and to add molecular signatures that will eventually allow a full understanding of this unique class of versatile cells.

ACKNOWLEDGMENTS

I thank Qin Shen and Karen Kirchofer for their invaluable help and advice in preparing this manuscript.

REFERENCES

1. Altman, J. (1970) Postnatal neurogenesis and the problem of neural plasticity, in *Developmental Neurobiology* (Himwich, W. A., ed.), C.C. Thomas, Springfield.
2. Temple, S. (1989) Division and differentiation of isolated CNS blast cells in microculture. *Nature* **340,** 471–473.
3. Reynolds, B. A., Tetzlaff, W., and Weiss, S. (1992) A multipotent EGF-responsive striatal embryonic progenitor cell produces neurons and astrocytes. *J. Neurosci.* **12,** 4565–4574.
4. Davis, A. A. and Temple, S. (1994) A self-renewing multipotential stem cell in embryonic rat cerebral cortex. *Nature* **372,** 263–266.
5. Johe, K. K., Hazel, T. G., Muller, T., Dugich-Djordjevic, M. M., and McKay, R. D. (1996) Single factors direct the differentiation of stem cells from the fetal and adult central nervous system. *Genes Dev.* **10,** 3129–3140.
6. Kalyani, A., Hobson, K., and Rao, M. S. (1997) Neuroepithelial stem cells from the embryonic spinal cord: isolation, characterization, and clonal analysis. *Dev. Biol.* **186,** 202–223.
7. Stemple, D. L. and Anderson, D. J. (1992) Isolation of a stem cell for neurons and glia from the mammalian neural crest. *Cell* **71,** 973–985.
8. Walsh, C. (1993) Cell lineage and regional specification in the mammalian neocortex. *Perspect Dev. Neurobiol.* **1,** 75–80.
9. Leber, S. M. and Sanes, J. R. (1991) Lineage analysis with a recombinant retrovirus: application to chick spinal motor neurons. *Adv. Neurol.* **56,** 27–36.
10. Sanes, J. R. (1989) Analysing cell lineage with a recombinant retrovirus. *Trends Neurosci.* **12,** 21–28.
11. Temple, S. and Alvarez-Buylla, A. (1999) Stem cells in the adult mammalian central nervous system. *Curr. Opin. Neurobiol.* **9,** 135–141.
12. Weiss, S., Dunne, C., Hewson, J., Wohl, C., Wheatley, M., Peterson, A. C., and Reynolds, B. A. (1996) Multipotent CNS stem cells are present in the adult mammalian spinal cord and ventricular neuroaxis. *J. Neurosci.* **16,** 7599–7609.
13. Shihabuddin, L. S., Ray, J., and Gage, F. H. (1997) FGF-2 is sufficient to isolate progenitors found in the adult mammalian spinal cord. *Exp. Neurol.* **148,** 577–586.
14. Gould, E., Reeves, A. J., Graziano, M. S., and Gross, C. G. (1999) Neurogenesis in the neocortex of adult primates. *Science* **286,** 548–552.
15. Marmur, R., Mabie, P. C., Gokhan, S., Song, Q., Kessler, J. A., and Mehler, M. F. (1998) Isolation and developmental characterization of cerebral cortical multipotent progenitors. *Dev. Biol.* **204,** 577–591.
16. Morrison, S. J., Shah, N. M., and Anderson, D. J. (1997) Regulatory mechanisms in stem cell biology. *Cell* **88,** 287–298.
17. Morrison, S. J., Uchida, N., and Weissman, I. L. (1995) The biology of hematopoietic stem cells. *Annu. Rev. Cell Dev. Biol.* **11,** 35–71.
18. Okabe, S., Forsberg-Nilsson, K., Spiro, A. C., Segal, M., and McKay, R. D. (1996) Development of neuronal precursor cells and functional postmitotic neurons from embryonic stem cells in vitro. *Mech. Dev.* **59,** 89–102.

19. Dinsmore, J., Ratliff, J., Deacon, T., Pakzaban, P., Jacoby, D., Galpern, W., and Isacson, O. (1996) Embryonic stem cells differentiated in vitro as a novel source of cells for transplantation. *Cell Transplant* **5**, 131–143.
20. Renoncourt, Y., Carroll, P., Filippi, P., Arce, V., and Alonso, S. (1998) Neurons derived in vitro from ES cells express homeoproteins characteristic of motoneurons and interneurons. *Mech. Dev.* **79**, 185–197.
21. Brustle, O., Jones, K. N., Learish, R. D., Karram, K., Choudhary, K., Wiestler, O. D., Duncan, I. D., and McKay, R. D. (1999) Embryonic stem cell-derived glial precursors: a source of myelinating transplants. *Science* **285**, 754–756.
22. Mujtaba, T., Piper, D. R., Kalyani, A., Groves, A. K., Lucero, M. T., and Rao, M. S. (1999) Lineage-restricted neural precursors can be isolated from both the mouse neural tube and cultured ES cells. *Dev. Biol.* **214**, 113–127.
23. Mujtaba, T., Mayer-Proschel, M., and Rao, M. S. (1998) A common neural progenitor for the CNS and PNS. *Dev. Biol.* **200**, 1–15.
24. Gage, F. H., Ray, J., and Fisher, L. J. (1995) Isolation, characterization, and use of stem cells from the CNS. *Annu. Rev. Neurosci* **18**, 159–192.
25. Gage, F. H. (1998) Stem cells of the central nervous system. *Curr. Opin. Neurobiol.* **8**, 671–676.
26. Cameron, H. A. and McKay, R. (1998) Stem cells and neurogenesis in the adult brain. *Curr. Opin. Neurobiol.* **8**, 677–680.
27. Kuhn, H. G. and Svendsen, C. N. (1999) Origins, functions, and potential of adult neural stem cells. *Bioessays* **21**, 625–630.
28. Rao, M. S. (1999) Multipotent and restricted precursors in the central nervous system. *Anat. Rec.* **257**, 137–148.
29. Tropepe, V., Sibilia, M., Ciruna, B. G., Rossant, J., Wagner, E. F., and van der Kooy, D. (1999) Distinct neural stem cells proliferate in response to EGF and FGF in the developing mouse telencephalon. *Dev. Biol.* **208**, 166–188.
30. Anderson, D. J. (1997) Cellular and molecular biology of neural crest cell lineage determination. *Trends Genet.* **13**, 276–280.
31. Anderson, D. J., Groves, A., Lo, L., Ma, Q., Rao, M., Shah, N. M., and Sommer, L. (1997) Cell lineage determination and the control of neuronal identity in the neural crest. *Cold Spring Harbor Symp. Quant. Biol.* **62**, 493–504.
32. LaBonne, C. and Bronner-Fraser, M. (1998) Induction and patterning of the neural crest, a stem cell-like precursor population. *J Neurobiol.* **36**, 175–189.
33. Gaiano, N. and Fishell, G. (1998) Transplantation as a tool to study progenitors within the vertebrate nervous system. *J. Neurobiol.* **36**, 152–161.
34. Na, E., McCarthy, M., Neyt, C., Lai, E., and Fishell, G. (1998) Telencephalic progenitors maintain anteroposterior identities cell autonomously. *Curr. Biol.* **8**, 987–990.
35. Takahashi, M., Palmer, T. D., Takahashi, J., and Gage, F. H. (1998) Widespread integration and survival of adult-derived neural progenitor cells in the developing optic retina. *Mol. Cell. Neurosci.* **12**, 340–348.
36. Campbell, K., Olsson, M., and Bjorklund, A. (1995) Regional incorporation and site-specific differentiation of striatal precursors transplanted to the embryonic forebrain ventricle. *Neuron* **15**, 1259–1273.

37. Frantz, G. D. and McConnell, S. K. (1996) Restriction of late cerebral cortical progenitors to an upper-layer fate. *Neuron* **17,** 55–61.
38. Olsson, M., Campbell, K., and Turnbull, D. H. (1997) Specification of mouse telencephalic and mid-hindbrain progenitors following heterotopic ultrasound-guided embryonic transplantation. *Neuron* **19,** 761–772.
39. Suhonen, J. O., Peterson, D. A., Ray, J., and Gage, F. H. (1996) Differentiation of adult hippocampus-derived progenitors into olfactory neurons in vivo. *Nature* **383,** 624–627.
40. Herrera, D. G., Garcia-Verdugo, J. M., and Alvarez-Buylla, A. (1999) Adult-derived neural precursors transplanted into multiple regions in the adult brain. *Ann. Neurol.* **46,** 867–877.
41. Quinn, S. M., Walters, W. M., Vescovi, A. L., and Whittemore, S. R. (1999) Lineage restriction of neuroepithelial precursor cells from fetal human spinal cord. *J. Neurosci. Res.* **57,** 590–602.
42. Bjornson, C. R., Rietze, R. L., Reynolds, B. A., Magli, M. C., and Vescovi, A. L. (1999) Turning brain into blood: a hematopoietic fate adopted by adult neural stem cells in vivo. *Science* **283,** 534–537.
43. Weissman, I. L. (2000) Stem cells: units of development, units of regeneration, and units in evolution. *Cell* **100,** 157–168.
44. Anderson, D. J. and Axel, R. (1985) Molecular probes for the development and plasticity of neural crest derivatives. *Cell* **42,** 649–662.
45. Hu, M., Krause, D., Greaves, M., Sharkis, S., Dexter, M., Heyworth, C., and Enver, T. (1997) Multilineage gene expression precedes commitment in the hemopoietic system. *Genes Dev.* **11,** 774–785.
46. Cao, Y., Wilcox, K. S., Martin, C. E., Rachinsky, T. L., Eberwine, J., and Dichter, M. A. (1996) Presence of mRNA for glutamic acid decarboxylase in both excitatory and inhibitory neurons. *Proc. Natl. Acad. Sci. USA* **93,** 9844–9849.
47. Arber, S., Han, B., Mendelsohn, M., Smith, M., Jessell, T. M., and Sockanathan, S. (1999) Requirement for the homeobox gene Hb9 in the consolidation of motor neuron identity. *Neuron* **23,** 659–674.
48. Thaler, J., Harrison, K., Sharma, K., Lettieri, K., Kehrl, J., and Pfaff, S. L. (1999) Active suppression of interneuron programs within developing motor neurons revealed by analysis of homeodomain factor HB9. *Neuron* **23,** 675–687.
49. Jan, Y. N. and Jan, L. Y. (1993) HLH proteins, fly neurogenesis, and vertebrate myogenesis. *Cell* **75,** 827–830.
50. Lo, L., Tiveron, M. C., and Anderson, D. J. (1998) MASH1 activates expression of the paired homeodomain transcription factor Phox2a, and couples pan-neuronal and subtype-specific components of autonomic neuronal identity. *Development* **125,** 609–620.
51. Lo, L., Morin, X., Brunet, J. F., and Anderson, D. J. (1999) Specification of neurotransmitter identity by Phox2 proteins in neural crest stem cells. *Neuron* **22,** 693–705.
52. Cau, E., Gradwohl, G., Fode, C., and Guillemot, F. (1997) Mash1 activates a cascade of bHLH regulators in olfactory neuron progenitors. *Development* **124,** 1611–1621.

53. Casarosa, S., Fode, C., and Guillemot, F. (1999) Mash1 regulates neurogenesis in the ventral telencephalon. *Development* **126,** 525–534.
54. Fode, C., Ma, Q., Casarosa, S., Ang, S.-L., Anderson, D. J., and Guillemot, F. (2000) A role for neural determination genes in specifying the dorsoventral identity of telencephalic neurons. *Genes Dev.* **14,** 67–80.
55. Chang, C. and Hemmati-Brivanlou, A. (1998) Cell fate determination in embryonic ectoderm. *J. Neurobiol.* **36,** 128–151.
56. Lemaitre, B., Nicolas, E., Michaut, L., Reichhart, J. M., and Hoffmann, J. A. (1996) The dorsoventral regulatory gene cassette spatzle/Toll/cactus controls the potent antifungal response in *Drosophila* adults. *Cell* **86,** 973–983.
57. Qian, X., Davis, A. A., Goderie, S. K., and Temple, S. (1997) FGF2 concentration regulates the generation of neurons and glia from multipotent cortical stem cells. *Neuron* **18,** 81–93.
58. Shen, Q., Qian, X., Capela, A., and Temple, S. (1998) Stem cells in the embryonic cerebral cortex: their role in histogenesis and patterning. *J. Neurobiol.* **36,** 162–174.
59. Doe, C. Q., Fuerstenberg, S., and Peng, C. Y. (1998) Neural stem cells: from fly to vertebrates. *J. Neurobiol.* **36,** 111–127.
60. Doe, C. Q. and Goodman, C. S. (1985) Early events in insect neurogenesis. I. Development and segmental differences in the pattern of neuronal precursor cells. *Dev. Biol.* **111,** 193–205.
61. Doe, C. Q. and Goodman, C. S. (1985) Early events in insect neurogenesis. II. The role of cell interactions and cell lineage in the determination of neuronal precursor cells. *Dev. Biol.* **111,** 206–219.
62. Hall, P. A. and Watt, F. M. (1989) Stem cells: the generation and maintenance of cellular diversity. *Development* **106,** 619–633.
63. Potten, C. S. and Loeffler, M. (1990) Stem cells: attributes, cycles, spirals, pitfalls and uncertainties. Lessons for and from the crypt. *Development* **110,** 1001–1020.
64. Lansdorp, P. M., Dragowska, W., and Mayani, H. (1993) Ontogeny-related changes in proliferative potential of human hematopoietic cells. *J. Exp. Med.* **178,** 787–791.
65. Martin, K., Kirkwood, T. B., and Potten, C. S. (1998) Age changes in stem cells of murine small intestinal crypts. *Exp. Cell Res.* **241,** 316–323.
66. Kuhn, H. G., Dickinson-Anson, H., and Gage, F. H. (1996) Neurogenesis in the dentate gyrus of the adult rat: age-related decrease of neuronal progenitor proliferation. *J. Neurosci.* **16,** 2027–2033.
67. Lin, H. and Schagat, T. (1997) Neuroblasts: a model for the asymmetric division of stem cells. *Trends Genet.* **13,** 33–39.
68. Bryan, T. M. and Cech, T. R. (1999) Telomerase and the maintenance of chromosome ends. *Curr. Opin. Cell Biol.* **11,** 318–324.
69. Langford, L. A., Piatyszek, M. A., Xu, R., Schold, S. C. J., and Shay, J. W. (1995) Telomerase activity in human brain tumours. *Lancet* **346,** 1267–1268.
70. Le, S., Zhu, J. J., Anthony, D. C., Greider, C. W., and Black, P. M. (1998) Telomerase activity in human gliomas. *Neurosurgery* **42,** 1120–1124.

71. Weil, R. J., Wu, Y. Y., Vortmeyer, A. O., Moon, Y. W., Delgado, R. M., Fuller, B. G., et al. (1999) Telomerase activity in microdissected human gliomas. *Mod. Pathol.* **12,** 41–46.

72. Counter, C. M., Hahn, W. C., Wei, W., Caddle, S. D., Beijersbergen, R. L., Lansdorp, P. M., et al. (1998) Dissociation among in vitro telomerase activity, telomere maintenance, and cellular immortalization. *Proc. Natl. Acad. Sci. USA* **95,** 14,723–14,728.

73. Blasco, M. A., Funk, W., Villeponteau, B., and Greider, C. W. (1995) Functional characterization and developmental regulation of mouse telomerase RNA. *Science* **269,** 1267–1270.

74. Lansdorp, P. M., Poon, S., Chavez, E., Dragowska, V., Zijlmans, M., Bryan, T., et al. (1997) Telomeres in the haemopoietic system. *Ciba Found. Symp.* **211,** 209–218.

75. Dye, C. A., Lee, J. K., Atkinson, R. C., Brewster, R., Han, P. L., and Bellen, H. J. (1998) The Drosophila sanpodo gene controls sibling cell fate and encodes a tropomodulin homolog, an actin/tropomyosin-associated protein. *Development* **125,** 1845–1856.

76. Herrera, E., Samper, E., and Blasco, M. A. (1999) Telomere shortening in mTR-/- embryos is associated with failure to close the neural tube. *EMBO J* **18,** 1172–1181.

77. Morrison, S. J., Wandycz, A. M., Hemmati, H. D., Wright, D. E., and Weissman, I. L. (1997) Identification of a lineage of multipotent hematopoietic progenitors. *Development* **124,** 1929–1939.

78. Bornemann, A., Maier, F., and Kuschel, R. (1999) Satellite cells as players and targets in normal and diseased muscle. *Neuropediatrics* **30,** 167–175.

79. Morrison, S. J. and Weissman, I. L. (1994) The long-term repopulating subset of hematopoietic stem cells is deterministic and isolatable by phenotype. *Immunity* **1,** 661–673.

80. Schultz, E. and McCormick, K. M. (1994) Skeletal muscle satellite cells. *Rev. Physiol. Biochem. Pharmacol.* **123,** 213–257.

81. Doerner, P. (1998) Root development: quiescent center not so mute after all. *Curr. Biol.* **8,** R42–R44.

82. Potten, C. S. (1998) Stem cells in gastrointestinal epithelium: numbers, characteristics and death. *Philos. Trans. R. Soc. Lond. B Biol. Sci* **353,** 821–830.

83. Whetton, A. D. and Graham, G. J. (1999) Homing and mobilization in the stem cell niche. *Trends Cell Biol.* **9,** 233–238.

83a. Jacobson, M. (1991) *Developmental Neurobiology* 3rd ed., Plenum Press, NY.

84. Takahashi, T., Nowakowski, R. S., and Caviness, V. S. J. (1995) The cell cycle of the pseudostratified ventricular epithelium of the embryonic murine cerebral wall. *J. Neurosci.* **15,** 6046–6057.

85. Morshead, C. M., Craig, C. G., and van der Kooy, D. (1998) In vivo clonal analyses reveal the properties of endogenous neural stem cell proliferation in the adult mammalian forebrain. *Development* **125,** 2251–2261.

86. Garcia-Verdugo, J. M., Doetsch, F., Wichterle, H., Lim, D. A., and Alvarez-Buylla, A. (1998) Architecture and cell types of the adult subventricular zone: in search of the stem cells. *J. Neurobiol.* **36,** 234–248.

87. Doetsch, F., Caille, I., Lim, D. A., Garcia-Verdugo, J. M., and Alvarez-Buylla, A. (1999) Subventricular zone astrocytes are neural stem cells in the adult mammalian brain. *Cell* **97,** 703–716.

88. Weiss, S. and van der Kooy, D. (1998) CNS stem cells: where's the biology (a.k.a. beef)? *J. Neurobiol.* **36,** 307–314.

89. Knoblich, J. A., Jan, L. Y., and Jan, Y. N. (1995) Asymmetric segregation of *Numb* and *Prospero* during cell division. *Nature* **377,** 624–627.

90. Hirata, J., Nakagoshi, H., Nabeshima, Y., and Matsuzaki, F. (1995) Asymmetric segregation of the homeodomain protein Prospero during *Drosophila* development. *Nature* **377,** 627–630.

91. Doe, C. Q., Chu-LaGraff, Q., Wright, D. M., and Scott, M. P. (1991) The prospero gene specifies cell fates in the *Drosophila* central nervous system. *Cell* **65,** 451–464.

92. Li, L. and Vaessin, H. (2000) Pan-neural *prospero* terminates cell proliferation during *Drosophila* neurogenesis. *Genes Dev.* **14,** 147–151.

93. Buescher, M., Yeo, S. L., Udolph, G., Zavortink, M., Yang, X., Tear, G., and Chia, W. (1998) Binary sibling neuronal cell fate decisions in the *Drosophila* embryonic central nervous system are nonstochastic and require inscuteable-mediated asymmetry of ganglion mother cells. *Genes Dev.* **12,** 1858–1870.

94. Wai, P., Truong, B., and Bhat, K. M. (1999) Cell division genes promote asymmetric interaction between *Numb* and *Notch* in the *Drosophila* CNS. *Development* **126,** 2759–2770.

95. Schober, M., Schaefer, M., and Knoblich, J. A. (1999) Bazooka recruits Inscuteable to orient asymmetric cell divisions in *Drosophila* neuroblasts. *Nature* **402,** 548–551.

96. Wodarz, A., Ramrath, A., Kuchinke, U., and Knust, E. (1999) Bazooka provides an apical cue for *Inscuteable* localization in *Drosophila* neuroblasts. *Nature* **402,** 544–547.

97. Chia, W., Kraut, R., Li, P., Yang, X., and Zavortink, M. (1997) On the roles of inscuteable in asymmetric cell divisions in *Drosophila. Cold Spring Harbor Symp. Quant. Biol.* **62,** 79–87.

98. Kraut, R. and Campos-Ortega, J. A. (1996) *inscuteable*, a neural precursor gene of *Drosophila*, encodes a candidate for a cytoskeleton adaptor protein. *Dev. Biol.* **174,** 65–81.

99. Broadus, J., Fuerstenberg, S., and Doe, C. Q. (1998) Staufen-dependent localization of prospero mRNA contributes to neuroblast daughter-cell fate. *Nature* **391,** 792–795.

100. Schuldt, A. J., Adams, J. H., Davidson, C. M., Micklem, D. R., Haseloff, J., St Johnston, D., and Brand, A. H. (1998) *Miranda* mediates asymmetric protein and RNA localization in the developing nervous system. *Genes Dev.* **12,** 1847–1857.

101. Shen, C. P., Knoblich, J. A., Chan, Y. M., Jiang, M. M., Jan, L. Y., and Jan, Y. N. (1998) Miranda as a multidomain adapter linking apically localized Inscuteable and basally localized Staufen and Prospero during asymmetric cell division in Drosophila. *Genes Dev.* **12,** 1837–1846.

102. Ikeshima-Kataoka, H., Skeath, J. B., Nabeshima, Y., Doe, C. Q., and Matsuzaki, F. (1997) *Miranda* directs *Prospero* to a daughter cell during *Drosophila* asymmetric divisions. *Nature* **390,** 625–629.

103. Lu, B., Rothenberg, M., Jan, L. Y., and Jan, Y. N. (1998) Partner of *Numb* colocalizes with *Numb* during mitosis and directs *Numb* asymmetric localization in *Drosophila* neural and muscle progenitors. *Cell* **95,** 225–235.

104. Chenn, A. and McConnell, S. K. (1995) Cleavage orientation and the asymmetric inheritance of *Notch1* immunoreactivity in mammalian neurogenesis. *Cell* **82,** 631–641.

105. Doerner, P. (1998) Root development: quiescent center not so mute after all. *Curr. Biol.* **8,** R42–R44.

106. Kornack, D. R. and Rakic, P. (1995) Radial and horizontal deployment of clonally related cells in the primate neocortex: relationship to distinct mitotic lineages. *Neuron* **15,** 311–321.

107. Reid, C. B., Tavazoie, S. F., and Walsh, C. A. (1997) Clonal dispersion and evidence for asymmetric cell division in ferret cortex. *Development* **124,** 2441–2450.

108. Tan, S. S., Kalloniatis, M., Sturm, K., Tam, P. P., Reese, B. E., and Faulkner-Jones, B. (1998) Separate progenitors for radial and tangential cell dispersion during development of the cerebral neocortex. *Neuron* **21,** 295–304.

109. Ware, M. L., Tavazoie, S. F., Reid, C. B., and Walsh, C. A. (1999) Coexistence of widespread clones and large radial clones in early embryonic ferret cortex. *Cereb. Cortex* **9,** 636–645.

110. Qian, X., Goderie, S. K., Shen, Q., Stern, J. H., and Temple, S. (1998) Intrinsic programs of patterned cell lineages in isolated vertebrate CNS ventricular zone cells. *Development* **125,** 3143–3152.

111. Shen, Q., Qian, X., Capela, A., and Temple, S. (1998) Stem cells in the embryonic cerebral cortex: their role in histogenesis and patterning. *J. Neurobiol.* **36,** 162–174.

112. Zhong, W., Feder, J. N., Jiang, M. M., Jan, L. Y., and Jan, Y. N. (1996) Asymmetric localization of a mammalian *numb* homolog during mouse cortical neurogenesis. *Neuron* **17,** 43–53.

113. Seecof, R. L., Donady, J. J., and Teplitz, R. L. (1973) Differentiation of *Drosophila* neuroblasts to form ganglion-like clusters of neurons in vitro. *Cell Differ.* **2,** 143–149.

114. Huff, R., Furst, A., and Mahowald, A. P. (1989) *Drosophila* embryonic neuroblasts in culture: autonomous differentiation of specific neurotransmitters. *Dev. Biol* **134,** 146–157.

115. Whetton, A. D. and Graham, G. J. (1999) Homing and mobilization in the stem cell niche. *Trends Cell. Biol.* **9,** 233–238.

116. Fuchs, E. and Segre, J. A. (2000) Stem cells: a new lease on life. *Cell* **100,** 143–155.

117. Bjornson, C.R., Rietze, R. L., Reynolds, B. A., Magli, M. C., and Vescovi, A. L. (1999) Turning brain into blood: a hematopoietic fate adopted by adult neural stem cells in vivo. *Science* **283,** 534–537.

118. Ferrari, G., Cusella-De, A. G., Coletta, M., Paolucci, E., Stornaiuolo, A., Cossu, G., and Mavilio, F. (1998) Muscle regeneration by bone marrow-derived myogenic progenitors [published erratum appears in *Science* 1998; 281:923]. *Science* **279,** 1528–1530.

119. Jackson, K. A., Mi, T., and Goodell, M. A. (1999) Hematopoietic potential of stem cells isolated from murine skeletal muscle. *Proc. Natl. Acad. Sci. USA* **96,** 14482–14486.

120. Steindler, D. A., Kadrie, T., Fillmore, H., and Thomas, L. B. (1996) The subependymal zone: "brain marrow." *Prog. Brain Res.* **108,** 349–363.

121. Murphy, M., Reid, K., Dutton, R., Brooker, G., and Bartlett, P. F. (1997) Neural stem cells. *J. Invest. Dermatol. Symp. Proc.* **2,** 8–13.

122. Lillien, L. (1997) Neural development: instructions for neural diversity. *Curr. Biol.* **7,** R168–R171.

123. Cameron, H. A., Hazel, T. G., and McKay, R. D. (1998) Regulation of neurogenesis by growth factors and neurotransmitters. *J. Neurobiol.* **36,** 287–306.

124. Mehler, M. F. and Gokhan, S. (1999) Postnatal cerebral cortical multipotent progenitors: regulatory mechanisms and potential role in the development of novel neural regenerative strategies. *Brain Pathol.* **9,** 515–526.

125. Olsson, M., Bjerregaard, K., Winkler, C., Gates, M., Bjorklund, A., and Campbell, K. (1998) Incorporation of mouse neural progenitors transplanted into the rat embryonic forebrain is developmentally regulated and dependent on regional and adhesive properties. *Eur. J. Neurosci.* **10,** 71–85.

126. Gage, F. H. (2000) Stem cells for the brain and spinal cord, in *Keystone Symposium on Stem Cells, Asymmetric Cell Division and Cell Fate* (abstract).

127. Morrison, S. J., Uchida, N., and Weissman, I. L. (1995) The biology of hematopoietic stem cells. *Annu. Rev. Cell Dev. Biol.* **11,** 35–71.

128. Watt, F. M. (1998) Epidermal stem cells: markers, patterning and the control of stem cell fate. *Philos. Trans. R. Soc. Lond. B Biol. Sci* **353,** 831–837.

129. Milner, L. A., Kopan, R., Martin, D. I., and Bernstein, I.D. (1994) A human homologue of the *Drosophila* developmental gene, *Notch*, is expressed in CD34+ hematopoietic precursors. *Blood* **83,** 2057–2062.

130. Varnum-Finney, B., Purton, L. E., Yu, M., Brashem-Stein, C., Flowers, D., Staats, S., et al. (1998) The *Notch* ligand, *Jagged-1*, influences the development of primitive hematopoietic precursor cells. *Blood* **91,** 4084–4091.

131. Jones, P., May, G., Healy, L., Brown, J., Hoyne, G., Delassus, S., and Enver, T. (1998) Stromal expression of *Jagged 1* promotes colony formation by fetal hematopoietic progenitor cells. *Blood* **92,** 1505–1511.

132. Li, L., Milner, L. A., Deng, Y., Iwata, M., Banta, A., Graf, L., et al. (1998) The human homolog of rat *Jagged1* expressed by marrow stroma inhibits differentiation of 32D cells through interaction with *Notch1*. *Immunity* **8,** 43–55.

133. Walker, L., Lynch, M., Silverman, S., Fraser, J., Boulter, J., Weinmaster, G., and Gasson, J. C. (1999) The *Notch/Jagged* pathway inhibits proliferation of human hematopoietic progenitors in vitro. *Stem Cells* **17,** 162–171.

134. Lewis, J. (1998) Notch signalling and the control of cell fate choices in vertebrates. *Semin. Cell Dev. Biol.* **9,** 583–589.

135. Campos-Ortega, J. A. (1995) Genetic mechanisms of early neurogenesis in *Drosophila melanogaster. Mol. Neurobiol.* **10,** 75–89.

136. Skeath, J. B. and Doe, C. Q. (1998) *Sanpodo* and *Notch* act in opposition to Numb to distinguish sibling neuron fates in the *Drosophila* CNS. *Development* **125,** 1857–1865.

137. Lindsell, C. E., Boulter, J., diSibio, G., Gossler, A., and Weinmaster, G. (1996) Expression patterns of *Jagged, Delta1, Notch1, Notch2,* and *Notch3* genes identify ligand-receptor pairs that may function in neural development. *Mol. Cell. Neurosci.* **8,** 14–27.

138. Nye, J. S., Kopan, R., and Axel, R. (1994) An activated *Notch* suppresses neurogenesis and myogenesis but not gliogenesis in mammalian cells. *Development* **120,** 2421–2430.

139. Foltz, D. R., Berechid, B. E., Nunzi, M.-G., Mugnani, E., and Nye, J. S. (1997) Expression of *Notch1* and its proteolytic fragments on neuronal dendrites and perikarya. *Soc. Neurosci. Abstr.* **120,**3 (abstract).

140. Berezovska, O., Xia, M. Q., and Hyman, B. T. (1998) *Notch* is expressed in adult brain, is coexpressed with presenilin-1, and is altered in Alzheimer disease. *J. Neuropathol. Exp. Neurol.* **57,** 738–745.

141. Sestan, N., Artavanis-Tsakonas, S., and Rakic, P. (1999) Contact-dependent inhibition of cortical neurite growth mediated by *notch* signaling. *Science* **286,** 741–746.

142. Johansson, C. B., Momma, S., Clarke, D. L., Risling, M., Lendahl, U., and Frisen, J. (1999) Identification of a neural stem cell in the adult mammalian central nervous system. *Cell* **96,** 25–34.

143. Wang, S., Sdrulla, A. D., diSibio, G., Bush, G., Nofziger, D., Hicks, C., Weinmaster, G. and Barres, B. A. (1998) *Notch* receptor activation inhibits oligodendrocyte differentiation. *Neuron* **21,** 63–75.

144. Shipley, G. D., Keeble, W. W., Hendrickson, J. E., Coffey, R. J. J., and Pittelkow, M. R. (1989) Growth of normal human keratinocytes and fibroblasts in serum-free medium is stimulated by acidic and basic fibroblast growth factor. *J. Cell. Physiol.* **138,** 511–518.

145. Reddi, A. H. and Cunningham, N. S. (1990) Bone induction by osteogenin and bone morphogenetic proteins. *Biomaterials* **11,** 33–34.

146. Fuchs, E. and Byrne, C. (1994) The epidermis: rising to the surface. *Curr. Opin. Genet. Dev.* **4,** 725–736.

147. Donovan, P. J. (1994) Growth factor regulation of mouse primordial germ cell development. *Curr. Top. Dev. Biol.* **29,** 189–225.

148. Allouche, M. and Bikfalvi, A. (1995) The role of fibroblast growth factor-2 (FGF-2) in hematopoiesis. *Prog. Growth Factor Res.* **6,** 35–48.

149. McKay, R. (1997) Stem cells in the central nervous system. *Science* **276,** 66–71.

150. Burgess, A. W. (1998) Growth control mechanisms in normal and transformed intestinal cells. *Philos. Trans. R. Soc. Lond. B. Biol. Sci.* **353,** 903–909.

151. Murphy, M. S. (1998) Growth factors and the gastrointestinal tract. *Nutrition* **14,** 771–774.

152. Benfey, P. N. (1999) Stem cells: a tale of two kingdoms. *Curr. Biol.* **9,** R171–R172.

153. Cox, D. N., Chao, A., and Lin, H. (2000) *piwi* encodes a nucleoplasmic factor whose activity modulates the number and division rate of germline stem cells [In Process Citation]. *Development* **127,** 503–514.

154. Nakamura, M., Okano, H., Blendy, J. A., and Montell, C. (1994) Musashi, a neural RNA-binding protein required for *Drosophila* adult external sensory organ development. *Neuron* **13,** 67–81.

155. Good, P., Yoda, A., Sakakibara, S., Yamamoto, A., Imai, T., Sawa, H., et al. (1998) The human *Musashi* homolog 1 (MSI1) gene encoding the homologue of *Musashi/Nrp*-1, a neural RNA-binding protein putatively expressed in CNS stem cells and neural progenitor cells. *Genomics* **52,** 382–384.

156. Kaneko, Y., Sakakibara, S., Imai, T., Suzuki, A., Nakamura, Y., Sawamoto, K., et al. (2000) Musashi1: an evolutionarily conserved marker for CNS progenitor cells including neural stem cells. *Dev. Neurosci.* **22,** 139–153.

157. Lendahl, U., Zimmerman, L. B., and McKay, R. D. (1990) CNS stem cells express a new class of intermediate filament protein. *Cell* **60,** 585–595.

158. Doetsch, F., Caille, I., Lim, D. A., Garcia-Verdugo, J. M., and Alvarez-Buylla, A. (1999) Subventricular zone astrocytes are neural stem cells in the adult mammalian brain. *Cell* **97,** 703–716.

159. Morrison, S. J., White, P. M., Zock, C., and Anderson, D. J. (1999) Prospective identification, isolation by flow cytometry, and in vivo self-renewal of multipotent mammalian neural crest stem cells. *Cell* **96,** 737–749.

160. Shah, N. M. and Anderson, D. J. (1997) Integration of multiple instructive cues by neural crest stem cells reveals cell-intrinsic biases in relative growth factor responsiveness. *Proc. Natl. Acad. Sci. USA* **94,** 11,369–11,374.

161. Park, J. K., Williams, B. P., Alberta, J. A., and Stiles, C. D. (1999) Bipotent cortical progenitor cells process conflicting cues for neurons and glia in a hierarchical manner. *J. Neurosci.* **19,** 10,383–10,389.

2

Multipotent Stem Cells in the Embryonic Nervous System

John A. Kessler, Mark F. Mehler, and Peter C. Mabie

Neural stem cells are defined by a number of properties, including their ability to proliferate, to maintain themselves (self-renew), to retain multilineage potential over time, and to generate large numbers of progeny, often through transient amplification of intermediate progenitor pools. Although self-renewal can occur through symmetric cell divisions that generate two identical daughter cells, asymmetric cell divisions that generate a renewable stem cell and a more lineage-restricted daughter cell are a hallmark of stem cells in many organ systems. Cells that do not self-renew but that nevertheless proliferate and have the capacity to generate multiple phenotypes are often referred to as multipotential progenitor cells, but they will be included in a broad definition of stem cells for the purposes of this review. Other stem cell-derived precursor populations that are able to proliferate but that have more restricted lineage potential (e.g., glial restricted or neuronal restricted cells) are discussed elsewhere (see Chapters 5 and 6; *1*).

At present there are no generally accepted markers that allow the unambiguous identification of stem cells in the developing nervous system in vivo. Expression of the intermediate filament proteins nestin and/or vimentin coupled with the lack of expression of markers of more differentiated progeny is frequently used to identify putative neural stem/progenitor cells in culture. Neural stem cells arise from generative zones, derived from the inner lining of the neural tube, that extend from periventricular regions of the telencephalon to the spinal cord within the mammalian central nervous system (CNS) *(2)*. These zones initially consist of pseudostratified ventral epithelium (ventricular zone [VZ]) that gives rise during late embryonic life to secondary subventricular zones (SVZs) that persist in an attenuated form

From: *Stem Cells and CNS Development*
Edited by: M. S. Rao © Humana Press Inc., Totowa, NJ

into the adult state. Neurons and radial glia are generated predominantly within the early embryonic VZ, whereas oligodendrocytes and astrocytes are largely generated during perinatal and early postnatal periods within regional cortical SVZs.

Patterns of labeling and growth of putative stem cells within the early embryonic cerebral cortical VZ suggest that the earliest cell divisions are symmetric, with the elaboration of equivalent daughter cells *(3,4)*. This process presumably allows exponential expansion of the resident progenitor population. Later, progenitor cell development and migration involve asymmetric cell divisions that cause the elaboration of apical and basal daughter cells, with neuronal differentiation of the basal cell and subsequent migration of the basal neuroblast to regions of the developing cortical plate. Following early neuroblast migration, an additional wave of symmetric cell divisions is essential for coordinating cortical laminar organization. Similarly, in slice cultures of developing ferret brain, early cell divisions are oriented primarily in a plane vertical to the ventricular surface and generate two apparently similar daughter cells. By contrast, later cell divisions occur predominantly in a horizontal plane and generate two different daughter cells by asymmetric cell division *(5)*. Interestingly VZ stem cells in low-density culture undergo stereotyped patterns of both symmetric and asymmetric cell divisions, suggesting that patterns of division are governed at least in part by cell-intrinsic programming *(6)*.

Although the mechanisms underlying asymmetric division of vertebrate stem cells remain unclear, important clues have emerged from studies of *Drosophila* development. A number of proteins have been identified that show a polarized distribution during asymmetric division of neural precursors, including Inscuteable, Miranda, Prospero, Staufen, Bazooka, and Numb proteins. Asymmetric localization of these proteins is microfilament dependent and coordinated with positioning of the mitotic spindle, leading to unequal distribution to daughter cells during cell division. Numb is required for the asymmetric cell division of at least some neural lineages *(7)*. Vertebrate homologs of Numb have been identified that are asymmetrically localized in cells in the developing mouse cortex and that appear to participate in the process of asymmetric cell division *(8)*. Interactions of Numb with Notch provide at least one way of integrating cell intrinsic and extrinsic mechanisms in determining the asymmetric fate of daughter cells *(9,10)*.

REGULATION OF STEM CELL PROLIFERATION AND SURVIVAL

Proliferation and survival of stem cells are regulated by a variety of factors including members of the fibroblast growth factor (FGF) and epidermal growth factor (EGF) families *(11–14)*. Studies of stem cells in culture have provided insight into some of the mechanisms governing progenitor cell proliferation and survival. Embryonic and postnatal stem cells do not survive well in culture in the absence of added growth factors, but they survive and proliferate when cultured in the presence of mitogens such as basic (b)FGF or EGF. In the absence of a culture substratum that promotes cell adherence, stem cells proliferate as clonal aggregates of cells ("neurospheres") ranging in size from a few cells to hundreds of cells. Stem cells that proliferate as neurospheres in the presence of these mitogens remain largely undifferentiated as judged by continued expression of nestin and vimentin and lack of expression of markers of more mature progeny, but they generate both neurons and glia when replated onto a culture substratum and upon withdrawal of the mitogen *(12,15,16)*. This facilitates uses of "neurosphere assays" in which the percent of neurospheres and the percent of cells within a neurosphere that commit to specific phenotypes are determined. By contrast, primary cultures of stem cells proliferate as a monolayer of cells when plated onto an adherent substratum in the presence of bFGF or EGF. Clonal analyses of low-density cultures have allowed examination of the developmental potential of single mitotic stem cells and the effects of defined epigenetic signals in altering cell fate *(17–19)*.

Stem cells exhibit differing requirements for EGF and bFGF during neural development. The preponderance of evidence suggests that the survival and proliferation of early embryonic progenitor species are regulated by bFGF *(11,14,20)*. Although early embryonic stem cells generated under the influence of bFGF are multipotential, they appear to be predisposed toward neuronal differentiation. Early embryonic VZ progenitor cells do not express the EGF receptor (EGFR), and the cells do not respond to ligand *(21,22)*. However, the receptor is progressively expressed during development by later SVZ progenitors and EGF and/or transforming growth factor-α (TGF-α) (which is another ligand for the EGFR) then regulate cellular proliferation and survival *(21–23a)*. Most evidence suggests that EGF-responsive progenitor cells are the predominant species present during the period of perinatal gliogenesis. The EGF-responsive cells appear to be derived from FGF-responsive stem cells, although they display different kinetics of proliferation *(20,24)*. EGF-responsive progenitor cells retain the capacity to generate all cellular phenotypes, but they appear to be predisposed to differentiate into glia.

There is substantial evidence that FGF and EGF receptor activation regulate stem cell proliferation and survival in vivo. Mice lacking functional bFGF have reduced tissue mass and reduced numbers of both neurons and glia in the cerebral cortex *(25,26)*, and injection of neonates with neutralizing antibodies to bFGF reduces DNA synthesis in several areas of brain *(27)*. Conversely, injection of bFGF into the cerebral ventricles of rat embryos increases the volume of cerebral cortex and the number of neurons generated *(26)*, and subcutaneous administration of bFGF to neonatal rats increases neuroblast proliferation in regions still undergoing neurogenesis *(28)*. Finally, ligands of the FGF family including bFGF are expressed contiguous to generative zones in the developing brain in vivo from early embryogenesis into adulthood *(29,30)*. Similarly, targeted deletion of the EGF receptor leads to defects in cortical neurogenesis *(23)*, and deletion of functional TGF-α (which activates EGF receptors) leads to diminished proliferation of precursors in the SVZ of mature animals. Additional evidence involving injection of EGF receptor ligands into brains of mature rats supports a role for these ligands in stem cell proliferation in adults (see Chapter 3). Finally, TGF-α is expressed in the developing brain in vivo from E13 into adulthood *(22)*.

A number of other factors have been implicated in the control of stem cell proliferation. *Sonic hedgehog* (Shh) is a member of the *hedgehog* (hh) multigene family that encodes signaling proteins involved in induction and patterning processes in vertebrate and invertebrate embryos (for review, see ref. *31*). However, in addition to its effects on axial patterning and cellular differentiation, Shh appears to regulate stem cell proliferation. Ectopic overexpression of Shh in the mouse dorsal neural tube increases rates of proliferation of embryonic spinal cord progenitor cells *(32)*. Although Shh increases proliferation of cultured neural stem cells, unlike bFGF or EGF it does not enhance cell survival *(33)*. The factor is a potent mitogen for cultured retinal progenitor cells, cerebellar granule cell precursors *(34)*, neuronal restricted precursors in the spinal cord *(35)*, and skeletal muscle cells, and overexpression of the Shh gene leads to basal cell carcinoma (for review, see ref. *31*). To activate target genes, the N-terminal signaling domain of Shh (Shh-N) binds to the Shh-binding protein, Patched (Ptc), which is complexed with Smoothened (Smo), to counteract the inhibition by Ptc of constitutive signaling activity mediated by Smo (for review, see ref. *36*). Disruption of the gene encoding Ptc leads to meduloblastoma and other primitive neuroectodermal tumors. These cumulative observations suggest that Shh is an important regulator of neural stem cell proliferation.

Stem cell proliferation may also be influenced by other types of regulatory signals including the wnt pathway, glutamate, γ-aminobutyric acid (GABA), biogenic amines, opioid peptides, and other peptides such as vasoactive intestinal peptide (VIP) and pituitary adenylate cyclase-activating peptide (PACAP). For example, injection of pregnant mice from E9 to

E11 with a VIP antagonist reduces bromodeoxyuridine (BrdU) labeling in germinal zones in the developing embryonic brains and reduces the subsequent size of the ventricular and intermediate zones *(37)*. Disruption of wnt and wnt-3a leads to deficits in expansion of dorsal neural prognitor cells *(38)*. However the precise role of these regulatory influences on stem cell proliferation in vivo are less well-characterized than the effects of bFGF, EGF, and Shh.

Stem cell proliferation also appears to be regulated by factors that actively promote exit of the cells from cell cycle. For example, treatment of cultured stem cells with members of the bone morphogenetic protein (BMP) family of factors promotes rapid exit of the cells from cell cycle, even in the presence of mitogens such as bFGF, EGF, or Shh *(33,39)*. Proliferation of cortical precursor cells cultured in the presence of bFGF is diminished by cotreatment with either neurotrophin 3 (NT3) *(40)* or GABA *(41)*. Glutamate treatment of embryonic cortical explants significantly reduces proliferation of putative progenitor cells *(42)*, and treatment with PACAP decreases proliferation of embryonic cortical precursors *(43)*. In most instances exit from cell cycle induced by these factors is associated with enhanced differentiation of the surviving cells.

Stem cell numbers appear to be regulated primarily by rates of proliferation that in turn reflect cell cycle duration, the length of time during which exponential expansion of cell numbers occurs, and the ratio of asymmetric to symmetric cell divisions *(4)*. However, there is now increasing recognition that stem cell numbers may also be influenced by apoptotic cell death within proliferative periventricular generative zones. Cell death is rare during early embryogenesis (E10), peaks during the E14–15 period, and begins to recede in late embryonic life *(44)* Targeted disruption of either caspase-9 or caspase-3 leads to decreased programmed cell death of cortical precursors, causing expansion and exencephaly of the forebrain as well as supernumerary neurons in the cerebral cortex *(45)*. By contrast, disruption of the c-Jun N-kinase signaling pathway leads to precocious degeneration of cerebral precursors *(45)*. Selective survival of different populations of progenitor cells may be important not only for regulation of cell numbers and specification of cellular phenotypes, but also for the morphogenesis of the brain. For example, the BMPs induce apoptosis of selected rhombencephalic neural crest-associated progenitor species, resulting in segmentation of the rhombencephalon *(46)*. Furthermore, the BMPs also induce apoptosis of VZ stem cells in culture *(19)* and of forebrain precursor cells in explants *(47)*. Stem cell numbers in different regions of the neuraxis thus appear to reflect regional rates of survival as well as proliferation.

DEVELOPMENTAL CHANGES IN STEM CELLS

Although stem cells retain the ability to generate neurons, oligodendroglia, and astrocytes throughout the embryonic and postnatal periods (and even in the adult — see Chapter 3), there are clearly developmental changes in both their bias toward differentiation into specific cell types and their responses to epigenetic signals. Early VZ stem cells in culture are predisposed to become neurons and to a lesser extent oligodendroglia, whereas SVZ stem cells are biased toward astrocytic differentiation. As noted previously, EGFRs are not expressed by VZ stem cells but are expressed by the cells in the SVZ. Although neural stem cells thus become progressively more biased toward a glial fate during development coincident with an increase in expression of EGFRs, the role of EGF signaling in this change is unclear. Retroviral introduction of extra EGFRs into VZ progenitor cells results in premature expression of traits characteristic of later SVZ progenitors including the bias toward astrocytic differentiation *(21)*. This suggests that developmental increases in levels of EGFRs expressed by progenitor cells may mediate changes in their responses to environmental signals and their tendency to differentiate into astrocytes. However, similar experiments involving introduction of extra copies of the EGFR into early embryonic retinal progenitor cells does not bias the cells toward a glial fate *(48)*. More importantly, pharmacologic blockade of EGFR signaling does not alter the developmentally increased bias of cultured progenitor cells to undergo astrocytic differentiation *(48a)*, suggesting that the competence to generate glia is temporally regulated by other mechanisms. There are also striking differences between VZ and SVZ stem cell responses to differentiating signals such as the BMPs *(49)* or leukemia inhibitory factor (LIF) *(50)*, so the same signals may induce different phenotypes at different developmental stages. Thus analysis of the factors regulating stem cell differentiation requires knowledge of the developmental stage and history of the cell.

MAINTENANCE OF STEM CELL FATE

The stem cell phenotype is maintained by both daughter cells during the period of symmetric cell divisions and rapid expansion of the stem cell pool in early embryos, and it is maintained by one daughter cell of each pair during later asymmetric cell divisions. Although the mechanisms underlying the maintenance of the stem cell phenotype in vertebrates are not yet well-described, it has become increasingly evident that there are active cell intrinsic as well as extrinsic mechanisms that inhibit lineage commitment by these cells. The most intensively studied example of such inhibitory signaling involves the Notch pathway. Notch and its ligands Delta and Serrate are integral membrane proteins that generally transmit signals only between cells in direct contact. Overexpression of Delta1 (i.e., activation of Notch) suppresses neurogenesis, whereas overexpression of a dominant negative inhibitor of Delta1 leads to premature commitment of stem cells to the neuronal fate *(51)*. Activation of Notch also regulates transcriptional activity, including inhibition of production of Notch ligands by that cell. Through a process termed lateral inhibition, cells that produce ligand force neighboring cells to produce less ligand, thereby enabling the ligand-producing cells to increase production even further. The effect of such a feedback loop is to amplify small differences between neighboring cells and to drive them into different developmental pathways. Delta1 is expressed by a scattered subset of cells (nascent neurons; *52*) in the outer part of the VZ zone, whereas Notch1 is expressed throughout the VZ *(53)*. Delta production by daughter cells undergoing neuronal differentiation activates Notch in their dividing partners, thereby inhibiting neuronal differentiation and maintaining a cohort of stem cells so that neurogenesis can continue. Notch1 signaling also inhibits differentiation into alternative fates such as oligodendroglial differentiation *(54)*. More recent studies suggest that notch signaling promotes the generation of radial glia *(54a)*.

Stem cells express a number of other proteins whose function appears to be related to the maintenance of the undifferentiated state. For example, HES1 *(55,56)* was originally isolated as a mammalian homolog of hairy and Enhancer of Split, which negatively regulate neurogenesis in *Drosophila*. HES1 negatively regulates transcriptional activation mediated by basic helix-loop-helix (bHLH) genes and normally functions to repress the commitment of multipotent progenitor cells to the neuronal lineage, thereby maintaining their self-renewing state *(57)*. Overexpression of Hes1 prevents both migration of neural stem cells out of the VZ and expression of neuronal markers *(58)*, whereas HES1-deficient brains prematurely express neurofilaments *(59)*. These observations suggest that the gene is required for the negative regulation of neuronal differentiation.

Stem cell fate may also be maintained by the four members of the ID (*i*nhibitor of *D*NA binding and *i*nhibitor of cell *d*ifferentiation) family of proteins that resemble bHLH factors but that lack a basic region necessary for DNA binding. The ID proteins act as dominant negative inhibitors by preferentially dimerizing with a subset of bHLH factors to form inactive complexes, thereby decreasing bHLH-mediated transcriptional activity (for review, see ref. *60*). Members of the ID family are expressed throughout the nervous system during neurogenesis with localization of ID transcripts within putative neural stem cells *(61,62)*. Targeted disruption of both ID1 and ID3 in the same animals results in premature withdrawal of neuroblasts from cell cycle and expression of neuron-specific differentiation markers

(63). These observations suggest that expression of ID proteins is necessary to maintain stem cells in the undifferentiated, proliferative state.

TRANSCRIPTION FACTORS THAT PROMOTE LINEAGE COMMITMENT BY STEM CELLS

The observation that maintenance of the stem cell phenotype requires inhibition of bHLH factors by ID proteins and HES1 suggests that bHLH factors are involved in directing stem cell differentiation. There is, in fact, a large body of evidence that regulatory cascades of bHLH and other transcription factors play essential roles in mammalian neurogenesis. Detailed discussions of the relationship of neurogenesis to neural induction and of the genes involved in neurogenesis are beyond the scope of this chapter but are available in a number of recent reviews *(64,65)*.

Briefly, bHLH neurogenic factors are thought to regulate neuronal development positively at the level of both commitment and of postmitotic differentiation. Overexpression of bHLH genes such as *Mash1, neurogenin (Ngn)*, or *neuroD* induces ectopic expression of neurons, whereas targeted deletion leads to deficits in the generation of neurons (for review, see refs. *66* and *67*). These factors act through stereotyped cascades; for example, *Ngn* expression precedes that of *neuroD*, and *Ngn* activates *neuroD* but not vice versa, suggesting that *Ngn* acts upstream of *neuroD* in neuron production.

The cascades may involve other types of transcription factors and homeobox genes. For example, the zinc finger transcription factor MyT1 is involved in neurogenesis, and blocking MyT1 function inhibits ectopic neurogenesis induced by *Ngn (68)*. This suggests that MyT1 activation is part of the cascade initiated by *Ngn*. Similarly, *Mash1* regulation of noradrenergic neuron differentiation depends in part on the homebox gene *Phox2a*, and targeted deletion of *Phox2a* abolishes the locus coeruleus, the major noradrenergic center in the brain *(69,70)*.

In summary, neuronal lineage commitment and progressive neuronal differentiation involve the coordinated interplay of positive and negative regulatory signals, including cascades of transcription factors that regulate lineage-specific gene expression. Furthermore, multiple signal cascades are involved in the generation of different populations of neurons, and activation of these cascades reflects the effects of both cell intrinsic and extrinsic factors that promote cell differentiation.

FACTORS REGULATING LINEAGE COMMITMENT BY STEM CELLS

Lineage commitment by stem cells results from the confluence of the effects of cell intrinsic and extrinsic signals, and there is substantial overlap among the factors involved in proliferation and survival and those that regulate lineage commitment and cellular differentiation. For example, in addition to effects on survival and proliferation, bFGF and other members of the FGF family influence lineage commitment by embryonic neural stem cells. Withdrawal of bFGF from stem cells in vitro promotes generation of neurons and glia, suggesting that the factor represses intrinsic programs of stem cell differentiation. Furthermore, exposure of stem cells to bFGF alters their subsequent developmental bias. Treatment of cultured stem cells with bFGF promotes expression of the EGF receptor *(20,71)* and enhances expression of differentiated traits such as the catecholamine biosynthetic enzyme tyrosine hydroxylase *(72)*. Furthermore, the concentrations of bFGF to which embryonic stem cells are exposed in vitro influences cell fate; low concentrations of bFGF favor neuronal differentiation, whereas higher threshold concentrations favor oligodendroglial differentiation *(14)*.

This may reflect preferential activation of different subtypes of FGF receptors by different concentrations of the factor, a conclusion supported by observations of the differential effects of other FGF family members. For example, treatment of cultured stem cells with FGF-1 in the presence of heparan sulfate proteoglycan preferentially promotes neuronal differentiation, whereas bFGF (FGF-2) treatment of sister cultures preferentially promotes proliferation *(73)*. FGF-8 collaborates with Shh to induce dopaminergic neurons in the mid/hindbrain, whereas FGF-4 in association with Shh induces a serotonergic cell fate *(74)*.

Shh also appears to be involved in the induction of neuronal phenotypes in the brain (for review, see ref. *31*) and in the induction of oligodendrocyte lineage commitment in the spinal cord *(75,76)*. Shh treatment of cultured neural stem cells promotes the elaboration of both neuronal and oligodendroglial lineage species *(33)*, suggesting that its differentiating effects reflect direct actions on neural stem cells. Furthermore, neural stem cells express smoothened *(33)*, the signaling component of the Shh receptor, and constitutively active forms of smoothened reproduce inductive effects of Shh *(77)*, suggesting that Shh exerts its inductive effects directly on stem cells. However, the final phenotype of cells induced by Shh depends on other inductive signals and on other genes expressed by progenitor cell populations. For example, mutation of the homeobox gene *Nkx2.2* in progenitor cells alters the inductive effects of Shh in specifying the neuronal identity (motor neuron vs interneuron) of the progeny *(78)*. As noted above, interactions between the effects of Shh and other growth factors including wnt, FGF-4, and FGF-8, are critical for specifying alternate cellular phenotypes in the brain *(74)* and for patterning of the dorsal compartment of the somite. Interactions between Shh and members of the BMP gene family are important for the specification of dorsal and intermediate dorsoventral cell types (for reviews, see *31* and *79*), and Shh inhibits BMP signaling, in part by inducing the endogenous BMP inhibitor noggin *(80)*. Shh and BMP signaling exert directly opposing effects on both proliferation and differentiation of cultured neural stem cells *(33)*.

Neuronal differentiation of stem cells is thus regulated by a diversity of factors including the Notch/delta pathway, NUMB family members *(10,81)*, FGF family members, Shh, BMP family members *(19,82,83)*, wnt family members *(84)*, retinoid-activated pathways *(85,86)*, and other signaling molecules (for review, see ref. *87*). The existence of so many pathways for neuronal lineage commitment and differentiation presumably reflects the diversity of neuronal phenotypes that must be generated. There is clearly diversity among stem/progenitor cell populations even at early developmental stages (for review, see refs. *11* and *88*), and there are developmental changes in stem cells that lead to markedly different cell fate decisions in response to the same factors at different developmental stages *(49)*. Commitment and differentiation of stem cells to specific neuronal lineages thus reflect complex patterns of events and parallel pathways for neurogenesis.

Just as there are multiple pathways of neuronal differentiation, there are several different pathways leading to astrocytic lineage commitment by neural stem cells. The peak period of gliogenesis occurs during late embryonic and early perinatal cerebral cortical development, and SVZ stem cells are biased toward astrocytic differentiation compared with VZ stem cells. Treatment with the BMPs promotes the elaboration of mature astrocytes from late embryonic SVZ-derived stem cells in culture as well as from early postnatal cerebral cortical multipotent and bipotent oligodendroglial-type 2 astroglial (O-2A) progenitor cells *(39,89)*. Ciliary neurotrophic factor (CNTF) and LIF also potentiate the generation of astrocytes from embryonic neural stem cells; genetic and developmental analyses confirm that a CNTF/LIF subgroup of factors that interacts with gp130/LIF-β receptors participates in astrogliogenesis *(90,91)*. However, BMP-2 treatment of progenitor cells cultured from animals that are deficient in the LIF-β receptor induces astrocytic lineage commitment, indicating that astrocytic differentiation does not require signaling through gp130/LIFRs *(91)*. CNTF and LIF signal through the JAK/STAT signaling pathway, whereas the BMPs signal through Smad-mediated pathways. Formation of a complex between STAT3 and Smad1, bridged by the transcriptional coactivator p300, may mediate cooperative effects of these two classes of factors on stem cell commitment to the astrocytic lineage *(92)*.

In view of the foregoing observations regarding multiple pathways of neuronal and astrocytic lineage commitment, it is not surprising that oligodendroglia (OLs) also appear to be generated from multiple lineages in response to a number of different epigenetic signals (for review, see ref. 93). During embryonic development in the spinal cord, the expression of Jagged, a Notch ligand, coincides with the elaboration of foci of OL precursors from paramedian generative zones. Shh, a notochord-derived signal, supports the generation of mature OL lineage species from caudal regions of the neuraxis (spinal cord), but its role as an obligate developmental signal for more anterior regions of the central nervous system is unclear. Oligodendroglia are first generated in the embryonic spinal cord in response to signals derived from the floor plate and notocord. Treatment of spinal cord explants with Shh induces both OLs and neurons *(76)*, and antibodies that neutralize Shh prevent OL lineage commitment *(75)*. Shh treatment of cultured embryonic stem cells derived from neurospheres also induces both oligodendroglial and neuronal differentiation *(33)*, suggesting that OL lineage commitment reflects direct effects of Shh on stem cells. However, other factors produced by neurons influence this process, and accumulating evidence suggests that members of the neuregulin family may be involved *(94)*. Other factors are capable of promoting OL lineage commitment by cultured neural stem cells; for example, increased concentrations of bFGF or brief exposure to thyroid hormone foster OL differentiation *(13,14)*. The regulation of later stages of OL differentiation from glial restricted precursors is described elsewhere (Chapter 6).

PATTERNS OF STEM CELL DIFFERENTIATION IN VIVO

Tracking the fate of stem cell progeny in vivo became possible after development of techniques for labeling individual VZ cells with replication-defective retroviral vectors that label all daughter cells with an inheritable marker such as β-galactosidase. In early experiments, injection of retroviruses into E14 murine lateral ventricles gave rise to small, scattered clones that largely consisted of a single cell type, neuron, oligodendrocyte, or astrocyte. Mixed clones of neurons and glia were uncommon (<1% of clones), and most neuronal clones were of a single cell type, pyramidal or nonpyramidal *(95–98)*. This gave rise to a concept that the VZ consisted largely of different progenitor cell types committed to specific lineages. However, there was a surprisingly large degree of scattering of cells, suggesting that unexpectedly large amounts of cell movement in vivo might be blurring clonal boundaries. Walsh and Cepko *(99)* addressed this problem by using a library of heterogeneous retroviral vectors with numerous genetic tags. Any daughter cells containing precisely the same mixture of tags were presumed to arise from the same progenitor, even if the progeny were scattered widely. Injection into E15–17 rat embryos generated some spread clones of clusters of different cell types and an equivalent number of smaller clones of a single cell type *(99,100)*. This is the pattern that would be expected if the retrovirus infected asymmetrically dividing cell pairs since one cell would display the multipotent stem cell phenotype, and the other would display the phenotype of a committed cell. In turn, this suggests that most cells generated by the VZ during this time period arise from multipotent progenitor cells undergoing active asymmetric division, a conclusion consistent with studies of stem cells in culture *(101,102)*. However, the precise proportions of stem cells, uncommitted progenitors, and committed progenitors at differing developmental stages and in different regions of the generative zones are unknown.

REFERENCES

1. Rao, M. S. (1999) Multipotent and restricted precursors in the central nervous system. *Anat. Rec.* **257**, 137–148.
2. Rakic, P. (1995) Radial versus tangential migration of neuronal clones in the developing cerebral cortex. *Proc. Nat. Acad. Sci. USA* **92**, 11,323–11,327.
3. Rakic, P. (1995) A small step for the cell, a giant leap for mankind: a hypothesis of neocortical expansion during evolution. *Trends Neurosci.* **8**, 383–388.
4. Caviness, V. S. Jr., and Takahashi, T. (1995) Proliferative events in the cerebral ventricular zone. *Brain Dev.* **17**, 159–163.
5. Chenn, A. and McConnell, S. K. (1995) Cleavage orientation and the asymmetric inheritance of Notch 1 immunoreactivity in mammalian neurogenesis. *Cell* **82**, 631–641.
6. Qian, X., Goderie, S., Shen, Q., Stern, J., and Temple, S. (1998) Intrinsic programs of patterned cell lineage in isolated vertebrate CNS ventricular zone cells. *Development* **125**, 3143–3152.
7. Spana, E. P., Kopczynski, C., Goodman, C. S., and Doe, C. Q. (1995) Asymmetric localization of Numb autonomously determines sibling neuron identity in *Drosophila* CNS *Development* **121**, 3489–3494.
8. Jan, Y. N. and Jan, L. Y. (1999) Asymmetry across species *Nature Cell Biol.* **1**, E42–44.
9. Jan, Y. N. and Jan, L. Y. (1998) Asymmetric cell division. *Nature* **392**, 775–778.
10. Wakamatsu, Y., Maynard, T. M., Jones, S. U., and Weston, J. A. (1999) NUMB localizes in the basal cortex of mitotic avian neuroepithelial cells and modulates neuronal differentiation by binding to NOTCH-1 *Neuron* **23**, 71–81.
11. Kilpatrick, T. J. and Bartlett, P. F. (1995) Cloned multipotential precursors from the mouse cerebrum require FGF-2, whereas glial restricted precursors are stimulated with either FGF-2 or EGF. *J. Neurosci.* **15**, 3653–3661.

12. Weiss, S., Reynolds, B. A., Vescovi, A., Morshead, C., Craig, C. G., and van der Kooy, D. (1996) Is there a neural stem cell in the mammalian forebrain? *Trends Neurosci.* **19**, 387–393.

13. Johe, K. K., Hazel, T. G., Muller, T., Dugich-Djordjevic, M., and McKay, R. (1996) Single factors direct the differentiation of stem cells from the fetal and adult central nervous system. *Genes Dev.* **10**, 3129–3140.

14. Qian, X., Davis, A. D., Goderie, S. K., and Temple, S. (1997) FGF2 concentration regulates the generation of neurons and glia from multipotent cortical stem cells. *Neuron* **18**, 81–93.

15. Reynolds, B. A., Tetzlaff, W., and Weiss, S. (1992) A multipotent EGF-responsive striatal embryonic progenitor cell produces neurons and astrocytes. *J. Neurosci.* **12**, 4565–4574.

16. Vescovi, A. L., Reynolds, B. A., Fraser, D. D., and Weiss, S. (1993) bFGF regulates the proliferative fate of unipotent (neuronal) and bipotent (neuronal/astroglial) EGF-generated CNS progenitor cells. *Neuron* **11**, 951–966.

17. Temple, S. (1989) Division and differentiation of isolated CNS blast cells in microculture *Nature* **340**, 471–473.

18. Sakakibara, S., Imai, T., Hamaguchi, K., Okabe, M., Aruga, J., Nakajima, K., et al. (1996) Mouse-Musashi-1, a neural RNA-binding protein highly enriched in the mammalian CNS stem cell. *Dev. Biol.* **176**, 230–242.

19. Mabie, P. C., Mehler, M. F., and Kessler, J. A. (1999) Multiple roles of bone morphogenetic protein signaling in the regulation of cortical cell number and phenotype. *J. Neurosci.* **19**, 7077–7088.

20. Ciccolini, F. and Svendsen, C. N. (1998) Fibroblast growth factor 2 (FGF-2) promotes acquisition of epidermal growth factor (EGF) responsiveness in mouse striatal precursor cells: identification of neural precursors responding to both EGF and FGF-2. *J. Neurosci.* **18**, 7869–7880.

21. Burrows, R. C., Wancio, D., Levitt, P., and Lillien, L. (1997) Response diversity and the timing of progenitor cell maturation are regulated by developmental changes in EGFR expression in the cortex. *Neuron* **19**, 251–267.

22. Kornblum, H. I., Hussain, R. J., Bronstein, J. M., Gall, C. M., Lee, D. C., and Seroogy, K. B. (1997) Prenatal ontogeny of the epidermal growth factor receptor and it ligand, transforming growth factor alpha, in the rat brain *J. Comp. Neurol* **380**, 243–261.

23. Threadgil, D. W., Flugosz, A. A., Hansen, A. A., Tennenbaum, T., Lichti, U., Yee, D., et al. (1995) Targeted disruption of mouse EGF receptor: effect of genetic background on mutant phenotype. *Science* **269**, 230–234.

23a. Kornblum, H. I., Hussain, R., Wiesen, J., Miettinen, P., Zurcher, S. D., Chow, K., Derynck, R., and Werb, Z. (1998) Abnormal astrocyte development and neuronal death in mice lacking the epidermal growth factor receptor *J Neurosci Res* **53**, 697–717.

24. Martens, D. J., Tropepe, V., and van der Kooy, D. (2000) Separate proliferation kinetics of fibroblast growth factor-responsive and epidermal growth factor-responsive neural stem cells within the embryonic forebraion germinal zone. *J. Neurosci.* **20**, 1085–1095.

25. Ortega, S., Ittmann, M., Tsang, S. H., Ehrlich, M., and Basilico, C. (1998) Neuronal defects and delayed wound healing in mice lacking fibroblast growth factor 2. *Proc. Natl. Acad. Sci. USA* **95,** 5672–5677.

26. Vaccarino, F. M., Schwartz, M. L., Raballo, R., Nilsen, J., Rhee, J., Zhou, M., et al. (1999) Changes in cerebral cortex size are governed by fibroblast growth factor during embryogenesis. *Nature Neurosci.* **2,** 246–253.

27. Tao, Y., Black, I. B., and DiCicco-Bloom, E. (1997) *In vivo* neurogenesis is prevented by neutralizing antibodies to basic fibroblast growth factor. *J. Neurobiol.* **33,** 289–296.

28. Tao, Y., Black, I. B., and DiCicco-Bloom, E. (1996) neurogenesis in neonatal rat brain is regulated by peripheral injection of basic fibroblast growth factor (bFGF). *J. Comp. Neurol.* **376,** 653–663.

29. Emoto, N., Gonzales, A. M., Walicke, P. A., Wada, E., Simmons, D. M., Shimasaki, S., and Baird, A. (1989) Basic fibroblast growth factor in the central nervous system: identification of specific loci of basic FGF expression in the rat brain. *Growth Factors* **2,** 21–29.

30. Nurcombe, V., Ford, M. D., Wildchut, J. A., and Bartlett, P. F. (1993) Developmental regulation of neural response to FGF-1 and FGF-2 by heparan sulfate proteoglycan. *Science* **260,** 103–106.

31. Goodrich, L. V. and Scott, M. P. (1998) Hedgehog and patched in neural development and disease. *Neuron* **21,** 1243–1257.

32. Rowitch, D. H., S-Jacques, B., Lee, S. M., Flax, J. D., Snyder, E. Y., and McMahon, A. P. (1999) Sonic hedgehog regulates proliferation and inhibits differentiation of CNS precursor cells. *J. Neurosci.* **19,** 8954–8965.

33. Zhu, G., Mehler, M. F., Zhao, J., Yu Yung, S., and Kessler, J. A. (1999) Sonic hedgehog and BMP2 exert opposing actions on proliferation and differentiation of embryonic neural progenitor cells. *Dev. Biol.* **215,** 118–129.

34. Wechsler-Reya, R. J. and Scott, M. P. (1999) Control of neuronal precursor proliferation in the cerebellum by sonic Hedgehog. *Neuron* **22,** 103–114.

35. Kalyani, A. J., Piper, D., Mujtaba, T., Lucero, M. T., and Rao, M. S. (1998) Spinal cord neuronal precursors generate multiple neuronal phenotypes in culture. *J. Neurosci.* **18,** 7856–7868.

36. McMahon, A. P. (2000) More surprises in the hedgehog signaling pathway. *Cell* **100,** 185–188.

37. Gressens, P., Hill, J. M., Paindaveine, B., Gozes, I., Fridkin, M., and Brenneman, D. E. (1994) Severe microcephaly induced by blockade of vasoactive intestinal peptide function in the primitive neuroepithelium of the mouse. *J. Clin. Invest.* **94,** 2020–2027.

38. Ikeya, M., Lee, S. M., Johnson, S. E., McMahon, A. P., and Takada, S. (1997) Wnt signaling required for expansion of neural crest and CNS progenitors. *Nature* **389,** 966–970.

39. Gross, R. E., Mehler, M. F., Mabie, P. C., Zang, Z., Santschi, L., and Kessler, J. A. (1996) Bone morphogenetic proteins promote astroglial lineage commitment by mammalian subventricular zone progenitor cells. *Neuron* **17,** 595–606.

40. Ghosh, A. and Greenberg, M. E. (1995) Distinct role for bFGF and NT3 in the regulation of cortical neurogenesis. *Neuron* **15,** 249–252.

41. Antonopoulos, J., Pappas, I. S., and Parnavelas, J. G. (1997) Activation of the GABAA receptor inhibits the proliferative effects of bFGF in cortical progenitor cells. *Eur. J. Neurosci.* **9**, 291–298.
42. LoTurco, J. J., Owens, D. F., Heath, M. J. S., Davis, M., and Kriegstein, A. R. (1995) GABA and glutamate depolarize cortical progenitor cells and inhibit DNA synthesis. *Neuron* **15**, 1287–1298.
43. Lu, N. and DiCicco-Bloom, E. (1997) Pituitary adenylate cyclase-activating polypeptide is an autocrine inhibitor of mitosis in cultured cortical precursor cells. *Proc. Natl. Acad. Sci USA* **94**, 3357–3362.
44. Blaschke, A. J., Staley, K., and Chun, J. (1996) Widespread programmed cell death in proliferative and postmitotic regions of the fetal cerebral cortex. *Development* **122**, 1165–1174.
45. Haydar, T. F., Kuan, C.-Y., Flavell, R. A., and Rakic, P. (1999) The role of cell death in regulating the size and shape of the mammalian forebrain. *Cerebral Cortex* **9**, 621–626.
46. Graham, A. and Lumsden, A. (1996) Patterning the cranial neural crest. *Biochem. Soc. Symp.* **62**, 77–83.
47. Furuta, Y., Piston, D. W., and Hogan, B. L. (1997) Bone morphogenetic proteins (BMPs) as regulators of dorsal forebrain development. *Development* **124**, 2203–2212.
48. Lillien, L. and Wancio, D. (1998) Changes in epidermal growth factor receptor expression and competence to generate glia regulate timing and choice of differentiation in the retina.
48a. Zhu, G., Mehler, M. F., Mabie, P. C., and Kessler, J. A. (2000) Developmental changes in neural progenitor cell lineage commitment do not depend on epidermal growth factor receptor signaling. *J. Neurosci. Res.* **59**, 312–320.
49. Zhu, G., Mehler, M. F., Mabie, P. C., and Kessler, J. A. (1999) Developmental changes in progenitor cells responsiveness to cytokines *J. Neurosci. Res.* **56**, 131–145.
50. Molne, M., Studer, L., Tabar, L., Ting, Y., Eiden, M., and McKay, R. (2000) Early cortical precursors do not undergo LIF-mediated astrocytic differentiation. *J. Neurosci. Res.* **59**, 301–311.
51. Austin, C. P., Feldman, D. E., Ida, J. A., and Cepko, C. L. (1995) Vertebrate retinal ganglion cells are selected from competent progenitors by the action of Notch. *Development* **121**, 3637–3650.
52. Henrique, D., Adam, J., Myat, A., Chitnis, A., Lewis, J., and Ish-Horowicz, D. (1995) Expression of a Delta homologue in prospective neurons in the chick. *Nature* **375**, 787–790.
53. Myat, A., Henrique, D., Ish-Horowicz, D., and Lewis, J. (1996) A chick homologue of Serrate and its relationship with Notch and Delta homologues during central neurogenesis. *Dev. Biol.* **174**, 233–247.
54. Wang, S., Sdrulla, A. D., diSibio, G., Bush, G., Nofziger, D., Hicks, C., Weinmaster, G., and Barres, B. A. (1998) Notch receptor activation inhibits oligodendroglial differentiation. *Neuron* **21**, 63–75.
54a. Galano, N., Nye, J. S., and Fishell, G. (2000) Radial glial identity is promoted by Notch1 signaling in the murine forebrain. *Neuron* **26**, 395–404.

55. Akazawa, C., Sasai, Y., Nakanishi, S., and Kageyama, R. (1992) Molecular characterization of a rat negative regulator with a basic helix-loop-helix structure predominately expressed in the developing system. *J. Biol. Chem.* **267,** 21,879–21,885.

56. Sasai, Y., Kageyama, R., Tagawa, Y., Shigemoto, R., and Nakanishi, S. (1992) Two mammolian helix-loop-helix factors structurally related to *Drosophilia* hairy and enhancor of split. *Gene Dev.* **6,** 2620–2634.

57. Nakamura, Y., Sakakibara, S., Miyata, T., Ogawa, M., Shimazaki, T., Weiss, S., Kageyama, R., and Okana, H. (2000) The bHLH gene hes1 as a repressor of the neuronal commitment of CNS stem cells. *J. Neurosci.* **20,** 283–293.

58. Ishibashi, M., Moriyoshi, K., Sasai, Y., Shiota, K., Nakanishi, S., and Kageyama, R. (1994) Persistent expression of helix-loop-helix factor HES-1 prevents mammalian neural differentiation in the central nervous system. *EMBO J.* **13,** 1799–1805.

59. Ishibashi, M., Siew-Lan, A., Shiota, K., Nakanishi, S., Kagyeyama, R., and Guillemot, F. (1995) Targeted disruption of mammalian hairy and Enhancer of split homolog-1 (HES-1) leads to upregulation of neural helix-loop-helix factors, premature neurogenesis, and severe neural tube defects *Genes Dev.* **9,** 3136–3148.

60. Norton, J. D., Deed, R. W., Craggs, G., and Sablitzky, F. (1998) ID helix-loop-helix proteins in cell growth and differentiation. *Trends Cell Biol.* **8,** 58–65.

61. Jen, Y., Manova, K., and Benezra, R. (1997) Each member of the ID gene family exhibits a unique expression pattern in mouse gastrulation and neurogenesis. *Dev. Dyn.* **208,** 92–106.

62. Riechmann, V. and Sablitzky, F. (1995) Mutually exclusive expression of two dominant-negative helix-loop-helix (dnHLH) genes, ID4 and ID3, in the developing brain of the mouse suggests distinct regulatory roles of these dnHLH proteins during cellular proliferation and differentiation of the nervous system. *Cell Growth Differ.* **6,** 837–843.

63. Lyden, D., Young, A. Z., Zagzag, D., Yan, W., Gerald, W., O'Reilly, R., Bader, B. L., Hynes, R. O., Zhuang, Y., Manova, K., and Benezra, R. (1999) ID1 and ID3 are required for neurogenesis, angiogenesis and vascularization of tumor xenografts. *Nature* **401,** 670–677.

64. Sasai, Y. (1998) Identifying the missing links: genes that connect neural induction and primary neurogenesis in vertebrate embryos. *Neuron* **21,** 455–458.

65. Weinstein, D. C. and Hemmati-Brivanlou, A. (1999) Neural induction. *Annu. Rev. Cell Dev. Biol.* **15,** 411–433.

66. Lee, J. E. (1997) Basic helix-loop-helix genes in neural development. *Curr. Opin. Neurobiol.* **7,** 13–20.

67. Kageyama, R. and Nakanishi, S. (1997) Helix-loop-helix factors in growth and differentiation of the vertebrate nervous system. *Curr. Opin. Genes Dev.* **7,** 659–665.

68. Bellefroid, E. J., Bourguignon, C., Hollemann, T., Ma, Q., Anderson, D. J., Kinter, C., and Pieler, T. (1996) X-MyT1, a *Xenopus* C2HC-type zinc finger protein with a regulatory function in neuronal differentiation. *Cell* **87,** 1191–1202.

69. Morin, X., Cremer, H., Hirsch, M. R., Kapur, R. P., Goridis, C., and Brunet, J. F. (1997) Defects in sensory and autonomic ganglia and absence of locus coeruleus in mice deficient for the homeobox gene Phox2a *Neuron* **18,** 411–423.

70. Hirsch, M. R., Tiveron, M. C., Guillemot, F., Brunet, J. F., and Goridis, C. (1998) Control of noradrenergic differentiation and Phox2a expression by MASH1 in the central and peripheral nervous system. *Development* **125,** 599–608.

71. Santa-Olalla, J. and Covarrubias, L. (1999) Basic fibroblast growth factor promotes epidermal growth factor responsiveness and survival of mesencephalic neural precursor cells *J. Neurobiol.* **40,** 14–27.

72. Daadi, M. M. and Weiss, S. (1999) Generation of tyrosine hydroxylase-producing neurons from precursors of the embryonic and adult forebrain. *J. Neurosci.* **19,** 4484–4497.

73. Bartlett, P. F., Brooker, G. J., Faux, C. H., Dutton, R., Murphy, M., Turnley, A., and Kilpatrick, T. J. (1998) Regulation of neural stem cell differentiation in the forebrain *Immunol. Cell. Biol.* **76,** 414–418.

74. Ye, W., Shimamura, K., Rubenstein, J. L., Hynes, M. A., and Rosenthal, A. (1998) FGF and Shh signals control dopaminergic and serotonergic cell fate in the anterior neural plate. *Cell* **93,** 755–766.

75. Orentas, D. M. and Miller, R. H. (1996) The origin of spinal cord oligodendrocytes is dependent on local influences from the notocord. *Dev. Biol.* **177,** 45–53.

76. Pringle, N. P., Yu, W. P., Guthrie, S., Roelink, H., Lumsden, A., Peterson, A. C., and Richardson, W. D. (1996) Determination of neuroepithelial cell fate: induction of the oligodendrocyte lineage by ventral midline cells and sonic hedgehog. *Dev. Biol.* **177,** 30–42.

77. Hynes, M., Ye, W., Wang, K., Stone, D., Murone, M., Sauvage, F. D., and Rosenthal, A. (2000) The seven-transmembrane receptor smoothened cell-autonomously induces multiple ventral cell types. *Nature Neurosci.* **3,** 41–46.

78. Briscoe, J., Sussel, L., Serup, P., Hartigan-O'Conner D., Jessell, T. M., Rubenstein, J. L., and Ericson, J. (1999) Homeobox gene *Nkx2.2* and specification of neuronal identity by graded sonic hedgehog signaling. *Nature* **398,** 622–627.

79. Roelink, H. (1996) Tripartite signaling of pattern: interactions between Hedgehogs, BMPs and Wnts in the control of vertebrate development. *Curr. Opin. Neurobiol.* **6,** 33–40.

80. Hirsinger, E., Duprez, D., JOuve, C., Malapert, P., Cooke, J., and Pourquie, O. (1997) Noggin acts downstream of Wnt and sonic hedgehog to antagonize BMPU in avian somite patterning. *Development* **124,** 4605–4614.

81. Verdi, J. M., Bashirullah, A., Goldhawk, D. E., Kubu, C. J., Jamali, M., Meakin, S. O., and Lipshitz, H. D. (1999) Distinct human NUMB isoforms regulate differentiation vs. proliferation in the neuronal lineage. *Proc. Natl. Acad. Sci. USA* **96,** 10,472–10,476.

82. Guo, S., Brush, J., Teraoka, H., Goddard, A., Wilson, S. W., Mullins, M. C., and Rosenthal, A. (1999) Development of noradrenergic neurons in the zebrafish hindbrain requires BMP, FGF8, and the homeodomain protein soulless/Phox2a. *Neuron* **24,** 555–566.

83. Alder, J., Lee, K. J., Jessell, T. M., and Hatten, M. E. (1999) Generation of cerebellar granule neurons in vivo by transplantation of BMP-treated neural progenitor cells. *Nature Neurosci.* **2,** 535–540.

84. Lumsden, A. and Krumlauf, R. (1996) Patterning the vertebrate neuraxis. *Science* **274,** 1109–1115.
85. Toresson, H., Mata de Urqiza, A., Fagerstrom, C., Perlmann, T., and Campbell, K. (1999) Retinoids are produced by glia in the lateral ganglionic eminence and regulate striatal neuron differentiation. *Development* **126,** 1317–1326.
86. Pierani, A., Brenner-Morton, S., Chiang, C., and Jessell, T. M. (1999) A sonic-hedgehog independent, retinoid-activated pathway of neurogenesis in the ventral spinal cord. *Cell* **97,** 903–915.
87. Cameron, H. A., Hazel, T. G., and McKay, R. (1998) Regulation of neurogenesis by growth factors and neurotransmitters. *J. Neurobiol.* **36,** 287–306.
88. Temple, S. and Qian, X. (1996) Vertebrate neural progenitor cells: subtypes and regulation. *Curr. Opin. Neurobiol.* **6,** 11–17.
89. Mabie, P. C., Mehler, M. F., Papavasiliou, A. Song, Q., and Kessler, J. A. (1997) Bone morphogenetic proteins induce astroglial differentiation of oligodendroglial-astroglial progenitor cells. *J. Neurosci.* **17,** 4112–4120.
90. McKay, R. (1997) Stem cells in the central nervous system. *Science* **276,** 66–71.
91. Koblar, S. A., Turnley, A. M., Classon, B. J., Reid, K. L., Ware, C. B., Cheema, S. S., Murphy, M., and Bartlett, P. F. (1998) Neural precursor differentiation into astrocytes requires signaling through leukemia inhibitory factor receptor. *Proc. Natl. Acad. Sci. USA* **95,** 3178–3181.
92. Nakashima, K., Yanagisawa, M., Arakawa, H., Kimura, N., Hisatsune, T., Kawabata, M., Miyazono, K., and Taga, T. (1999) Synergistic signaling in fetal brain by STAT3-Smad1 complex bridged by p300. *Science* **16,** 479–482.
93. Rogister, B., Ben-Hur, T., and Dubois-Dalcq, M. (1999) From neural stem cells to myelinating oligodendrocytes. *Mol. Cell. Neurosci.* **14,** 287–300.
94. Vartanian, T., Fischbach, G., and Miller, R. H. (1999) Failure of spinal cord oligodendrocyte development in mice lacking neuregulin. *Proc. Natl. Acad. Sci. USA* **96,** 731–735.
95. Luskin, M. B., Pearlman, A. L., and Sanes, J. R. (1988) Cell lineage in the cerebral cortex of the mouse studied *in vivo* and *in vitro* with a recombinant retrovirus. *Neuron* **1,** 635–647.
96. Price, J. and Thurlow, L. (1988) Cell lineage in the rat cerebral cortex: a study using retroviral-mediated gene transfer. *Development* **104,** 473–482.
97. Walsh, C. and Cepko, C. L. (1988) Clonally related cortical cells show several migration patterns. *Science* **241,** 1342–1345.
98. Parnavelas, J. G., Barfield, J. A., Franke, E., and Luskin, M. B. (1991) Separate progenitor cells give rise to pyramidal and nonpyramidal neurons in the rat telencephalon. *Cerebral Cortex* **1,** 463–468.
99. Walsh, C. and Cepko, C. (1992) Widespread dispersion of neuronal clones across functional regions of the cerebral cortex. *Science* **255,** 434–440.
100. Reid, C. B., Liang, I., and Walsh, C. (1995) Systematic widespread clonal organization in cerebral cortex. *Neuron* **15,** 299–310.
101. Davis, A. A. and Temple, S. (1994) A self-renewing multipotential stem cell in embryonic rat cerebral cortex. *Nature* **372,** 263–266.
102. Williams, B. P. and Price, J. (1995) Evidence for multiple precursor cell types in the embryonic rat cerebral cortex. *Neuron* **14,** 1181–1188.

3

Multipotent Stem Cells in the Adult Central Nervous System

Luca Bonfanti, Angela Gritti, Rossella Galli, and Angelo L. Vescovi

INTRODUCTION

A stem cell is an undifferentiated cell capable of extensive proliferation that can give rise to more stem cells (self-renewal) as well as to a progeny that will terminally differentiate and integrate into the tissue of residence, to replace cells lost to physiological turnover or injury. These features make stem cells an essential element both in the embryonic development and in the maintenance of adult tissue cell renewal. For example, a constant number of terminally differentiated cells in tissues and organs like the intestinal crypts, the skin, and blood are maintained through stem cell activity throughout life. However, not all tissues are endowed with physiologic continuous cell renewal, and, when this is the case, they display this property at very different rates. In this context, the mature nervous tissue has long been considered incapable of cell renewal and structural remodeling, especially in mammals. Studies carried out in the last decades have uncovered unexpected potential concerning the existence and extent of postnatal and adult neurogenesis in the mammalian central nervous system (CNS). As a consequence, questions have been raised regarding the existence and location of true stem cells in the CNS, which may be able to sustain adult neurogenesis continuously and may even provide an endogenous source of new neural cells in the course of reparative processes.

Since very recent findings have enriched what was previously known on this topic, a brief survey of postnatal neurogenesis in mammals is given, before discussing stem cells in the adult nervous system.

From: *Stem Cells and CNS Development*
Edited by: M. S. Rao © Humana Press Inc., Totowa, NJ

GENESIS OF NEW CELLS IN THE POSTNATAL AND ADULT NERVOUS SYSTEM

Location

A protracted neurogenesis characterizes some regions of the mammalian brain during the postnatal period. A well-studied, delayed neurogenesis does occur in a transient cell layer of the assembling cerebellar cortex, called the external granular layer. The external granular layer of the cerebellum persists through the first 3 postnatal weeks in rodents, as an actively mitotic layer that generates the granule cells for the internal granular layer *(1)*. In most mammalian species, the events that characterize early postnatal structural plasticity have disappeared in the adult CNS, wherein a persistent neurogenesis is considered to be restricted to three sites: the olfactory neuroepithelium (in the olfactory mucosa, which lines the nasal cavities), the olfactory bulb, and the hippocampus. In primates, an addition of new neurons has recently been described to occur also in neocortical association areas (prefrontal, inferior temporal, and posterior parietal cortex; *2*).

The olfactory neuroepithelium contains only one neuronal cell type, the olfactory receptor neurons, which are continuously renewed through differentiation of newly generated basal cells, regarded as unipotent, neuronal stem cells *(3)*. These neurons enter the brain at the level of the olfactory bulb exclusively with their axonal processes *(4)*; these processes are enwrapped by a specialized type of glial cells called ensheathing glia, which are permissive to axonal growth *(5)*.

In the hippocampus of several mammalian species *(6–9)* including humans *(10)*, cell proliferation leading to neurogenesis has been described in the dentate gyrus granular layer. In the rat, newly generated granule cells can be detected up to 11 mo of age *(11)* and extend dendrites and axons through the hilus and the CA3 region of Ammon's horn *(12)*.

The olfactory bulb is characterized by massive neurogenesis both into its main (linked to olfaction) and accessory (linked to pheromones) parts *(13–15)*. In this case, the sites of ultimate cell differentiation appear to be clearly distinct from the area within which cell proliferation occurs, namely, a remnant of the primitive forebrain subventricular zone persisting during adulthood as an actively mitotic layer called the subependymal layer (SEL; for review, see ref. *16*). Unlike other neurogenic sites of the mature nervous system, and in a way that somehow recapitulates the embryonic neurogenetic process, the SEL also represents a pathway for long-distance cell migration of undifferentiated precursors (see Chapter *4*).

Place of Birth, Life, and Death

In thinking about the highly established neuronal circuitries of an adult brain, one can imagine that the stem/progenitor cells and their progeny must reside close to each other. By contrast, it is now well-accepted that the final location of the newly generated cells, at least in rodents and primates, can be either very close or very far with respect to their site of birth, this separation even spanning different neuroanatomical regions (reviewed in ref. *16*).

During embryonic neurogenesis, the whole CNS originates from cells that proliferate within the germinal layers lining the primitive ventricular cavities, whose progeny promptly engage in widespread and massive migration *(17)*. In the adult, with the only exception of the olfactory receptor neurons, originating from the olfactory placodes *(18)*, the newly generated cell populations also have their origin in actively proliferating cell layers that surround the brain ventricular cavities. This is clearly the case of the olfactory bulb and prefrontal cortex, whose progenitors came from the SEL of the lateral ventricles after long-distance migration *(2,19,20)*.

Other examples of postnatal (cerebellum) and adult (hippocampus) neurogenesis, although seemingly independent from the ventricular germinal layers, are indirectly linked to them. The external granular layer of the cerebellum originates after a massive tangential cell migration from the wall of the fourth ventricle (see references in ref. *1*), followed by a centripetal migration of neuroblasts along persisting radial glial cells, called Bergmann glia *(21)*. The real origin of the newly generated hippocampal granule cells is still a matter of debate, since they are visualized in the subgranular zone of the dentate gyrus, but an extension of the SEL in the caudalward part of the ventricles can be seen between the hippocampal cortex and fornix on the one side and the corpus callosum on the other, marking an obliterated part of the ventricle because of a "rolling in" of the hippocampus during development *(22)*. Thus, it has been proposed that hippocampal granule cells can originate during adulthood both from local proliferation in the granule cell layer and after short migration from the hilus *(7,23)*, in a manner similar to that described during postnatal development *(24)*.

Signifincance of Cross-Species Specificity

Adult neurogenesis has been discovered and studied extensively in rodents *(4,6,7,11,19,20,22,25)*, but recent reports have demonstrated that it also persists in the brain of adult primates *(2,9,26)* and humans *(10)*.

At present, the function(s) of adult mammalian neurogenesis remains enigmatic. Apart from the olfactory receptor neurons, which undergo cell renewal into an environment not comparable to the mature brain nervous tissue (see above paragraph), all newly generated cells of the postnatal and adult brain are small-sized local circuit neurons (interneurons), with tightly packed cell bodies and short processes. This feature probably reflects a physiologic requirement for cells whose fate is to occupy a space in the highly established environment of the mature neuropil, possibly having to be functionally and cytoarchitecturally integrated therein.

The genesis of new cells preferentially occurs in brain regions that are implicated in different types of learning and memory. So far, adult neurogenesis has been consistently demonstrated across a wide range of species only in the hippocampus *(6,7,9,10,26)*, a region implicated in the encoding of current and the retention of past experience, with particular reference to spatial memory *(27,28)*. The production of new neurons in the hippocampus is regulated by environmental cues *(8,26,29–32)* and can also be manipulated experimentally *(33,34)*, suggesting that it does represent a significant functional process. In particular, it has been shown that hippocampal neurogenesis is specifically affected by hippocampus-dependent associative learning, but not by hippocampus-independent learning tasks *(30)*.

Interestingly, when considering other brain regions, some interspecies differences can be observed. For example, in certain primates the addition of new neurons has been observed in the prefrontal cortex *(2)*, an association area involved in behavioral plasticity that does not appear to be endowed with adult neurogenesis in rodents. By contrast, most neurogenesis occurring in rodents takes place in the olfactory bulb, a region that subserves a role in learning and memory in these species *(35)*. Thus, the difference(s) observed across species might be related to the importance of each brain region in sustaining memory and learning processes that are linked to specific behavioral tasks and that may be relevant to the survival and thriving of the different species — similarly to what is observed in adult neurogenesis in songbirds (for review, see ref. *36*). Nevertheless, due to obvious technical difficulties, a complete mapping of the sites of neurogenesis in primates, as well as in other species, cannot be considered fully acquired at present. For example, besides the indication that the human forebrain SEL may contain actively proliferating neural precursors *(37,38)*, very little is known about the fate and destination of these cells in vivo.

STEM CELLS IN THE CENTRAL NERVOUS SYSTEM

During the last 10 years, studies on neurogenesis had been characterized by the intensive search for "stem" or "stem-like" cells that shared some of the critical stem cell features and potentials observed in similar cells from other, less quiescent tissues. A functional parallelism between CNS and blood lineages has been recently proposed in an attempt to approach the daunting task of studying the development of neural precursor cells by adopting the theoretical framework of a system like the hemopoietic, in which stem cells have been most rigorously and directly identified *(39,40)*. This has recently led some to propose the use of terms such as "neuropoiesis" for the process of persistent neurogenesis and "brain marrow" for the anatomical regions that support it and that are likely to embody *bona fide stem* cells *(41)*.

The abundant literature now available on neural stem cell biology has grown very rapidly in the last few years. In spite of a continuous delivery of striking results, leading to substantial advancement in this field, it remains rather difficult to integrate and compare results from studies conducted at different developmental stages or on different CNS regions, and in which different in vivo or in vitro approaches have been used. This is particularly true if we consider how the adoption of different experimental paradigms can dramatically affect the evaluation of critical functional properties and eventually the identification of stem or precursor cells from a given tissue *(42)*. Therefore, the current terminology regarding what is to be considered a stem cell, a stem-like cell, a stem/progenitor cell, or progenitor in the nervous system remains rather confusing, particularly with respect to the meaning that the different research teams working in this field attribute to each of these definitions (for further details, see ref. *41* and Chapter 1).

From Developing to Adult Brain

As mentioned in the introduction, an evident shift takes place in development that leads from the situation of high plasticity existing during prenatal and early postnatal CNS formation to the high cytoarchitectural and functional stability observed throughout adulthood. Thus, working with neural stem cells, two major questions can arise. The first, which is addressed in Chapter 2, concerns how the overall complexity of the mature brain is progressively achieved. This issue can be approached on one level by isolating and purifying the precursor (stem) cell types and by studying the progressive modification of their developmental potential that eventually leads to the genesis of terminally differentiated neural cells *(43)*. The second, which receives more attention in the present chapter, concerns where, how, and why functionally active multipotent stem cells do persist within the adult brain.

Both of the above questions are strictly related to the processes of fate commitment and differentiation, which in the nervous system, as well as in other tissues, are the result of the integration between an intrinsic genetic program and the response to environmental cues in uncommitted precursor cells. Thus, the path of development that a stem cell enters might depend on the environmental milieu in which it is found, which, in turn, strictly depends on when (among different developmental stages) and where (among different CNS areas) the stem cell is located. Most of the present evidence concerning the role(s) of the effector molecules that affect proliferation, differentiation, and survival of the stem cells and their progeny comes from in vitro studies in which the stem cells have been removed from their original environment. Although these studies have revealed a great deal of information about some of the basic functional characteristics of adult neural stem cells, they allow us only to determine the potential attributes that are part of a stem cell repertoire, some of which may never be adopted in an in vivo environment, whose complexity cannot be fully reproduced in culture. Thus, a wide gap still exists between evidence for the occurrence of true CNS stem cells in vitro and in vivo.

Furthermore, the functional and antigenic characterization of embryonic and adult stem cells is currently rather inadequate and leaves many important questions open. Among these, the question stands out of whether multiple kinds of age- and region-specific neural stem cells exist or whether a single "basic" stem cell type is present at different stages of development and in different brain areas that then displays diverse functional attributes when challenged in different experimental contexts in vivo and ex vivo. A plethora of studies carried out by explanting, culturing, and transplanting stem cells in animals at different developmental stages strongly suggest that embryonic and adult neural stem cells share the same features when removed from their environmental milieu. For example, it has been shown that embryonic and adult stem cells respond to different growth factors in a similar manner *(44)*. On the other hand, other reports suggest that differences do exist when subsequent developmental stages or different CNS regions are considered *(40,45,46)*, thus leaving open the possibility that more than one type of multipotent neural stem cells does exist.

The following chapter focuses on adult CNS stem cells, yet we ought to consider that the apparent diversity between the embryonic and the adult stem cells may in part result from the lack of molecular, cellular, and functional characterization of these peculiar brain cell types.

IDENTIFICATION OF STEM CELLS IN THE ADULT CENTRAL NERVOUS SYSTEM

As a result of the absence of specific molecular or antigenic markers capable of distinguishing stem cells from the wide populations of lineage-restricted progenitor/precursor cells, the present knowledge is largely based on culturing experiments from CNS tissue explants and, to a significantly lesser extent, on in vivo studies. As described in the following paragraphs, the topographic localization of stem cells in different areas of the CNS (and sometimes even their existence) is still a matter of discussion depending on which type of cell (stem cell, stem-like cell, progenitor cell) we are considering and what experimental paradigm is being used.

At present, it is generally accepted that "true" stem cells do indeed exist in restricted areas of the adult forebrain, which are remnants of the embryonic subventricular zone and directly linked to the sites of protracted neurogenesis *(47–53,41)*. Nevertheless, stem-like cells can also be isolated from nonneurogenic areas *(54)*, raising the issue of the possible existence of both "potential" and "actual" stem cells *(42)* within the adult CNS.

Stem Cells in Neurogenic Areas

A series of observations that led to the revisitation of the stem cell concept as applied to the nervous system originated from the work of Richards et al. *(55)* and Reynolds and Weiss *(56)*. These teams reported the generation of new neuronal and glial cells from explants of the forebrain of the adult rodent in vitro. Subsequent studies ascertained that the site harboring the dividing progenitor cells is the periventricular region of the lateral ventricles *(57,58)*, an area in which a progressive and essentially complete turnover of the proliferating cell population appeared to occur every 12–28 d *(59)*.

If different regions of the adult nervous system do contain actively dividing neuronal progenitors, those contained in the SEL have been characterized as *bona fide* stem cells *(45,57,60–65)*. At least in vitro, they actually fulfill most of the critical criteria accepted to characterize stem cells *(42)*, namely, an extensive capacity to self-renew, the ability to undergo asymmetric cell division and to generate a multilineage progeny (multipotentiality), extensive flexibility in the use of these options, and the capacity to exist in a mitotically quiescent state (reviewed in refs. *40* and *66*). For these reasons, the SEL is now considered as a major reservoir of multipotent stem cells in the adult rodent brain.

The stem properties of the cells isolated from the SEL can be observed by culturing them in serum-free medium, with the addition of polypeptide growth factors such as epidermal growth factor (EGF) or basic fibroblast growth factor-2 (FGF-2; Table 1). Treatment with EGF induces these cells to extend their proliferative potential, forming the so-called neurospheres, spheres of cells among which stem cells can be found *(66)*. Similarly, the culturing of SEL cells in the presence of FGF-2 also results in the expansion of a stem cell population *(45)*. The same growth factor has also been proven to exert a role in modulating the commitment of the SEL stem cell progeny to a differentiated fate *(62)*, thus suggesting that the generation of different cell types from a neural stem cell could be regulated through a sequential

Table 1
Characteristics of Stem/Progenitor Cells Isolated from Neurogenic (N) and Nonneurogenic (Non-N) Regions of the Adult Mammalian CNS

Region		Growth Factor(s)	References
SEL (lateral ventricle)	N	EGF	*56,57,62*
		FGF-2	*45*
		EGF + FGF-2	*46,63*
Hippocampus	N	FGF-2	*61,69,70*
Striatum and Septum	Non-N	FGF-2	*54,61*
III and IV ventricles	Non-N	EGF + FGF-2	*46*
Spinal cord	Non-N	FGF-2	*82*
		EGF + FGF-2	*46*

a The growth factors used to isolate the cells and the relevant references are indicated.

action of growth factors *(45,67)*. Although in vivo these periventricular SEL-derived cells give rise mainly to neurons *(19,20*; reviewed in ref. *16)*, under appropriate conditions in vitro, the neurospheres can generate a large number of progeny that can cover all the major cell types of the CNS, including neurons, astrocytes, and oligodendrocytes (reviewed in refs. *40* and *66*).

Until recently SEL-derived neural stem cells have been considered to belong to different classes, such as EGF-responsive cells (also called neurosphere stem cells) and FGF-2-dependent cells (called neuroepithelial stem cells; reviewed in ref. *68*). However, in recent in vitro work, it has been demonstrated that these two types of adult CNS stem cells are indeed the same cell type that has the capacity to respond to both EGF and FGF-2 (Table 1). Depending on the pattern of epigenetic stimulation (i.e., the growth factor combination) to which they are exposed, SEL stem cells can modify their proliferation mode and may vary the ratio of symmetric vs asymmetric cell division that takes place within the stem cell population *(63)*. The finding showing that the vast majority of adult SEL-resident stem cells are simultaneously EGF and FGF-2 responsive does not completely rule out the possibility that an extremely small, hardly detectable subset of SEL stem cells may indeed respond selectively to only one of these growth factors.

Multipotent cells similar to the SEL stem cells have been isolated from the adult rat hippocampus *(47,69)*, but they differ from the former in several aspects. First, their proliferative expansion in vitro does not require the addition of EGF, as they are regulated mainly by FGF-2 (Table 1). Second, they do not form neurospheres but rather adhere to the substratum. Finally, unlike the newly generated cells of the SEL, those of the dentate gyrus do not undergo long-distance migration in vivo, and about 15% of them differentiate into glial cells *(70)*. The self-renewal properties of these cells have not been formally proved by serial subcloning analysis, and thus they have frequently been referred to as hippocampus-derived progenitor cells or stem-like cells *(70,71)*. Nevertheless, the fact that these cells can steadily be expanded and that they retain multipotentiality over extended subculturing demonstrate the existence of *bona fide* stem cells in these cultures. Furthermore, as discussed in the next paragraph in more detail, stem cells from different CNS regions are likely to require different culture/environmental conditions to display their stem cell properties. In this view the multipotent progenitors isolated from the dentate gyrus should also be considered true neural stem cells *(53)*.

It is worth mentioning that hippocampus-derived progenitor cells can contribute to the elucidation of adult neurogenesis regulation. Indeed, a series of combined behavioral and anatomic studies, other than showing the influence of environmental factors on the production of new hippocampal neurons *(8,26,29–32)*, has demonstrated that hormones (glucocorticoids) and excitatory neurotransmitters (glutamate) actively regulate hippocampal neurogenesis, inhibiting cell production in the dentate gyrus *(29,33,72)*. These observations contributed to build the molecular basis for the current idea that stress may indeed reduce the rate of adult neurogenesis *(26)*.

Given the current state of the art in stem cell research, it is clear that the relationship among the SEL stem cells, the dentate gyrus stem cells, and the neural stem cells found in different brain regions at different ages cannot be elucidated conclusively. A strict distinction of stem cells in classes, on the basis of either their growth factor responsiveness or their behavior in culture, is probably misleading and is unlikely to increase our knowledge of the basic biology of these peculiar cells. As far as in vitro studies are concerned, some differences between the various cells could be related to the alternative harvesting and culture conditions adopted in different laboratories, which could act in a different manner on cells at different, progressive stages of differentiation (i.e., from stem to precursor cell). In other words, multipotent cells that respond to EGF or FGF-2, or both, might correspond to different stages of a single cell lineage *(45,53,73,74)*.

On the other hand, they could also correspond to inherently distinct cell types, if we accept the idea that intrinsic differences exist among cells isolated at different developmental stages and/or residing in different CNS regions. For example, EGF-dependent stem cells of the adult ventricular neuroaxis are not detectable during early development *(46)*, when FGF-responsive cells can be isolated from the same site (reviewed in ref. *68*). Indeed, progenitor cells in the ventricular zone at early stages of cortical development do not respond to EGF by dividing *(60)*, whereas in the subventricular zone during later developmental stages, cortical progenitor cells acquire the competence to divide in response to EGF *(75)*. One possible mechanism of such a regulation has been indicated in different EGF receptor/ligand levels, which could modulate the responsiveness to the environmental signals controlling proliferation, cell fate, and migration properties *(76–78)*. For example, after the introduction of extra EGF receptors into early progenitor cells, both in vitro and in vivo, several features proper to later stages were also acquired *(76)*.

Although help in clarifying conflicting findings concerning the neural stem cell identification, their growth capacity, fate potential, and lineage relationship(s) may also come from refined ultrastructural and functional in vivo studies (see Chapter *4*), a final resolution of the many outstanding issues regarding stem cells and their identification will be possible when, as in the hemopoietic field, neural stem cell-specific antigenic markers become available.

Stem-Like Cells in Nonneurogenic Areas

It has been reported that multipotent cells can also be isolated from nonneurogenic regions of the adult mammalian CNS, such as the walls of the third and fourth ventricles and the spinal cord *(46)* and, to a lesser extent, from the septum and the striatum *(61)*. Multipotent cells can also be obtained from white matter-rich areas *(54)*, suggesting that a latent neurogenic program may persist in glial progenitors. Other evidence, such as the existence of oligodendrocyte precursor cells in the white matter of the adult human brain *(79,80)* and the migration of oligodendrocytes precursors from the SEL to the corpus callosum of adult mice after demyelination *(81)*, are of fundamental importance for studies on spontaneous myelin regeneration.

Most of the cells isolated from nonneurogenic brain parenchyma do not fulfill the formal criteria to be considered true stem cells. They may rather be viewed as neural precursors or "stem-like cells" *(53,61,74,82)*. Among them, those residing close to the neuroaxis appear to possess a higher "degree of stemness" *(46)*. Interestingly, stem-like cells from the lateral ventricle (neurogenic in the adult) can be induced to proliferate, self-renew, and expand in the presence of EGF alone, whereas those from the third/fourth ventricles and the spinal cord (nonneurogenic in the adult) require a combination of EGF and FGF-2 *(46)*. However, other reports suggest that FGF-2 alone is sufficient for the isolation of multipotent stem-like cells from the adult spinal cord *(82)*, from the nonneurogenic parenchyma of the adult brain *(61)*, and from the SEL *(45)*. At present, as previously mentioned, it remains unclear whether different stem cell types exist or whether several stages of a single cell type can be affected differently by the adoption of specific culture conditions (Table 1).

Because of its prominent functional implications, another brain region—the cerebral cortex—has been the object of intense investigation with respect to neural precursors. At present no convincing evidence shows that precursor/stem-like cells can be obtained from the adult neocortex. However, the isolation of multipotent progenitors has been described in the cerebral cortex of 2-d-old postnatal rats *(83)*. Although neurogenesis is considered concluded at this stage in several species, the assembly and maturation of the cortex has not yet been fully accomplished in terms of connections and gliogenesis, its microenvironment is not comparable to that found in the adult, and residual neural precursors may persist for a limited time span. On the other hand, several observations suggest that cortical areas of certain mammalian species may also be the site of structural plasticity throughout adulthood. As recently elegantly demonstrated *(2)*, adult neurogenesis does occur in some neocortical areas of primates. In the adult rat brain, although neurogenesis has not been conclusively demonstrated, numerous polysialylated neural cell adhesion molecule (PSA-NCAM)-positive neurons (see next section) are present in the piriform cortex *(84)*; moreover, PSA-NCAM-positive, spindle-shaped cells oriented along the white matter of the external capsule, thus occupying a position between the ventricles and the piriform cortex, are also detectable *(16)*. Thus, cortical neurogenesis could occur in restricted regions, depending on the species, probably as a result of cell migration from the SEL of the lateral ventricles. Nevertheless, further studies are required to determine whether progenitors/stem-like cells do exist along these migration routes or within the adult cortical parenchyma itself.

On the whole, the results indicate the possibility that stem cells of the adult CNS are not restricted to the sites of persistent neurogenesis and suggest that specific environmental conditions are required to expand or "wake" these cells in order for them to adopt stem cell behavior.

Stem Cell Quiescence

To explain the existence of stem-like cells in brain regions devoid of cell proliferation, one would admit that "potential," or quiescent, stem cells could be present throughout the CNS, and that they can actually enter a proliferation state and become "actual" stem cells only at well-defined sites or under particular conditions *(42)*. According to a view borrowed from the hematopoietic system *(39,42,85)*, some stem cells would be noncycling "quiescent" stem cells locked in G0, while other stem cells are actively engaged in the cell cycle. Although mitotic quiescence is a property shared by some, but not all, stem cells (for more details, see ref. *39*), several studies now indicate that this scenario may apply to neural stem cells.

An attractive hypothesis has been advanced suggesting that, even in the highly neurogenic area of the SEL, stem cells belong to two different cell populations: quiescent and actively proliferating *(57,65)*. Initial studies had argued that the SEL stem cells that were quiescent in vivo specifically corresponded to the EGF-responsive neurosphere-forming cells observed in vitro *(56)*, whereas constitutively proliferating cells did not *(57)*. Thus, one could argue that SEL precursors displaying distinct kinetic properties in vitro might indeed represent distinct cell subtypes, characterized by selective responsiveness to specific growth factors. However, this does not appear to be the case. In fact, more recent studies have proved that large numbers (>50%) of the EGF-responsive stem cells isolated in vitro are indeed constitutively proliferating in vivo *(63,86)* and that the vast majority, if not all, of these cells are also simultaneously responsive to FGF-2 *(63)*. Thus, it is very unlikely that distinct subtypes of SEL stem cells, which display distinct in vivo kinetics, can be distinguished merely on the basis of their responsiveness to specific growth factors. Furthermore, it remains unclear whether the quiescent and constitutively proliferating kinetically active SEL precursors observed in vivo are indeed distinct cell types or rather represent different functional states of the same cell types, which become exposed to different environmental niches. The latter hypothesis may be confirmed through finding that the activity of SEL stem cells can be varied and modulated epigenetically in vivo, as demonstrated by the dramatic increase in the number of newborn cells that can be observed in the adult brain after intracerebroventricular administration of EGF, FGF-2 *(86,87)*, or brain-derived neurotrophic factor (BDNF) *(88)*. Interestingly, whereas FGF-2 and BDNF induce an increase in the number of neurons, EGF substantially decreases this number, enhancing differentiation in the glial lineage *(86,87)*. Moreover, EGF has an inhibitory effect on the progression of cells through their usual migratory routes and induces their displacement from the SEL into the adjacent brain parenchyma *(86)*. This strongly suggests that different cell functions, including cell proliferation, survival, commitment, and migration, can be brought about under different environmental conditions in a rather complex fashion, somehow mimicking what is observed in CNS stem cell cultures.

The existence of "potential" stem cells could explain the isolation of stem-like cells at levels of the ventricular neuroaxis, which is devoid of a subventricular layer *(46,58)*. Johansson and colleagues *(58)* have recently proposed that slowly dividing stem cells exist in the ependymal monolayer lining the ventricular cavities in the adult. These cells would generate more rapidly dividing "transit" cells (progenitors) where a subependymal layer persists beneath the ependyma (i.e., the forebrain lateral ventricles); they would remain quiescent where the ependyma is adjacent to the mature parenchyma (i.e., the brainstem and the spinal cord). Identification of ependymal cells as stem elements has recently been questioned *(89–91)*. However, a working hypothesis can be proposed, which would be independent from the exact identity of the adult neural stem cell: the persistence of a subependymal "no man's land" area could provide the environmental conditions required by the neural stem cells to proliferate and generate a multipotent progeny in the otherwise nonpermissive adult brain neuropil. The absence of such a region, either in other parts of the ventricular system or in the brain parenchyma could, conversely, prevent the expansion of stem cells, which would adopt a more quiescent behavior.

To explain the inability of a large part of potential stem cells to generate progeny in vivo, a failure of most adult CNS areas to support cell migration actively, other than cell survival and proliferation, has also been invoked. During development, a permissive environment made of extracellular matrix and cell surface molecules, along with special structures such as radial glia, allows the displacement of neuronal precursors from the site of genesis to the site of their full differentiation. Some of these features are retained in neurogenic areas of the adult brain. For example, the polysialylated, "embryonic" isoform of the neural cell adhesion molecule NCAM (PSA-NCAM), which plays an important, although not essential *(92)*, role in cell migration and cell shaping *(93;* reviewed in ref. *16)*, is selectively expressed in regions of the adult brain endowed with structural plasticity *(84)*, on the membrane of newly generated cells in the SEL *(94)* and in the hippocampus *(95)*. Also, glial cells appear to retain specific features in neurogenic areas: short radial glia-like cells are present in the hippocampal dentate gyrus *(23)*, and "glial tubes" formed by a particular type of astrocytic cells, expressing embryonic cytosceleton proteins such as vimentin and nestin, surround the cells migrating through the SEL toward the olfactory bulb *(16,96–98)*. Finally, recent findings concerning synaptic plasticity in the hippocampus *(99,100)* suggest the implication of PSA-NCAM in sensitizing neurons to growth factors, thus opening new perspectives for the presence of this protein in neurogenetic regions of the adult CNS.

On the whole, the examples given support the view of an adult mammalian CNS endowed with a striking capacity for structural plasticity, revealed as an intrinsic ability to undergo continuous, spatially restricted neurogenesis, as well as a cohort of "potentialities" that can become apparent under particular conditions in vitro or in vivo. Since physiologically occurring plasticity appears to be linked to the existence of specific anatomic and molecular features, it has gradually emerged that stem cells must obey a finely regulated set of extrinsic signals, strictly depending on the microenviroment in which these cells reside.

THE EMERGING ROLE OF THE LOCAL ENVIRONMENT

It is known that the context in which the stem cells live, both temporally and spatially, has a significant impact on their developmental potential and their terminal differentiation (reviewed in ref. 78). The successful differentiation of adult neural stem cells into cells of the hematopoietic lineage, obtained after transplantation into the bone marrow of irradiated animals, is a striking example of how a different tissue environment can affect the behavior of a stem cell *(101)*. This approach strongly suggests that the neural fate adopted by adult SEL stem cells is the result of a default differentiation program and indicates that these cells may be pluripotent in nature, possessing the appropriate machinery to respond to unusual patterns of epigenetic stimulation that may trigger their differentiation along lineages other than the neuroectodermal *(101)*.

As pointed out in a recent review article *(41)*, the bone marrow and the SEL share several important features in addition to their content in stem/progenitor cells, including the presence of common extracellular matrix proteins (e.g., tenascin; *38,97,102*) and cells with supportive functions (bone marrow stromal cells and brain astrocytes). Many studies in the hematopoietic area (reviewed in ref. *103*) strongly indicate that stromal cells and extracellular matrix proteins play a crucial role in affecting the proliferative capacity, fate, adhesivity, and migration of stem/progenitor cells. In a recent report, extensive neurogenesis was observed by culturing cells explanted from the SEL onto astrocyte monolayers, even in the absence of serum or growth factors *(104)*, thus suggesting that astrocytes of the SEL could be the "stromal" cells of the brain marrow.

The assumption that local "environmental" signals, rather than some intrinsic features of different stem cell populations, are important for neuronal differentiation is strengthened by transplantation approaches in different neurogenic or nonneurogenic areas. For example, adult rat hippocampus-derived progenitor cells grafted into the SEL (neurogenic) or the cerebellum (nonneurogenic) of adult hosts actually give rise to neurons in a site-specific manner only in the former *(105)*. Interestingly, some of the neuronal phenotypes that appear later in the olfactory bulb after the graft into the SEL, such as tyrosine hydroxylase, are not present in the hippocampus *(105)*. The same cells, when transplanted into the retina (nonneurogenic) at different developmental stages, display a partial integration only in newborn animals (in which the retina has not completed its development) but not in adults *(106)*.

CONCLUSIONS

The CNS is a complex tissue embodying numerous cell types that interact in a highly coordinated and cytoarchitecturally appropriate fashion. Such a complex tissue is formed over a short time during embryogenesis, starting from embryonic neural stem cells through the progressive proliferation, migration, and differentiation of neuronal and glial precursor cells. Despite the striking progress accomplished in this field in the last few years, a complete understanding of the molecular mechanisms underlying embryonic neurogenesis is far from being achieved. At the same time, the existence of stem/progenitor cells has also been ascertained in the mature CNS, with particular reference to restricted brain areas, where neurogenesis is protracted after birth, and a recapitulation of several crucial events that characterize neural development can be observed throughout life. An outstanding challenge for future research in this field will be the unraveling of mechanisms that allow striking structural plasticity to persist into (and adapt to) an adult mammalian brain. In particular, a deeper understanding of the phenomena underlying the state of "mitotic quiescence," as applied to the nervous system, should reveal the extent to which the technique of manipulating neural stem cells can be harnessed to develop novel therapeutic paradigms, concepts, and techniques for restoring the damaged brain.

As discussed in the above paragraphs, the gap still existing between in vitro and in vivo approaches to neural stem cell biology certainly represents an important hindrance to our understanding of the basic biology of neural stem cells, probably not of immediate solution. In this context, two major points remain obscure at both the beginning and the end of the entire process leading from a stem cell to a fully differentiated cell in the mature CNS. The first point concerns the evidence of self-renewal activity for neural stem cells in vivo, a task that has been addressed in the hematopoietic system by transplantation of stem cells that can be subsequently harvested, reisolated, and transplanted into fresh recipients (39,42). The second point faces the issue of conclusive demonstration that the stem cell progeny can actually be incorporated into functional neural circuits. At present, many investigators agree that most of the newly generated stem cell differentiated progeny in the adult mammalian brain are neuronal precursors, which subsequently differentiate into local circuit neurons (interneurons). It is not still clear whether all these cells give rise to neurons and whether all these neurons actually survive for long times, becoming incorporated into neural networks. Then, if some cells are incorporated into established functional circuits, the question arises of whether they either add to the preexisting neuronal populations, as suggested for the hippocampal dentate gyrus, at least into midlife *(25,31)*, or replace older cells that die; this might be a regulated process aimed at the maintenance of a cell turnover that, in turn, implies a certain rate of cell death, such as in the olfactory neuroepithelium for example *(3)*. Answering these questions will provide more insight into the role that neural stem cells may actually play within the adult mammalian CNS and the potential use of these particular cells for repairing the damaged nervous tissue.

REFERENCES

1. Altman, J. and Bayer, S. A. (1997) Transformation of the paired cerebellar primordia into the unified primitive cerebellum, in *Development of the Cerebellar System: In Relation to its Evolution, Structure and Function* (Altman, J. and Bayer, S. A., eds.), CRC Press, Boca Raton, FL, pp. 138–169.
2. Gould, E., Reeves, A. I., Graziano, M. S. A., and Gross, C. G. (1999a) Neurogenesis in the neocortex of adult primates. *Science* **286,** 548–552.
3. Calof, A. L., Mumm, J. S., Rim, P. C., and Shou, J. (1998) The neuronal stem cell of the olfactory epithelium. *J. Neurobiol.* **36,** 190–205.
4. Graziadei, P. P. C. and Monti Graziadei, G. A. (1979) Neurogenesis and neuron regeneration in the olfactory system of mammals. I. Morphological aspects of differentiation and structural organization of the olfactory neurons. *J. Neurocytol.* **8,** 1–18.
5. Ramon-Cueto, A. and Valverde, F. (1995) Olfactory bulb ensheathing glia: a unique cell type with axonal growth-promoting properties. *Glia* **14,** 163–173.
6. Kaplan, M. S. and Hinds, J. W. (1977) Neurogenesis in the adult rat: electron microscopic analysis of light radioautographs. *Science* **197,** 1092–1094.
7. Kuhn, H. G., Dickinson-Anson, H., and Gage, F. H. (1996) Neurogenesis in the dentate gyrus of the adult rat: age-related decrease of neuronal progenitor proliferation. *J. Neurosci.* **16,** 2027–2033.
8. Gould, E., McEwen, B. S., Tanapat, P., Galea, L. A. M., and Fuchs, E. (1997) Neurogenesis in the dentate gyrus of the adult tree shrew is regulated by psychosocial stress and NMDA receptor activation. *J. Neurosci.* **17,** 2492–2498.
9. Gould, E., Reeves, A. I., Fallah, M., Tanapat, P., and Gross, C. G. (1999b) Hippocampal neurogenesis in adult Old World primates. *Proc. Natl. Acad. Sci. USA* **96,** 5263–5267.
10. Eriksson, P. S., Perfilieva, E., Biork-Eriksson, T., Alborn, A. M., Nordborg, C., Peterson, D. A., and Gage, F. H. (1998) Neurogenesis in the adult human hippocampus. *Nature Med.* **4,** 1313–1317.
11. Kaplan, M. S. and Bell, D. H. (1984) Mitotic neuroblasts in the 9-day-old and 11-month-old rodent hippocampus. *J. Neurosci.* **4,** 1429–1441.
12. Stanfield, B. B. and Trice, J. E. (1988) Evidence that granule cells generated in the dentate gyrus of adult rats extend axonal projections. *Exp. Brain Res.* **72,** 399–406.
13. Altman, J. (1969) Autoradiographic and histological studies of postnatal neurogenesis. IV. Cell proliferation and migration in the anterior forebrain, with special reference to persisting neurogenesis in the olfactory bulb. *J. Comp. Neurol.* **137,** 433–458.

14. Hinds, J. W. (1968) Autoradiographic study of histogenesis in the mouse olfactory bulb II. Cell proliferation and migration. *J. Comp. Neurol.* **134,** 305–322.

15. Bonfanti, L., Peretto, P., Merighi, A., and Fasolo, A. (1997) Newly generated cells from the rostral migratory stream in the accessory olfactory bulb of the adult rat. *Neuroscience* **81,** 489–502.

16. Peretto, P., Merighi, A., Fasolo, A., and Bonfanti, L. (1999) The subependymal layer in rodents: a site of structural plasticity and cell migration in the adult mammalian brain. *Brain Res. Bull.* **49,** 221–243.

17. Jacobson, M. (ed.) (1991) Histosenesis and morphogenesis of cortical structures, in *Developmental Neurobiology.* Plenum Press, New York, pp. 401–451.

18. Farbman, A. I. (1991) Developmental neurobiology of the olfactory system, in *Smell and Taste in Health and Disease* (Getchell, T. V., Doty, R. L., Bartoshuk, L. M., and Snow Jr, J. B. eds.), Raven Press, New York, pp. 9–33.

19. Luskin, M. B. (1993) Restricted proliferation and migration of postnatally generated neurons derived from the forebrain subventricular zone. *Neuron* **11,** 173–189.

20. Lois, C. and Alvarez-Buylla, A. (1994) Long-distance neuronal migration in the adult mammalian brain. *Science* **264,** 1145–1148.

21. Rakic, P. (1971) Neuron-glia relationship during granule cell migration in developing cerebellar cortex. A Golgi and electronmicroscopic study in Macacus rhesus. *J. Comp. Neurol.* **141,** 283–312.

22. Smart, I. (1961) The subependymal layer of the mouse brain and its cell production as shown by radioautography after thymidine-H3 injection. *J. Comp. Neurol.* **116,** 325–338.

23. Cameron, H. A., Woolley, C. S., McEwen, B. S., and Gould, E. (1993) Differentiation of newly born neurons and glia in the dentate gyrus of the adult rat. *Neuroscience* **56,** 337–344.

24. Schlessinger, A. R., Cowan, W. M., and Gottlieb, D. I. (1975) An autoradiographic study of the time of origin and the pattern of granule cell migration in the dentate gyrus of the rat. *J. Comp. Neurol.* **159,** 149–176.

25. Bayer, S. A., Yackel, J. W., and Puri, P. S. (1982) Neurons in the rat dentate gyrus granular layer substantially increase during juvenile and adult life. *Science* **216,** 890–892.

26. Gould, E., Tanapat, P., McEwen, B. S., Flugge, G., and Fuchs, E. (1998) Proliferation of granule cell precursors in the dentate gyrus of adult monkeys is diminished by stress. *Proc. Natl. Acad. Sci. USA* **95,** 3168–3171.

27. Hampson, R. E., Simeral, J. D., and Deadwylir, S. A. (1999) Distribution of spatial and nonspatial information in dorsal hippocampus. *Nature* **402,** 610–614.

28. Wallenstein, G. V., Eichenbaum, H., and Hasselmo, M. E. (1998) The hippocampus as an associator of discontiguous events. *Trends Neurosci.* **21,** 317–323.

29. Cameron, H. A. and Gould, E. (1994) Adult neurogenesis is regulated by adrenal steroids in the dentate gyrus. *Neuroscience* **61,** 303–209.

30. Gould, E., Beylin, A., Tanapat, P., Reeves, A., and Shors, T. J. (1999c) Learning enhances adult neurogenesis in the hippocampal formation. *Nature Neurosci.* **2,** 260–265.

31. Kempermann, G., Kuhn, H. G., and Gage, F. H. (1998) Experience-induced neurogenesis in the senescent dentate gyrus. *J. Neurosci.* **18,** 3206–3212.

32. van Praag, H., Kempermann, G., and Gage, F.H. (1999) Running increases cell proliferation and neurogenesis in the adult mouse dentate gyrus. *Nature Neurosci.* **2,** 266–270.

33. Cameron, H. A., McEwen, B. S., and Gould, E. (1995) Regulation of adult neurogenesis by NMDA receptor activation. *J. Neurosci.* **15,** 4687–4692.

34. Parent, J. M., Yu, T. V., Leibowitz, R., Geschwind, D. H., Sloviter, R. S., and Lowenstein, D. H. (1997) Dentate granule cell neurogenesis is increased by seizures and contributes to aberrant network reorganization in the adult hippocampus. *J. Neurosci.* **17,** 3727–3738.

35. Brennan, P., Kaba, H., and Keverne, E. B. (1990) Olfactory recognition: a simple memory system. *Science* **250,** 1223–1226.

36. Alvarez-Buylla, A. and Kirn, J. R. (1997) Birth, migration, incorporation and death of vocal control neurons in adult songbirds. *J. Neurobiol.* **33,** 585–601.

37. Kirschenbaum, B., Nedergaard, M., Preuss, A., Barami, K., Fraser, R. A. R., and Goldman, S. A. (1994) In vitro neuronal production and differentiation by precursor cells derived from the adult human forebrain. *Cereb. Cortex* **6,** 576–589.

38. Kukekov, V. G., Laywell, E. D., Suslov, O., Scheffler, B., Thomas, L. B., O'Brian, T. F., et al. (1999) Multipotent stem/progenitor cells with similar properties arise from two neurogenic regions of adult human brain. *Exp. Neurol.* **156,** 333–344.

39. Morrison, S. J., Shah, N. M., and Anderson, D. J. (1997) Regulatory mechanisms in stem cell biology. *Cell* **88,** 287–298.

40. Weiss, S. and van der Kooy, D. (1998) CNS stem cells: where's the biology (a.k.a. Beef)? *J. Neurobiol.* **36,** 307–314.

41. Scheffler, B., Horn, M., Blumcke, I., Laywell, E. D., Coomes, D., Kukekov, V. G., and Steindler, D. A. (1999) Marrow-mindedness: a perspective on neuropoiesis. *Trends Neurosci.* **22,** 348–357.

42. Loeffler, M. and Potten, C. S. (1997) Stem cells and cellular pedigrees — a conceptual introduction, in *Stem Cells.* (Potten, C. S., ed.), Academic, London, pp. 1–27.

43. McKay, R. (1997) Stem cells in the central nervous system. *Science* **276,** 66–71.

44. Johe, K. K., Hazel, T. G., Muller, T., Dugich-Djordjevic, M. M., and McKay, R. D. (1996) Single factors direct the differentiation of stem cells from the fetal and adult central nervous system. *Genes Dev.* **10,** 3129–3140.

45. Gritti, A., Parati, E. A., Cova, L., Frolichsthal, P., Galli, R., Wanke, E., et al. (1996) Multipotential stem cells from the adult mouse brain proliferate and self renew in response to basic fibroblast growth factor. *J. Neurosci.* **16,** 1091–1100.

46. Weiss, S., Dunne, C., Hewson, J., Wohl, C., Wheatley, M., Peterson, A. C., and Reynolds, B. A. (1996b) Multipotent CNS stem cells are present in the adult mammalian spinal cord and ventricular neuroaxis. *J. Neurosci.* **16,** 7599–7609.

47. Gage, F. H., Ray, J., and Fisher, L. J. (1995) Isolation, characterization, and use of stem cells from the CNS. *Annu. Rev. Neurosci.* **18,** 159–192.

48. Alvarez-Buylla, A. and Temple, S. (1998) Stem cells in the developing and adult nervous system. *J. Neurobiol.* **36,** 105–110.
49. Cameron, H. A. and McKay, R. (1998) Stem cells and neurogenesis in the adult brain. *Curr. Opin. Neurobiol.* **8,** 677–680.
50. Frisen, J., Johansson, C. B., Lothian, C., and Lendahl, U. (1998) Central nervous system stem cells in the embryo and adult. *Cell. Mol. Life Sci.* **54,** 935–945.
51. Gage, F. H. (1998) Stem cells of the central nervous system. *Curr. Opin. Neurobiol.* **8,** 671–676.
52. Kuhn, H. G. and Svendsen, C. N. (1999) Origin, functions, and potential of adult neural stem cells. *Bioessays* **21,** 625–630.
53. Shihabuddin, L. S., Palmer, T. D., and Gage, F. H. (1999) The search for neural progenitor cells: prospects for the therapy of neurodegenerative disease. *Mol. Med. Today* **5,** 474–480.
54. Palmer, T. D., Markakis, E. A., Willhoite, A. R., Safar, F., and Gage, F. H. (1999) Fibroblast growth factor-2 activates a latent neurogenic program in neural stem cells from diverse regions of the adult CNS. *J. Neurosci.* **19,** 8487–8497.
55. Richards, K. J., Kilpatrick, T. J., and Bartlett, P. F. (1992) De novo generation of neuronal cells from the adult mouse brain. *Proc. Natl. Acad. Sci. USA* **9,** 8591–8595.
56. Reynolds, B. A. and Weiss, S. (1992) Generation of neurons and astrocytes from isolated cells of the adult mammalian central nervous system. *Science* **255,** 1707–1710.
57. Morshead, C. M., Reynolds, B. A., Craig, C. G., McBurney, M. W., Staines, W. A., and Morassutti D. (1994) Neural stem cells in the adult mammalian forebrain: a relatively quiescent subpopulation of subependymal cells. *Neuron* **13,** 1071–1082.
58. Johansson, C. B., Momma, S., Clarke, D. L., Risling, M., Lendahl, U., and Frisén, J. (1999) Identification of a neural stem cell in the adult mammalian central nervous system. *Cell* **96,** 25–34.
59. Craig, C. G., D'Sa, R., Morshead, C. M., Roach, A., and van der Kooy, D. (1999) Migrational analysis of the constitutively proliferating subependyma population in adult mouse forebrain. *Neuroscience* **93,** 1197–1206.
60. Kilpatrick, T. J. and Bartlett, P. F. (1993) Cloning and growth of multipotential neural precursors: requirements for proliferation and differentiation. *Neuron* **10,** 255–265.
61. Palmer, T. D., Ray, J., and Gage, F. H. (1995) FGF-2 responsive neuronal progenitors reside in proliferative and quiescent regions of the adult rodent brain. *Mol. Cell. Neurosci.* **6,** 474–486.
62. Gritti, A., Cova, L., Galli, R., Parati, E. A., and Vescovi, A. L.(1995) BFGF supports the proliferation of EGF-generated neuronal precursor cells of the adult mouse CNS. *Neurosci. Lett.* **185,** 151–154.
63. Gritti, A., Frolichsthal, P., Galli, R., Parati, E. A., Cova, L., and Pagano, S. F. (1999) Epidermal and fibroblast growth factors behave as mitogenic regulators for a single multipotent stem cell-like population from the subventricular region of the adult mouse forebrain. *J. Neurosci.* **19,** 3287–3297.

64. Reynolds, B. A. and Weiss, S. (1996) Clonal and population analyses demonstrate that an EGF-responsive mammalian embryonic CNS precursor is a stem cell. *Dev. Biol.* **175,** 1–13.

65. Morshead, C. M., Craig, C. G., and van der Koy, D. (1998) In vivo clonal analyses reveal the properties of endogenous neural stem cells in the adult mammalian forebrain. *Development* **125,** 2251–2261.

66. Weiss, S., Reynolds, B. A., Vescovi, A. L., Morshead, C., Craig, C. G., and van der Kooy, D. (1996a) Is there a neural stem cell in the mammalian forebrain? *Trends Neurosci.* **19,** 387–393.

67. Vescovi, A. L., Reynolds, B. A., Fraser, D. D., and Weiss, S. (1993) bFGF regulates the proliferative fate of unipotent (neuronal) and bipotent (neuronal/astroglial) EGF-generated CNS progenitor cells. *Neuron* **11,** 951–966.

68. Rao, M. S. (1999) Multipotent and restricted precursors in the central nervous system. *Anat. Rec.* **257,** 137–148.

69. Palmer, T. D., Takahashi, J., and Gage, F. H. (1997) The adult rat hippocampus contains primordial neural stem cells. *Mol. Cell. Neurosci.* **8,** 389–404.

70. Gage, F. H., Kempermann, G., Palmer, T. D., Peterson, D. A., and Ray, J. (1998) Multipotent progenitor cells in the adult dentate gyrus. *J. Neurobiol.* **36,** 249–266.

71. Ray, J., Palmer, T. D., Suhonen, J., Takahashi, J., and Gage, F. H. (1997) Neurogenesis in the adult brain: lesson learned from the studies of progenitor cells from the embryonic and adult central nervous system in *Isolation, Characterization and Utilization of CNS Stem Cells* (Gage F. H. and Christen Y., eds.), Springer-Verlag, Berlin, pp. 128–148.

72. Gould, E., Cameron, H. A., Daniels, D. C., Woolley, C. S., and McEwen, B. S. (1992) Adrenal hormones suppress cell division in the adult rat dentate gyrus. *J. Neurosci.* **12,** 3642–3650.

73. Ciccolini, F. and Svendsen, C. N. (1998) Fibroblast growth factor 2 (FGF-2) promotes acquisition of epidermal growth factor (EGF) responsiveness in mouse striatal precursor cells: identification of neural precursors responding to both EGF and FGF-2. *J. Neurosci.* **18,** 7869–7880.

74. Temple, S. and Alvarez-Buylla, A. (1999) Stem cells in the adult mammalian central nervous system. *Curr. Opin. Neurobiol.* **9,** 135–141.

75. Kilpatrick, T. J. and Bartlett, P. F. (1995) Cloned multipotential precursors from the mouse cerebrum require FGF-2, whereas glial restricted precursors are stimulated with either FGF-2 or EGF. *J. Neurosci.* **15,** 3653–3661.

76. Burrows, R. C., Wancio, D., Levitt, P., and Lillien, L. (1997) Response diversity and the timing of progenitor cell maturation are regulated by developmental changes in EGFR expression in the cortex. *Neuron* **19,** 251–267.

77. Qian, X., Davis, A. A., Goderie, S. K., and Temple, S. (1997) FGF2 concentration regulates the generation of neurons and glia from multipotent cortical stem cells. *Neuron* **18,** 81–93.

78. Lillien, L. (1998) Neural progenitors and stem cells: mechanisms of progenitor heterogeneity. *Curr. Opin. Neurobiol.* **8,** 37–44.

79. Armstrong, R. C., Dorn, H. H., Kufta, C. V., Friedman, E., and Dubois-Dalcq, M. E. (1992) Pre-oligodendrocytes from adult human CNS. *J. Neurosci.* **12,** 1538–1547.

80. Scolding, N., Franklin, R., Stevens, S., Heldin, C-H., Compston, A., and Newcombe, J. (1998) Oligodendrocyte progenitors are present in the normal adult human CNS and in the lesions of multiple sclerosis. *Brain* **121**, 2221–2228.

81. Nait-Ouesmar, B., Decker, L., Lachapelle, F., Avellana-Adalid, V., Bachelin, C., Baron-VanEvercooren, A. (1999) Progenitor cells of the adult mouse subventricular zone proliferate, migrate and differentiate into oligodendrocytes after demyelination. *Eur. J. Neurosci.* **11**, 4357–4366.

82. Shihabuddin, L. S., Ray, J., and Gage, F. H. (1997) FGF-2 is sufficient to isolate progenitors found in the adult mammalian spinal cord. *Exp. Neurol.* **148**, 577–586.

83. Marmur, R., Mabie, P. C., Gokhan, S., Song, Q., Kessler, J. A., and Mehler, M. F. (1998) Isolation and developmental characterization of cerebral cortex multipotent progenitors. *Dev. Biol.* **204**, 577–591.

84. Bonfanti, L., Olive, S., Poulain, D. A., and Theodosis, D. T. (1992) Mapping of the distribution of polysialylated neural cell adhesion molecule throughout the central nervous system of the adult rat: an immunohistochemical study. *Neuroscience* **49**, 419–436.

85. Potten, C. S. and Loeffler, M. (1990) Stem cells: attributes, cycles, spirals, pitfalls and uncertainties. Lesson for and from the crypt. *Development* **110**, 1001–1020.

86. Craig, C. G., Tropepe, V., Morshead, C. M., Reynolds, B. A., Weiss, S., and van der Kooy, D. (1996) In vivo growth factor expansion of endogenous subependymal neural precursor cell populations in the adult mouse brain. *J. Neurosci.* **16**, 2649–2658.

87. Kuhn, H. G., Winkler, J., Kempermann, G., Thal, L. J., and Gage, F. H. (1997) Epidermal growth factor and fibroblast growth factor-2 have different effects of neural progenitors in the adult rat brain. *J. Neurosci.* **17**, 5820–5829.

88. Zigova, T., Pencea, V., Wiegand, S. J., and Luskin, M. B. (1998) Intraventricular administration of BDNF increases the number of newly generated neurons in the adult olfactory bulb. *Mol. Cell. Neurosci.* **11**, 234–245.

89. Chiasson, B. J., Tropepe, V., Morshead, C. M., and van der Kooy, D. (1999) Adult mammalian forebrain ependymal and subependymal cells demonstrate proliferative potential, but only subependymal cells have neural stem cell characteristics. *J. Neurosci.* **19**, 4462–4472.

90. Temple, S. (1999) The obscure origins of adult stem cells. *Curr. Biol.* **9**, 397–399.

91. Doetsch, F., Caille, I., Lim, D. A., Garcia-Verdugo, J. M., and Alvarez-Buylla, A. (1999) Subventricular zone astrocytes are neural stem cells in the adult mammalian brain. *Cell* **97**, 703–716.

92. Chazal, G., Durbec, P., Jankovsky, A., Rougon, G., and Cremer, H. (2000) Consequences of neural cell adhesion molecule deficiency on cell migration in the rostral migratory stream of the mouse. *J. Neurosci.* **20**, 1446–1457.

93. Hu, H., Tomasiewicz, H., Magnuson, T., and Rutishauser, U. (1996) The role of polysialic acid in migration of olfactory bulb interneuron precursors in the subventricular zone. *Neuron* **16**, 735–743.

94. Bonfanti, L. and Theodosis, D. T. (1994) Expression of polysialylated neural cell adhesion molecule by proliferating cells in the subependymal layer of the adult rat, in its rostral extension and in the olfactory bulb. *Neuroscience* **62,** 291–305.

95. Seki, T. and Arai, Y. (1993) Highly polysialylated neural cell adhesion molecule (NCAM-H) is expressed by newly-generated granule cells in the dentate gyrus of the adult rat. *J. Neurosci.* **13,** 2351–2358.

96. Lois, C., Garcia-Verdugo, J., and Alvarez-Buylla, A. (1996) Chain migration of neuronal precursors. *Science* **271,** 978–981.

97. Jankovski, A. and Sotelo, C. (1996) Subventricular zone-olfactory bulb migratory pathway in the adult mouse: cellular composition and specificity as determined by heterochronic and heterotopic transplantation. *J. Comp. Neurol.* **371,** 376–396.

98. Peretto, P., Merighi, A., Fasolo, A., and Bonfanti, L. (1997) Glial tubes in the rostral migratory stream of the adult rat. *Brain Res. Bull.* **42,** 9–21.

99. Vutskitis, L., Paccaud, J. P., Djebbara-Hannas, Z., Durbec, P., Rougon, G., and Kiss, J. Z. (1999) PSA-NCAM modulates the responsiveness of cultured cortical neurons to BDNF. *Soc. Neurosci. Abstr.* **25,** 510.

100. Muller, D., Djebbara-Hannas, Z., Jourdain, P., Vutskits, L., Durbec, P., Rougon, G., and Kiss, J. Z. (2000) BDNF restores LTP in PSA-NCAM deficient hippocampus. *Proc. Natl. Acad. Sci. USA* **8,** 4315–4320.

101. Bjornson, C. R. R., Rietze, R. L., Reynolds, B. A., Magli, M. C., and Vescovi, A. L. (1999) Turning brain into blood: a hematopoietic fate adopted by adult neural stem cells in vivo. *Science* **283,** 534–537.

102. Thomas, L. B., Gates, M. A., and Steindler, D. A. (1996) Young neurons from the adult subependymal zone proliferate and migrate along an astrocyte, extracellular matrix-rich pathway. *Glia* **17,** 1–14.

103. Weissmann, I. L. (2000) Stem cells: units of regenaration and units in evolution. *Cell* **100,** 157–168.

104. Lim, D. A. and Alvarez-Buylla, A. (1999) Interaction between astrocytes and adult subventricular zone precursors stimulates neurogenesis. *Proc. Natl. Acad. Sci. USA* **96,** 7526–7531.

105. Suhonen, J. O., Peterson, D. A., Ray, J., and Gage, F. H. (1996) Differentiation of adult hippocampus-derived progenitors into olfactory neurons in vivo. *Nature* **383,** 624–627.

106. Takahashi, M., Palmer, T. D., Takahashi, J., and Gage, F. H. (1998) Widespread integration and survival of adult-derived neural progenitor cells in the developing optic retina. *Mol. Cell. Neurosci.* **12,** 340–348.

4

Glial Characteristics of Adult Subventricular Zone Stem Cells

Daniel A. Lim and Arturo Alvarez-Buylla

INTRODUCTION

It is now widely accepted that new neurons are added continuously to select regions of the adult mammalian brain. Over 30 years of reports describing neurogenesis in the adult brains of fish, frogs, reptiles, birds, and rodents *(1–6)* have recently culminated in studies demonstrating the birth of new central nervous system (CNS) neurons in primates including humans *(7,8)*. Hence, the century-old, dogmatic proposition of a fixed, ended, immutable adult brain has been refuted, spurring new investigations into the regenerative capacity of the CNS.

The dentate gyrus of the hippocampus *(9)* and the lateral ventricle subventricular zone (SVZ) *(10)* are two brain regions in which neurons are born in the adult. The SVZ is the larger of these two germinal zones and consists of a layer of cells adjacent to the ependyma along the entire length of the lateral ventricular wall. In postnatal *(11)* and adult rodents *(12)*, cells born in the SVZ migrate from the ventricular wall into the olfactory bulb (OB), where they differentiate into interneurons. Interestingly, in the monkey brain, the SVZ has been found to generate new neurons for the prefrontal, inferior temporal, and posterior parietal cortex *(7)*. In the adult human SVZ, investigators have also found proliferating cells *(8,13)*, raising the intriguing possibility that we — like rodents and other primates — harbor a neurogenic population of cells within the walls of our brain ventricles *(14)*.

The proliferation of SVZ cells continues throughout life *(15,16)*. It has been estimated that at least 30,000 new OB neurons are born in the mouse every day to replace those that are dying *(12)*. This profound level of continuous neurogenesis argues for the presence of a self-renewing cell type within the SVZ. Self-renewing cells from the SVZ have been propagated in

From: *Stem Cells and CNS Development*
Edited by: M. S. Rao © Humana Press Inc., Totowa, NJ

vitro in both adherent and nonadherent cultures, and these cells can differentiate into neurons, astrocytes, and oligodendrocytes *(17–19)*. Self-renewal and multilineage differentiation are two generic attributes of stem cells. A population of cells in the SVZ satisfies these two criteria and can thus be described as a neural stem cell. However, the precise definition of stem cells is a matter of debate *(20–22)*. The SVZ stem cell is perhaps most analogous to stem cells found in the skin, intestine, and blood. Stem cells of the SVZ and these other regions generate new cells for their respective organ systems throughout the life of the animal. The constant production of new cells complements normal cell turnover, maintaining the tissue cell population. It is not clear how similar adult brain stem cells are to those of the embryo. We define the adult mouse SVZ stem cell as the self-renewing cell type responsible for maintaining the constant production of OB neurons in vivo.

Perhaps the most misleading notion about stem cells is that they should be undifferentiated or primitive, lacking expression of markers attributed to more mature cells. This perception has led many researchers to ignore the "mature-looking" cell as potential stem cells. However, it is becoming increasingly clear that stem cells can bear what were thought to be the biochemical hallmarks of differentiated cells. For instance, skin stem cells express intermediate filament keratins found in mature keratinocytes *(23,24)*. Intestinal crypt stem cells, which continuously replace the epithelial lining of the digestive tract, have been described as being more epithelial than primitive *(25)*. Hematopoietic stem cells (HSC), perhaps the best studied of all stem cells, express what have been considered to be lineage-restricted factors *(26)*.

In this review, we incorporate our understanding of the cellular composition of the SVZ with recent experimental results that identify the neural stem cell. Surprisingly, the stem cell candidate possesses attributes of mature glial cells. Given the prevailing view that glial cells represent an end point in neural development, a glial-like stem cell seems extraordinary. We therefore review the data concerning the SVZ stem cell identity. We then discuss the possibility that glial-like cells might be stem cells at other developmental times. Considering the accumulating evidence of glial-like stem cells, we propose a revision to our current understanding of developmental neural cell lineages.

CELLULAR COMPOSITION AND ORGANIZATION OF THE ADULT MOUSE SVZ

In the adult mouse, neuroblasts born along the entire length of the SVZ migrate anteriorly to the OB. The migratory neuroblasts (type A cells) travel as chains of themselves *(27,28)* along many interconnecting paths widely distributed throughout the lateral ventricle wall *(29)*. These paths converge at the anterior SVZ, where the confluence of type A cells continues along a restricted path into the OB. In neonatal rat brain, it has been suggested that SVZ neuroblasts originate exclusively in the anterior SVZ, the so-called SVZa *(30)*. This is not observed in the adult. The high concentration of neuroblasts that converge at the anterior SVZ may give the impression that this is the site of origin of these cells. Further work is required to describe potential differences between neonatal and adult SVZ and to determine the nature of cells in the neonatal caudal SVZ.

Chains of migrating type A cells in the adult mouse brain are ensheathed by the processes of slowly dividing SVZ astrocytes (type B cells) *(27)*. Scattered along the type A cell chains are clusters of rapidly dividing immature cells (type C cells). Type C cell clusters are often interposed between type B and A cells *(31)*. (See Fig. 1 for a schematic cross-section of the SVZ.)

SVZ cell types were assigned by their morphologic, immunocytochemical, and ultrastructural characteristics *(31)*. Type A cells are immunopositive for a neuron-specific β-tubulin revealed by the monoclonal antibody Tuj1 and express a polysialated form of neuronal cell adhesion molecule (PSA-NCAM). Type B cells contain intermediate filament bundles with glial fibrillary acidic protein (GFAP), a marker assigned to mature astrocytes. Type C cells are ultrastructurally immature and do not stain for markers of mature brain cells. Based on their immature cellular appearance, proximity to the type A cells, and high mitotic activity, type C cells have been proposed to be precursors of type A cells. Adjacent to the SVZ is the layer of multiciliated ependymal cells. Interestingly, all SVZ cell types and the ependyma express nestin *(31)*, an intermediate filament protein found in neuroepithelial stem cells *(32)*.

Ependymal cells line the luminal surface of the brain ventricle and appear highly differentiated, bearing multiple beating cilia that move cerebrospinal fluid through the ventricular system. The lateral ventricle ependyma is generally described as a layer of multiciliated epithelial cells that separate the SVZ from the ventricular lumen. However, upon closer examination using electron microscopy (EM), the ependymal layer does not appear entirely contiguous. In normal mice, a small number of type B cells make direct contact with the ventricle *(33)*. Some of these type B cells contact the ventricle by extending a thin cellular process between ependymal cells, whereas a few have a larger luminal surface (see "x"-marked cell in Fig. 1). Thus, the boundary between the ependymal layer and the SVZ is somewhat blurred by the small number of type B cells that are interdigitated with the ependymal cells. In addition to their unusual cellular location, some of the ventricle-contacting type B cells possess a single, thin cilium lacking the central pair of microtubules. Similar single cilium with this 9 + 0 microtubule arrangement have been described in embryonic neuroepithelial cells *(34,35)* and adult

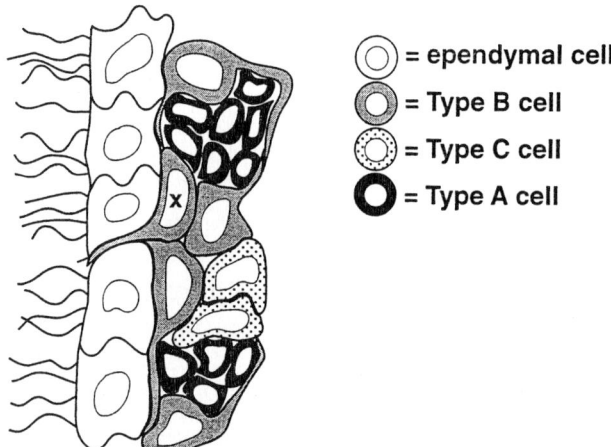

Fig. 1. Schematic cross section of the adult SVZ. Ependymal cells (white) are multiciliated and are closely apposed to the underlying SVZ cells. The ventricular lumen is to the left. Type B cells (gray) are slowly dividing astrocyte-like cells that ensheath chains of migrating type A cells (black). In this cross-section, type A cells would be migrating out of the plane of the paper. Type C cells (stippled) are highly mitotic and are found as clusters along the chains of type A cells. See Cellular Composition and Organization of the Adult Mouse SVZ section for details. Some type B cells (marked with "x") extend a cellular process between ependymal cells and contact the ventricle. Many of these ventricle-contacting type B cells have a short, single cilium lacking the central pair of microtubules (9 + 0 arrangement). Ventricle-contacting type B cells may be an actively dividing SVZ stem cell. See The Ventricle-Contacting Type B Cell: Interkinetic Nuclear Movement? section for details.

avian brain neuronal precursors *(36)*. As we discuss subsequently, this cilium may be an indicator of progression through the cell cycle.

WHICH ARE THE SVZ STEM CELLS?

As mentioned in the introduction, stem cells can express markers of differentiated cells. Perhaps, then, it should not be surprising that candidates for the adult SVZ stem cell have been found to possess attributes of mature glia. Johansson et al. *(37)* recently suggested that ependymal cells are multipotent neural stem cells. Independently, Doetsch et al. *(38)* from our laboratory identified the type B cell as the SVZ stem cell. Although the results of Johansson et al. and Dotesch et al. agree in that they both describe a neural stem cell with mature cellular characteristics, they disagree on the exact identity of the SVZ stem cell. The experimental foundation of the controversial claims about SVZ stem cell identity thus deserves careful appraisal.

First, we evaluate the data concerning which cells remain in the SVZ for extended periods after division. Second, we review the in vivo lineage analysis of ependymal and type B cells. Third, we review the evidence that type B cells can regenerate the SVZ after ablation of rapidly dividing cells. Finally, we consider the in vitro experiments that identify the stem cell.

The Label-Retaining Cell of the SVZ

A traditional view of adult stem cells is that they divide very slowly. For instance, to maintain hematopoiesis, HSCs enter the cell cycle every 1–3 mo *(39,40)*, and the slowest cycling cell in the skin has stem cell behavior *(41)*. Accordingly, data from two studies suggest that SVZ stem cells are the most slowly dividing cells of this region *(33,42)*. As a result of their slow cell cycle, stem cells are labeled infrequently by a single pulse of a nucleotide analog such as [^{3}H]thymidine or bromodeoxyuridine (BrdU). Efficient labeling of stem cells requires continuous or repeated administration of [^{3}H]thymidine or BrdU for a prolonged duration. Once having incorporated the label, the stem cells retain the mitotic marker for an extended period and can thus be identified as label-retaining cells (LRCs). Rapidly dividing progenitor cells dilute out the label and/or migrate from the region.

The label-retaining experiment has been performed in the SVZ by both Johannson et al. *(37)* and Doetsch et al. *(38)*. Both groups administered BrdU for at least 2 wk in the drinking water. One week after the end of BrdU administration, brain sections were processed with BrdU immunohistochemistry. Because some labeled nuclei appear to be very close to or within the ependymal layer, Johannson et al. concluded that ependymal cells are the LRCs. Although the BrdU-positive nuclei are very clearly labeled, the resolution of the light microscope is not sufficient to distinguish ependymal cells from the closely apposed SVZ cells. Type B cells sometimes have their nuclei separated from the ventricular lumen by only a thin process of an adjacent ependymal cell, and such a nuclei could be easily mistaken as belonging to the ependymal layer *(33,38)*. In addition, as described earlier, some type B cells are interposed between ependymal cells and actually contact the ventricle; these type B cells may have their nuclei directly in the ependymal layer.

The precision of LRC identification can be improved with double immunohistochemistry for BrdU and a cell-specific marker. Doetsch et al. *(38)* double-immunostained brain sections for BrdU and an ependymal cell marker, either CD24 or S100β. Although many BrdU-positive nuclei are close to the ventricle, none of the labeled nuclei belong to CD24-positive or S100β-positive ependymal cells. Given the close association of ependymal cells and type B cells, EM was required to confirm the membrane boundaries of LRCs. [³H]Thymidine was continuously infused for 12 d into the lateral ventricle and [³H]thymidine-labeled cells analyzed at the EM. No ependymal cells were labeled by [³H]thymidine.

Do ependymal cells ever divide in vivo? Previous reports are also conflicting. A survey of the literature reveals reports concluding that the lateral ventricle ependyma do not divide and others that claim to have evidence of ependymal proliferation in the same area *(43)*. It is difficult to come to any conclusion from the earlier reports, as EM was not used to confirm ependymal cell identity in the lateral ventricle wall. There are also reports of ependymal cell proliferation in the fourth ventricle and central canal of the spinal cord. However, more detailed EM analysis is required to confirm that the proliferating cells in these caudal regions correspond to multiciliated ependymal cells. Johannson et al. show by EM a ventricle-contacting cell in mitosis in the central canal; however, this cell appears to be unciliated. Perhaps the ependymal cells along the neuraxis are not all equivalent. In children, ependymal tumors occur most frequently in the fourth ventricle, and this may reflect such intrinsic differences *(44)*.

Epithelial layers with a stem cell component often display a remarkable regenerative capacity in pathologic conditions. If the ependyma contains a stem cell, then one might expect this epithelium to regenerate after injury. However, there is presently no convincing evidence of ependymal regeneration after injury *(45)*. Interestingly, injury to the ependymal cells stimulates the subependymal astrocytes to proliferate and form a gliotic scar, which appears to substitute for the missing ependyma *(45,46)*.

The present data indicate that the majority of SVZ LRCs are type B cells and not ependymal. LRCs, of course, are not necessarily stem cells. Labeled type B cells might simply represent endogenous local glial cell turnover. Furthermore, the SVZ stem cells may enter the cell cycle so rarely that a 2-wk period of labeling would not identify them.

Tracing the Fate of Ependymal Cells

To determine whether ependymal cells can give rise to OB neurons, Johannsen et al. performed experiments that were designed to follow the fate of ependymal cells in vivo. To label ependymal cells, they injected either the fluorescent lipophilic label DiI or adenovirus carrying the β-galactosidase marker into the lateral ventricular lumen. One day after injection, labeled cells appeared to be restricted to the ependymal layer. Ten days after injection, a large number of labeled cells were found either en route to the OB or in the OB. Interpretation of these data is difficult. Intraventricular injections of tracer substances label all cells that contact the ventricle. Since some type B cells contact the ventricle, it is possible that the DiI injection labeled B cells as well as the ciliated ependyma. Transfer of DiI from the ependyma to other cell types is also difficult to rule out. Johannsen et al. employed the adenoviral vector in an attempt to exclude these possibilities. They found that only ependymal cells express the adenovirus receptor CXADR and thus assumed that this would restrict adenoviral infection to the ependyma. This assumption is, however, not correct, as adenoviral vectors have been reported to infect multiple brain cell types *(47)* including type B astrocytes and other SVZ cells *(48)*. Furthermore, the fact that type B cells contact the ventricle complicates the adenovirus result in the same way it renders the DiI experiment difficult to interpret. The present data support a more general conclusion that cells closely associated with the ependymal layer can generate OB neurons.

Tracing the Fate of Type B Cells

Injections of ecotropic retroviruses encoding markers into the SVZ result in labeled OB neurons. Since type A, B, and C cells are all mitotic, the ecotropic retrovirus traces the fate of all SVZ cells and does not identify any one cell type as being in the OB neuron lineage. To follow the fate of type B cells specifically, Doetsch et al. *(38)* injected an avian leukosis retroviral vector (RCAS) encoding alkaline phosphatase (AP) into the SVZ of transgenic mice. The transgene in the recipient mice directs expression of the avian retrovirus receptor to glial fibrillary acidic protein (GFAP)-positive cells *(49)*. Hence, only mitotic type B cells are labeled by the RCAS vector. One day after injection, only type B cells express the RCAS marker AP gene, confirming the specificity of the initial infection. Then, 3.5 d later, AP-positive cells are found en route to the OB, and by 14 d, many AP-positive neurons integrate into the OB. Although this experiment clearly demonstrates that type B cells can produce OB neurons, it does not exclude ependymal cells from this lineage. One could argue that ependymal cells produce OB neurons directly or through a type B cell intermediate. Also, the data do not demonstrate that type B cells self-renew in vivo. However, the presence of AP-positive type B cells in the SVZ 14 d after infection suggests that stem cells were originally infected.

Regeneration of SVZ after Elimination of Rapidly Dividing Cells

Because they divide more slowly than other cell types, adult stem cells are thought to be more resistant to antimitotic agents. Thus, treatment with certain types of antimitotic drugs should be able to eliminate rapidly dividing progenitor cells while sparing a population of stem cells capable of regenerating the killed cells. Infusion of the antimitotic cytosine-β-D-arabinofuranoside (Ara-C) into the SVZ for 6 d eliminates all type A and C cells *(33)*. The only cell types remaining are type B and ependyma. At the end of Ara-C treatment, no BrdU incorporation is observed in the SVZ. However, 12 h after Ara-C removal, type B cells begin incorporating BrdU. No ependymal cells incorporate the mitotic marker. Two days later, the first type C cells appear, and by 14 d, the entire cellular and architectural composition of the SVZ is regenerated. The appearance of type C cells followed by type A cells suggests a developmental lineage of B to C to A. (See Fig. 2 for a lineage schematic.)

Type B cells incorporate BrdU almost immediately after Ara-C removal. This rapid appearance of mitotic cells suggests that stem cells are recruited into cell division by the absence of progenitor cells (negative feedback loop). If this notion of stem cell induction were true, then one would predict the stem cells themselves to be slowly killed off with continued Ara-C administration. This appears to be the case. Increasing the duration of Ara-C treatment decreases the number of type B cells remaining in the SVZ (Doetsch and Alvarez-Buylla, unpublished observations). With continued Ara-C treatment or local irradiation *(50)*, it may be possible to deplete the SVZ of stem cells.

Since both type B and ependymal cells remain after Ara-C treatment, it is still possible that ependymal cells are neural stem cells. Ependymal cells may have remained quiescent during the pulse of BrdU and were thus not observed to divide. However, ependymal cell mitoses or [^{3}H]thymidine incorporation into ependymal cells were never seen at any of the different survivals studied after termination of the Ara-C treatment. The data indicate that dividing type B cells are sufficient to regenerate the SVZ. The complete elimination of all type B cells would possibly test the ependymal cell potential as a stem cell. Targeting cell death to type B cells or ependymal cells would also be useful to address stem cell identity.

Neural Stem Cells In Vitro

Neural stem cells isolated from the adult brain SVZ can be propagated as nonadherent clusters of cells called neurospheres *(42,51,52)*. Cell proliferation is maintained by high concentrations of epidermal growth factor (EGF). Upon removal of EGF, these cultured cells are capable of differentiating into neurons, astrocytes, and oligodendrocytes. Johannsen et al. *(37)* tested ependymal cells for their ability to produce multipotent neurospheres. From a population of dissociated SVZ cells, multiciliated ependymal cells were micromanipulated into single culture wells. About 6% of these cells form neurospheres. Johannsen et al. *(37)* also show that cells isolated by their expression of Notch-1 give rise to multipotent neurospheres. However, it is not clear that only ependymal cells express Notch-1.

The provocative in vitro results of Johannsen et al. *(37)* have not yet been confirmed by other groups. In similar experiments, Doetsch et al. *(38)* were

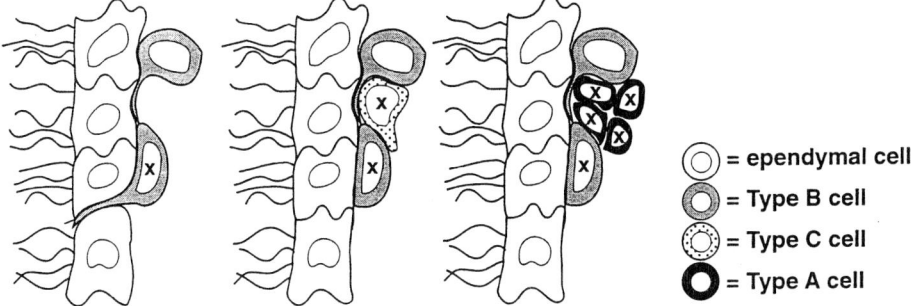

Fig. 2. Proposed SVZ cell lineage. In this model, a type B cell (left, marked with "x") divides asymmetrically to produce another type B cell and a type C cell (middle, marked with "x"). The type C cell may be a transit-amplifying cell type, which divides multiple times before generating type A cells (right, black cells marked with "x"). The remaining type B cell can later reenter the cell cycle to produce more neurons. Type B cells may also divide symmetrically.

not able to grow neurospheres from ependymal cells. Chiasson et al. *(53)* report that whereas ependymal cells can divide in culture, the proliferation is independent of EGF, and the EGF-grown ependymal cells cannot be passaged and do not differentiate into neurons. It is possible that the differences are simply a result of differences in the way the experiments were performed. Johannsen et al. *(37)* used neurosphere-conditioned medium in their cultures; Doetsch et al. *(38)* and Chiasson et al. *(53)* did not. In addition, Chiasson et al. *(53)* used ependyma from the medial ventricular wall, and ependymal cells in different regions of the brain might not be equivalent.

Although Doetsch et al. *(38)* were not able to confirm the results of Johannsen et al. *(37)*, they were able to show that type B cells function as an in vitro neural stem cell precursor. An adenovirus engineered to direct green-fluorescent protein (GFP) expression to infected GFAP-positive cells was injected into the SVZ. The restriction of GFP expression to type B cells was confirmed by both confocal microscopy and EM. GFP co-localizes to cells that are GFAP positive, as revealed by monoclonal antibodies, and is completely excluded from type A and ependymal cells. Although ependymal cells are stained by polyclonal antibodies to GFAP *(31)*, monoclonal antibodies to GFAP do not seem to stain ependyma *(38)*.

Importantly, the adenovirus expressing GFP under the control of the GFAP promoter does not seem to label ependymal cells. In fact, EM analysis demonstrates GFP-positive type B cells immediately adjacent to GFP-negative ependyma. SVZ dissections from injected animals were dissociated to single cells and plated onto clonal isolation dishes. GFP-positive single cells were identified in culture and their locations mapped. The fate of GFP-positive cells was followed over the course of several days. About 10% of GFP-positive cells give rise to neurospheres that can be passaged and differentiated into glial and neurons. Thus, type B cells can give rise to neurospheres.

Although it is tempting to relate EGF-responsive cells in vitro to the in vivo stem cell, a more conservative viewpoint is that the neurosphere assay reveals cell types that can self-renew and become multipotent in response to EGF signaling. The caveats of in vitro stem cell study are perhaps obvious but should be reiterated. Stem cells in vivo reside in niches that are likely to provide a microenvironment critical for their behavior, and cultured stem cells are removed from their normal cellular context. Furthermore, stem cells in culture are exposed to nonphysiological concentrations of mitogenic factors, which may alter their "normal" developmental potential. For example, hippocampal precursors grown in the presence of fibroblast growth factor-2 (FGF2) can differentiate into neurons phenotypically distinct from those of the hippocampus *(54)*. More intriguingly, neurospheres grown from the adult CNS have recently been found capable of producing several hematopoietic lineages when injected into the circulation of irradiated recipient mice *(55)*. In vitro cultures might remove transcriptional silencing and in such a way "deprogram" a cell, making the transcriptional profile more "generic" and allowing a wider diversity of final cell fates.

Thus, any demonstration of stem cell behavior in vitro must be cautiously interpreted. In vitro manipulations may be necessary for stem cell behavior to be unveiled in a particular cell. Although ciliated ependymal cells may divide in vitro in response to EGF, the evidence for ependymal cell division in vivo is not conclusive. Likewise, it is not clear that EGF is the primary mitogen for type B cells in vivo *(56)*, and so the multipotentiality of neurospheres derived from B cells may be a consequence of high levels of EGF signaling. Discovering the molecular signals present in the SVZ is critical for future in vitro studies.

Clues about the molecular signals critical for stem cell biology may come from the intercellular interactions observed in vivo. For instance, hematopoietic stem cells are best maintained in vitro upon cultures of bone marrow stromal cell monolayers *(57)*. Skin stem cells are similarly clonogenic when cultured in contact with fibroblasts, their in vivo cellular neighbors *(58)*. In the SVZ, all cell types are in contact with astrocytes. Reconstituting the interaction between astrocytes and SVZ stem cells in vitro recapitulates the extensive production of type A cells observed in vivo *(59)*.

Neurogenesis in these clonogenic cultures is not dependent on exogenously added growth factors or serum. Understanding the molecular nature of the astrocyte-stem cell interaction may allow for the design of culture assays that fully reproduce in vivo stem cell behavior. Furthermore, the clonogenic in vitro assay may prove to recapitulate the biology of SVZ stem cells more faithfully than high concentrations of EGF or FGF-2.

The SVZ Stem Cell: Ependymal Cell or Astrocyte?

Without having good evidence of ependymal cell division in vivo, it is difficult to conclude that ependymal cells are the SVZ stem cells. However, the evidence of Johannson et al. *(37)* demonstrating that ependymal cells can form multipotent neurospheres in vitro is intriguing although not yet demonstrated by other groups. Perhaps this remarkable in vitro activity of ependymal cells is a reflection of a more generalized phenomenon of in vitro manipulations conferring a wider spectrum of activities on cells. More direct evidence of ependymal cell behavior in vivo is needed to corroborate the in vitro results.

Type B cells have been confirmed by EM to be the in vivo LRC. Type B cells produce OB neurons and can form multipotent neurospheres in vitro. Type B cells are also sufficient to regenerate the SVZ after the elimination of rapidly dividing cells. This body of evidence demonstrating stem cell behavior of type B cells should alter our perception of cells with glial characteristics in the brain. The expression of GFAP can no longer be only ascribed to cells committed to a glial lineage. Cells with morphologic, ultrastructural, and antigenic features of astrocytes may very well have the ability to serve as stem cells. It remains to be determined whether all brain astrocytes retain the ability to become stem cells. In the SVZ, only 10% of the GFAP-positive type B cells can form multipotent neurospheres. Given these data, it is likely that at any one time, only a subset of type B cells can serve as stem cells. If only a subset of type B cells are stem cells, markers specific to those cells would be useful for identification and isolation.

NEURAL STEM CELLS: FROM THE EMBRYO TO THE ADULT

The Ventricle-Contacting Type B Cell: Interkinetic Nuclear Movement?

As described earlier and reviewed above, some type B cells contact the ventricle. In some of these ventricle-contacting type B cells the centriole projects a single 9 + 0 cilium, similar to those on neuroepithelial cells and neuronal precursors of the avian brain. In cultured cells, the appearance of a 9 + 0 cilium has been correlated with cell cycle progression *(60)*. About 6 h before S-phase, one of the interphase centrioles of fibroblasts becomes ciliated, differentiating it from quiescent cells. Perhaps, then, the 9 + 0 cilium on type B cells indicates their progression through the cell cycle. If this is correct, the ventricle-contacting type B cells could be activated SVZ stem cells just hours away from DNA replication.

Could it be that all dividing type B cells transiently contact the ventricle at some point during the cell cycle? Such a mechanism would be reminiscent of the interkinetic nuclear movement observed in the ventricular zone of embryos *(61,62)* and the adult avian brain *(36)*. In the ventricular zone, the nucleus of an actively dividing cell migrates to and from the ventricular lumen at different points in the cell cycle, with mitosis occurring at the ventricular wall. Similarly, a dividing type B cell may actually push aside neighboring ependymal cells and contact the ventricle as it progresses through the cell cycle. It is also possible that the ventricle-contacting type B cells represent dedifferentiating ependymal cells. If such an event actually occurs in vivo, one would expect to observe some transitional cell types representing stages between ependymal and type B cells. However, no such transitional types have been found.

Nomenclature: Is There an Adult Ventricular Zone?

It is perhaps appropriate at this point to raise the problem of nomenclature and its inherent conceptual influences. First, the SVZ, sometimes called the subependymal layer (SEL) or zone (SEZ), suggests that this germinal center functions underneath the ependymal covering without contacting ventricular fluid. In fact, it is widely accepted that the so-called ventricular zone (VZ) *(63)* disappears during development and is no longer present in the adult. Note that these influential statements are purely based on gross anatomic observations and not on testing for whether a VZ-like cell is present in the adult. Clearly, the recent finding that some type B cells may transiently come into contact with the cerebrospinal fluid both challenges this view and contradicts the prefix "sub-" in SVZ, SEL, or SEZ. Second, using the terms SEL or SEZ may not only be inaccurate in terms of the localization of the germinal cells but would also suggest that the adult cell genesis in this region is fundamentally different from that of the embryo. When is the SVZ no longer an SVZ? When does it become a SEZ? Based on the anatomic localization of the embryonic SVZ, and recent transplantation experiments *(64)*, it is evident that the SVZ of the lateral ganglionic eminence (LGE) has cells of similar properties to those present in the adult SVZ. Until an updated terminology becomes available, it is perhaps best to keep one term, SVZ, to describe both the developing and adult germinal zone that underlies the VZ and ependyma, respectively.

Do the Lineages of Neurons and Glia Separate Early in Development?

The recent identification of glial-like neural stem cells raises an important long-standing controversial question concerning the developmental origin of neurons and glia in general. Do neurons and glia arise from a multipotential cell type, or are there specific types of cells devoted to one lineage or the other?

The controversy of multipotent stem cells vs lineage-restricted precursors can be traced back to the earliest studies of the neural tube (65). The CNS arises from a sheet of cells called the neural epithelium. Early in development, the neural epithelium invaginates from the rest of the embryo, forming the neural tube. In 1887, Wilhelm His founded the concept of subclasses of neuroepithelial cells that are consigned to becoming either neurons or glia. His described two neural tube cell types based on their appearance. Germinal cells were the rounded mitotic cells near the lumen, and he proposed these to be precursors of neurons. He also described a columnar matrix of cells, called them spongioblasts, and proposed that they were committed to giving rise to glial cells. His was probably misled by artifacts of histology, as he improperly described spongioblasts as a syncytium rather than separate cells. The theories of His were countered by Schaper in 1894 and 1897. Schaper concluded that the germinal cells and spongioblasts are essentially the same cell type at different stages of cell division. However, it was not until 1935 that F.C. Sauer produced new evidence in favor of Schaper's theory. Later, more cell cycle studies confirmed that all neuroepithelial cells are the same and that the differences in their location and appearance simply represent a different stage of the cell cycle.

Multipotent neuroepithelial stem cells are believed to exist in later stages of brain development. Retroviral lineage analysis and in vitro studies demonstrate that at least some VZ cells are capable of producing both neurons and glia (14,66–71). However, it is not clear whether most of the neurons and glia in the brain are generated from such multipotent VZ cells. The multipotent stem cell of the cortical VZ appears to be rarely infected by retroviruses at these late stages of development (72,73). The VZ stem cell might be either rare or slowly dividing during the formation of cortical structures. It seems that proliferating progenitors committed to specific lineages are preponderant at these developmental stages (74,75).

Radial Glial Cells as Neural Stem Cells

Radial glial cells arise during VZ development and are unique in that they extend processes from the ventricular lumen to the pial surface. Radial glial processes are commonly thought to serve as guides that neuroblasts migrate on to reach their final destination (76). After neuronal production ceases, radial glial are believed to retract from the ventricular and pial surfaces and differentiate into brain astrocytes *(77–80)*. It has been suggested that radial glia do not divide during the period of neurogenesis *(81)*.

It is, however, interesting to compare the morphology of the neuroepithelial stem cell with that of radial glial cells. In the primitive neural tube, the neuroepithelial stem cells contact both the ventricular and pial surfaces (Fig. 3, left). As development progresses, the wall of the neural tube thickens as layers of cells are added. If multipotent neuroepithelial cells are to maintain their ventricular and pial contacts, then they must elongate to accommodate the thickening of the neural tube wall. Radial glial cells are elongated cells, many extending processes to both the ventricular and pial surfaces (Fig. 3, middle). Is it possible, then, that what we call radial glial cells are really just neuroepithelial stem cells with an elongated morphology?

Several lines of evidence support the idea that radial glia are neural stem cells. Mammalian radial glial cells express nestin *(32,82)*, an intermediate filament that is found in neuroepithelial cells *(83)* as well as some cultured neural stem cells *(19)*. Radial glial cells were also found to be mitotic in vivo *(84,85)*, suggesting that radial glial cells have neuroepithelial characteristics *(86)*. In the avian brain, radial glia persist into adult life *(87)*. These cells continue to divide in the adult avian brain. Their division correlates spatially and temporally with the appearance of new neurons, leading to the proposition that these radial cells are neuronal precursors *(88)*. Furthermore, the primary precursors in the adult avian brain undergo interkinetic nuclear migration *(36)*, a phenomenon typified by neuroepithelial cells. Retroviral lineage studies of the adult avian brain have described clones containing radial glia and multiple young neurons *(89)*. In retroviral lineage studies of the developing mammalian striatum and chick optic tectum, clones of cells containing a single radial glial with neurons have also been found *(67,90)*. Although the initially infected cell cannot be defined, such a clonal composition could be explained by an asymmetrically dividing radial cell that gives rise to both neuronal and glial lineages.

Origin and Nature of SVZ Type B cells

There is good evidence that radial glia become astrocytes late in development. Transitional forms between radial glia and astrocytes have been observed in vivo *(77–80)* and in vitro *(91)*. Furthermore, radial glial vitally labeled by injections of tracers onto the surface of the brain can differentiate into astrocytes in vitro *(92)*. It is interesting to consider that SVZ type B cells might be derived from radial glial cells and retain some neuroepithelial stem cell characteristics into adulthood (Fig. 3, right). The microenvironment of the SVZ might provide signals that program type B cells for

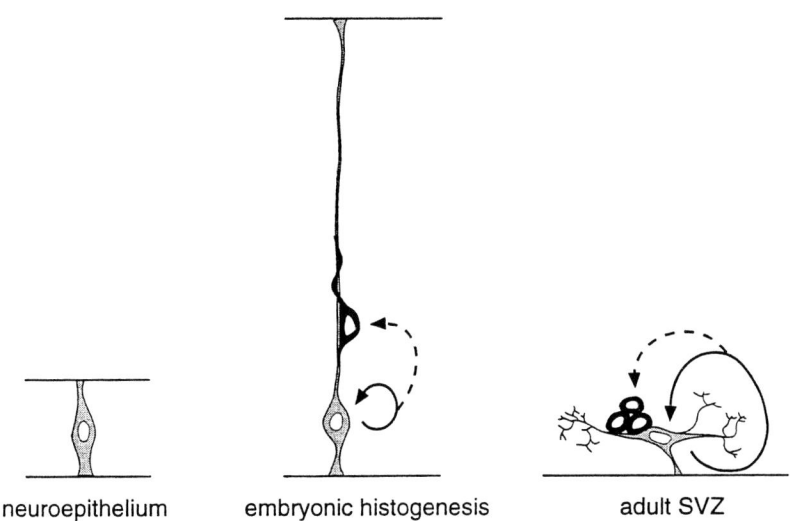

Fig. 3. Hypothetical relationship of neural stem cells from the early embryo to the adult SVZ. Left: Neural stem cells (gray) in the early neuroepithelium extend from the ventricular (bottom) to the pial (top) surfaces. Middle: Like neuroepithelial stem cells, many radial glial cells (gray) also contact both the ventricular and pial surfaces. Radial glia may be neural stem cells, perhaps an elongated form of the stem cell of the early neuroepithelium. Radial glia are known to divide and may self-renew (solid arrow) and produce neurons (black), possibly through intermediate cell types (dotted arrow). Right: Radial glial give rise to astrocytes later in development. Cells derived from radial glial may come to reside in the adult SVZ, where they are identified as type B cells (gray). Like radial glia and neuroepithelial cells, some SVZ type B cells contact the ventricle. These astrocyte-like cells behave as stem cells in that they self-renew (solid arrow) and produce neurons (black), possibly through intermediate cell types (dotted arrow).

continuous OB neuron production. Many astrocytes throughout the brain are also thought to be derived from radial glia, raising the intriguing possibility that some of these cells may also behave as neural stem cells under appropriate conditions. Compelling evidence in support of this idea comes from the demonstrations that in vitro stem cells can be propagated from regions other than the SVZ. Neurosphere-generating cells are found all along the entire ventricular neuroaxis *(93)*. In addition, multipotent stem cells can also be isolated in vitro from the cortex, septum, hippocampus, and SVZ *(94,95)*. It will be interesting to determine whether precursors similar to type B cells exist in these diverse brain regions.

Are Glial Tumors Neural Stem Cell Tumors?

Most brain tumors are glial. However, based on the above discussion, we should consider the possibility that tumors may arise from stem cells. In fact, EGF receptor overexpression in nestin-positive postnatal brain cells lacking the INK4a-ARF locus leads to a high incidence of gliomas *(96)*, suggesting that a genetic alteration in a stem-like cell can generate a tumor that would be histologically classified as a glioma. Interestingly, the SVZ in some mammals is the most common site of gliomas induced by chemical carcinogens *(97)*. Although it is not a frequent site of tumors in humans, the SVZ is perhaps a site where stem cells acquire the initial genetic alterations in tumor cell progression. Certain mutations may enhance the migration of stem cells, leading to the subsequent formation of tumors at a distance from the SVZ. In fact, the overexpression of FGF-2 in glial cells stimulates their migration *(49)*. Also, infusion of FGF-2 or EGF into the lateral ventricle causes SVZ cells to migrate deeper into the striatum *(98,99)*.

THE PROBLEM WITH CALLING A CELL A GLIAL CELL

Some of the unavoidable, historical misconceptions about glial cells appear to have persisted to the present, and it is now important to reexamine the cells that that we call glial in a new light. The concept of neuroglia, meaning "nerve glue," was originally based on an assumption by Rudolf Virchow in the mid-1800s that there must be a mesoderm-derived connective tissue-like component to the nervous system *(65)*. Virchow's theory has since been refuted; however, his ideas about the derivation and nature of glial cells seemingly instilled the field with the idea that glial cells should be distantly related to neurons. Perhaps the shadow of Virchow's conjectures extends to the present day, making it difficult to consider the possibility that glial-like cells are neural stem cells. As proposed in an earlier study, perhaps the radial glial cell should be called simply a radial cell to remove the influential aspect of the word "glial" *(88)*. Alternatively, some of the historical weight of the word "glial" needs to be lightened. In the brain, the term "glia" may be taken to encompass both fully differentiated supportive cells and others that are capable of behaving as neural stem cells. Neural stem cells may have roles often assigned to glia. The cellular anatomy of both the adult SVZ and the developing VZ suggests that stem cells play important structural roles as the scaffold upon which neurogenesis and neuronal migration occur. Hence, it appears that in some cases, glial cells and stem cells are one in the same.

ACKNOWLEDGMENTS

We thank Amy Newman for assistance in preparing this manuscript. This work was supported by NIH grants HD32116 and NS28478. D.A.L. is supported by NIH grant GM07739.

REFERENCES

1. Altman, J. (1970) Postnatal neurogenesis and the problems of neural plasticity, in *Developmental Neurobiology.* (Himwich, W. A., ed.), Charles C.Thomas, Springfield, pp. 197–237.
2. Lopez-Garcia, C., Molowny, A., Garcia-Verdugo, J. M., Martinez-Guijarro, F. J., and Bernabeu, A. (1990) Late generated neurons in the medial cortex of adult lizards send axons that reach the Timm-reactive zones *Dev. Brain Res.* **57,** 249–254.
3. Straznicky, A. and Gaze, R. M. (1971) The growth of the retina in *Xenopus laevis*: an autoradiographic analysis *J. Embryol. Exp. Morph.* **26,** 67–79.
4. Birse, S. C., Leonard, R. B., and Coggeshall, R. E. (1980) Neuronal increase in various areas of the nervous system of the guppy, Lebistes *J. Comp. Neurol.* **194,** 291–301.
5. Goldman, S. A. and Nottebohm, F. (1983) Neuronal production, migration, and differentiation in a vocal control nucleus of the adult female canary brain. *Proc. Natl. Acad. Sci.USA* **80,** 2390–2394.
6. Alvarez-Buylla, A. and Lois, C. (1995) Neuronal stem cells in the brain of adult vertebrates *Stem Cells* **13,** 263–272.
7. Gould, E., Reeves, A. J., Graziano, M. S. A., and Gross, C. G. (1999) Neurogenesis in the neocortex of adult primates *Science* **286,** 548–552.
8. Eriksson, P. S., Perfilieva, E., Bjork-Eriksson, T., Alborn, A., Nordborg, C., Peterson, D. A., and Gage, F. H. (1998) Neurogenesis in the adult human hippocampus. *Nature Med.* **4,** 1313–1317.
9. Gage, F. H., Kempermann, G., Palmer, T., Peterson, D. A., and Ray, J. (1998) Multipotent progenitor cells in the adult dentate gyrus. *J. Neurobiol.* **36,** 249–266.
10. García-Verdugo, J. M., Doetsch, F., Wichterle, H., Lim, D. A., and Alvarez-Buylla, A. (1998) Architecture and cell types of the adult subventricular zone: in search of the stem cells. *J. Neurobiol.* **36,** 234–248.
11. Luskin, M. B. (1993) Restricted proliferation and migration of postnatally generated neurons derived from the forebrain subventricular zone *Neuron* **11,** 173–189.
12. Lois, C. and Alvarez-Buylla, A. (1994) Long-distance neuronal migration in the adult mammalian brain *Science* **264,** 1145–1148.
13. Globus, J. H. and Kuhlenbeck, H. (1944) The subependymal cell plate (matrix) and its relationship to brain tumors of the ependymal type *J. Neuropathol. Exp. Neurol.* **3,** 1–35.
14. Goldman, S. A. (1998) Adult neurogenesis: from canaries to the clinic. *J. Neurobiol.* **36,** 267–286.
15. Goldman, S. A., Kirschenbaum, B., Harrison-Restelli, C., and Thaler, H. T. (1997) Neuronal precursors of the adult rat subependymal zone persist into

senescence, with no decline in spatial extent or response to BDNF. *J. Neurobiol.* **32**, 554–566.

16. Kuhn, H. G., Dickinson-Anson, H., and Gage, F. H. (1996) Neurogenesis in the dentate gyrus of the adult rat: age-related decrease of neuronal progenitor proliferation. *J. Neurosci.* **16**, 2027–2033.

17. Gage, F. H., Ray, J., and Fisher, L. J. (1995) Isolation, characterization, and use of stem cells from the CNS *Annu. Rev. Neurosci.* **18**, 159–192.

18. Weiss, S., Reynolds, B. A., Vescovi, A. L., Morshead, C., Craig, C. G., and Van der Kooy, D. (1996) Is there a neural stem cell in the mammalian forebrain? *Trends Neurosci.* **19**, 387–393.

19. McKay, R. (1997) Stem cells in the central nervous system. *Science* **276**, 66–71.

20. Morrison, S. J., Shah, N. M., and Anderson, D. J. (1997) Regulatory mechanisms in stem cell biology. *Cell* **88**, 287–298.

21. Gage, F. H. (1998) Discussion point: stem cells of the central nervous system *Curr. Opin. Neurobiol.* **8**, 671–675.

22. Alvarez-Buylla, A. and Temple, S. (1998) Stem cells in the developing and adult nervous system. *J. Neurobiol.* **36**, 105–110.

23. Coulombe, P. A., Kopan, R., and Fuchs, E. (1989) Expression of keratin K14 in the epidermis and hair follicle: insights into complex programs of differentiation *J. Cell Biol.* **109**, 2295–2312.

24. Vasioukhin, V., Degenstein, L., Wise, B., and Fuchs, E. (1999) The magical touch: genome targeting in epidermal stem cells induced by tamoxifen application to mouse skin *Proc. Natl. Acad. Sci.USA* **96**, 8551–8556.

25. Fuchs, E. and Segre, J. A. (2000) Stem cells: a new lease on life. *Cell* **100**, 143–155.

26. Hu, M., Krause, D., Greaves, M., Sharkis, S., Dexter, M., Heyworth, C., and Enver, T. (1997) Multilineage gene expression precedes commitment in the hemopoietic system *Genes Dev.* **11**, 774–785.

27. Lois, C., Garcia-Verdugo, J. M., and Alvarez-Buylla, A. (1996) Chain migration of neuronal precursors *Science* **271**, 978–981.

28. Wichterle, H., Garcia-Verdugo, J. M., and Alvarez-Buylla, A. (1997) Direct evidence for homotypic, glia-independent neuronal migration. *Neuron* **18**, 779–791.

29. Doetsch, F. and Alvarez-Buylla, A. (1996) Network of tangential pathways for neuronal migration in adult mammalian brain *Proc. Natl. Acad. Sci.USA* **93**, 14,895–14,900.

30. Luskin, M. B. (1998) Neuroblasts of the postnatal mammalian forebrain: their phenotype and fate. *J. Neurobiol.* **36**, 221–233.

31. Doetsch, F., Garcia-Verdugo, J. M., and Alvarez-Buylla, A. (1997) Cellular composition and three-dimensional organization of the subventricular germinal zone in the adult mammalian brain. *J. Neurosci.* **17**, 5046–5061.

32. Lendahl, U., Zimmerman, L. B., and McKay, R. D. G. (1990) CNS stem cells express a new class of intermediate filament protein. *Cell* **60**, 585–595.

33. Doetsch, F., Garcia-Verdugo, J. M., and Alvarez-Buylla, A. (1999) Regeneration of a germinal layer in the adult mammalian brain. *Proc. Natl. Acad. Sci.USA* **96**, 11,619–11,624.

34. Sotelo, J. R. and Trujillo-Cenóz, O. (1958) Electron microscope study on the development of ciliary components of the neural epithelium of the chick embryo. *Z. Zellforsch.* **49,** 1–12.

35. Stensaas, L. J. and Stensass, S. S. (1968) Light microscopy of glial cells in turtles and birds. *Z. Zellforsch.* **91,** 315–340.

36. Alvarez-Buylla, A., García-Verdugo, J. M., Mateo, A., and Merchant-Larios, H. (1998) Primary neural precursors and intermitotic nuclear migration in the ventricular zone of adult canaries. *J.Neurosci.* **18,** 1020–1037.

37. Johansson, C. B., Momma, S., Clarke, D. L., Risling, M., Lendahl, U., and Frisén, J. (1999) Identification of a neural stem cell in the adult mammalian central nervous system. *Cell* **96,** 25–34.

38. Doetsch, F., Caille, I., Lim, D. A., García-Verdugo, J. M., and Alvarez-Buylla, A. (1999) Subventricular zone astrocytes are neural stem cells in the adult mammalian brain. *Cell* **97,** 1–20.

39. Bradford, G. B., Williams, B., Rossi, R., and Bertoncello, I. (1997) Quiescence, cycling, and turnover in the primitive hematopoietic stem cell compartment *Exp. Hematol.* **25,** 445–453.

40. Cheshier, S. H., Morrison, S. J., Liao, X., and Weissman, I. L. (1999) In vivo proliferation and cell cycle kinetics of long-term self-renewing hematopoietic stem cells. *Proc. Natl. Acad. Sci. USA* **96,** 3120–3125.

41. Morris, R. J. and Potten, C. S. (1994) Slowly cycling (label-retaining) epidermal cells behave like clonogenic stem cells in vitro. *Cell Prolif.* **27,** 279–289.

42. Morshead, C. M., Reynolds, B. A., Craig, C. G., McBurney, M. W., Staines, W. A., Morassutti, D., Weiss, S., and Van der Kooy, D. (1994) Neural stem cells in the adult mammalian forebrain: a relatively quiescent subpopulation of subependymal cells. *Neuron* **13,** 1071–1082.

43. Bruni, J. E. (1998) Ependymal development, proliferation, and functions: a review. *Microsc. Res. Tech.* **41,** 2–13.

44. Bigner, D. D., McLendon, R. E., and Bruner, J. M. (1998) in Tumors of neuroglial cells, *Russell & Rubinstein's Pathology of Tumors of the Nervous System,* vol. I. Oxford University Press, Oxford, pp. 392–419.

45. Bruni, J. E., Del Bigio, M. R., and Clattenburg, R. E. (1985) Ependyma: normal and pathological. A review of the literature *Brain Res. Rev.* **9,** 1–19.

46. Grondona, J. M., Pérez-Martín, M., Cifuentes, M., Pérez, J., Jiménez, A. J., Pérez-Fígares, J. M., and Fernández-LLebrez, P. (1996) Ependymal denudation, aqueductal obliteration and hydrocephalus after a single injection of neuraminidase into the lateral ventricle of adult rats *J. Neuropathol. Exp. Neurol.* **55,** 999–1008.

47. Davidson, B. L., Allen, E. D., Kozarsky, K. F., Wilson, J. M., and Roessler, B. J. (1993) A model system for in vivo gene transfer into the central nervous system using an adenoviral vector *Nature Genet.* **3,** 219–223.

48. Yoon, S. O., Lois, C., Alvirez, M., Alvarez-Buylla, A., Falck-Pederson, E., and Chao, M. V. (1996) Adenovirus-mediated gene delivery into neuronal precursors of the adult mouse brain *Proc. Natl. Acad. Sci. USA* **93,** 11,974–11,979.

49. Holland, E. C. and Varmus, H. E. (1998) Basic fibroblast growth factor induces cell migration and proliferation after glia-specific gene transfer in mice. *Proc. Natl. Acad. Sci. USA* **95,** 1218–1223.

50. Bellinzona, M., Gobbel, G. T., Shinohara, C., and Fike, J. R. (1996) Apoptosis is induced in the subependyma of young adult rats by ionizing irradiation. *Neurosci. Lett.* **208,** 163–166.
51. Reynolds, B. and Weiss, S. (1992) Generation of neurons and astrocytes from isolated cells of the adult mammalian central nervous system. *Science* **255,** 1707–1710.
52. Gritti, A., Parati, E. A., Cova, L., Frolichsthal, P., Galii, R., and Wanke, E. (1996) Multipotential stem cells from the adult mouse brain proliferate and self-renew in response to basic fibroblast growth factor. *J. Neurosci.* **16,** 1091–1100.
53. Chiasson, B. J., Tropepe, V., Morshead, C. M., and Van der Kooy, D. (1999) Adult mammalian forebrain ependymal and subependymal cells demonstrate proliferative potential, but only subependymal cells have neural stem cell characteristics. *J. Neurosci.* **19,** 4462–4471.
54. Suhonen, J. O., Peterson, D. A., Ray, J., and Gage, F. H. (1996) Differentiation of adult hippocampus-derived progenitors into olfactory neurons *in vivo*. *Nature* **383,** 624–627.
55. Bjornson, C. R. R., Rietze, R. L., Reynolds, B., Magli, M. C., and Vescovi, A. L. (1999) Turning Brain into blood: a hematopoietic fate adopted by adult neural stem cells *in vivo*. *Science* **283,** 534–471.
56. Tropepe, V., Craig, C. G., Morshead, C. M., and Van der Kooy, D. (1997) Transforming growth factor-a null and senescent mice show decreased neural progenitor cell proliferation in the forebrain subependyma. *J. Neurosci.* **17,** 7850–7859.
57. Deryugina, E. I. and Muller-Sieburg, C. E. (1993) Stromal cells in long-term cultures: keys to the elucidation of hematopoietic development? *Crit. Rev. Immunol.* **13,** 115–150.
58. Rheinwald, J. G. and Green, H. (1975) Serial cultivation of strains of human epidermal keratinocytes: the formation of keratinizing colonies from single cells. *Cell* **6,** 331–337.
59. Lim, D. A. and Alvarez-Buylla, A. (1999) Interaction between astrocytes and adult subventricular zone precursors stimulates neurogenesis. *Proc. Natl. Acad. Sci.USA* **96,** 7526–7531.
60. Ho, P. T. C. and Tucker, R. W. (1989) Centriole ciliation and cell cycle variability during g1 phase of BALB/c 3T3 Cells. *J. Cell. Physiol.* **139,** 398–406.
61. Sauer, F. C. (1935) Mitosis in the neural tube. *J. Comp. Neurol.* **62,** 377–405.
62. Takahashi, T., Nowakowski, R. S., and Caviness, Jr., V. S. (1993) Cell cycle parameters and patterns of nuclear movement in the neocortical proliferative zone of the fetal mouse. *J. Neurosci.* **13,** 820–833.
63. The Boulder Committee (1970) Embryonic vertebrate central nervous system: revised terminology. *Anat. Rec.* **166,** 257–262.
64. Wichterle, H., Garcia-Verdugo, J. M., Herrera, D. G., and Alvarez-Buylla. A. (1999) Young neurons from medial ganglionic eminence disperse in adult and embryonic brain *Nature Neurosci.* **2,** 461–466.
65. Jacobson, M. (1991) The germinal cell, histogenesis, and lineages of nerve cells, in *Developmental Neurobiology,* Plenum, New York, pp. 91–93.
66. Temple, S. (1989) Division and differentiation of isolated CNS blast cells in microculture. *Nature* **340,** 471–473.

67. Gray, G. E. and Sanes, J. R. (1992) Lineage of radial glia in the chicken optic tectum. *Development* **114,** 271–283.
68. Qian, X., Goderie, S. K., Shen, G., Stern, J. H., and Temple, S. (1998) Intrinsic programs of patterned cell lineages in isolated vertebrate CNS ventricular zone cells. *Development* **125,** 3143–3152.
69. Cepko, C. L., Austin, C. P., Walsh, C., Ryder, E. F., Halliday, A., and Fields-Berry, S. C. (1990) Studies of cortical development using retrovirus vectors. Cold Spring Harbor Symp. Quant. Biol. **LV,** 265–278.
70. Galileo, D. S., Gray, G. E., Owens, G. C., Majors, J., and Sanes, J. R. (1990) Neurons and glia arise from a common progenitor in chicken optic tectum: demonstration with two retroviruses and cell type-specific antibodies. *Proc. Natl. Acad. Sci. USA* **87,** 458–462.
71. Gray, G. E., Clover, J. C., Majors, J., and Sanes, J. R. (1988) Radial arrangement of clonally related cells in the chicken optic tectum: lineage analysis with a recombinant retrovirus *Proc. Natl. Acad. Sci. USA* **85,** 7356–7360.
72. Luskin, M. B., Pearlman, A. L., and Sanes, J. R. (1988) Cell lineage in the cerebral cortex of the mouse studied in vivo and in vitro with a recombinant retrovirus. *Neuron* **1,** 635–647.
73. Walsh, C. and Cepko, C. L. (1988) Clonally related cortical cells show several migration patterns. *Science* **241,** 1342–1345.
74. Luskin, M. B., Parnavelas, J. G., and Barfield, J. A. (1993) Neurons, astrocytes, and oligodendrocytes of the rat cerebral cortex originate from separate progenitor cells: an ultrastructural analysis of clonally related cells. *J. Neurosci.* **13,** 1730–1750.
75. Mayer-Proschel, M., Kalyani, A. J., Mujtaba, T., and Rao, M. S. (1997) Isolation of lineage-restricted neuronal precursors from multipotent neuroepithelial stem cells. *Neuron* **19,** 773–785.
76. Rakic, P. (1972) Mode of cell migration to the superficial layers of fetal monkey neocortex. *J. Comp. Neurol.* **145,** 61–84.
77. Ramón y Cajal, S. (1911) Histogenesis of the spinal cord and spinal ganglia. *Histologie du Système Nerveux de l'Homme et des Vertébrés,* Maloine, Paris, **Vol. I,** pp. 509–514.
78. Schmechel, D. E. and Rakic, P. (1979) A Golgi study of radial glia cells in developing monkey telencephalon: morphogenesis and transformation into astrocytes. *Anat. Embryol.* **156,** 115–152.
79. Levitt, P. R., Cooper, M. L., and Rakic, P. (1981) Coexistence of neuronal and glial precursor cells in the cerebral ventricular zone of the fetal monkey: an ultrastructural immunoperoxidase analysis. *J. Neurosci.* **1,** 27–39.
80. Pixley, S. K. R. and De Vellis, J. (1984) Transition between immature radial glia and mature astrocytes studied with a monoclonal antibody to vimentin. *Dev. Brain Res.* **15,** 201–209.
81. Schmechel, D. E. and Rakic, P. (1979) Arrested proliferation of radial glial cells during midgestation in rhesus monkey. *Nature* **277,** 303–305.
82. Hockfield, S. and McKay, R. D. G. (1985) Identification of major cell classes in the developing mammalian nervous system *J. Neurosci.* **5,** 3310–3328.

83. Zimmerman, L., Parr, B., Lendahl, U., Cunningham, M., McKay, R. G. B., and Mann, J. (1994) Independent regulatory elements in the nestin gene direct transgene expression to neural stem cells or muscle precursors. *Neuron* **12,** 11–24.

84. Misson, J. P., Edwards, M. A., Yamamoto, M., and Caviness, Jr., V. S. (1988) Mitotic cycling of radial glial cells of the fetal murine cerebral wall: a combined autoradiographic and immunohistochemical study. *Dev. Brain Res.* **38,** 183–190.

85. Frederiksen, K. and McKay, R. D. G. (1988) Proliferation and differentiation of rat neuroepithelial precursor cells in vivo. *J. Neurosci.* **8,** 1144–1151.

86. McKay, R. D. G. (1989) The origins of cellular diversity in the mammalian central nervous system. *Cell* **58,** 815–821.

87. Alvarez-Buylla, A., Theelen, M., and Nottebohm, F. (1988) Mapping of radial glia and of a new cell type in adult canary brain. *J. Neurosci.* **8,** 2707–2712.

88. Alvarez-Buylla, A., Theelen, M., and Nottebohm, F. (1990) Proliferation "hot spots" in adult avian ventricular zone reveal radial cell division. *Neuron* **5,** 101–109.

89. Goldman, J. E. (1995) Lineage, migration, and fate determination of postnatal subventricular zone cells in the mammalian CNS. *J. Neurooncol.* **24,** 61–64.

90. Halliday, A. L. and Cepko, C. L. (1992) Generation and migration of cells in the developing striatum. *Neuron* **9,** 15–26.

91. Culican, S. M., Baumrind, N. L., Yamamoto, M., and Pearlman, A. L. (1990) Cortical radial glia: Identification in tissue culture and evidence for their transformation to astrocytes. *J. Neurosci.* **10,** 684–692.

92. Voigt, T. (1989) Development of glial cells in the cerebral wall of ferrets: direct tracing of their transformation from radial glia into astrocytes. *J. Comp. Neurol.* **289,** 74–88.

93. Weiss, S., Dunne, C., Hewson, J., Wohl, C., Wheatley, M., Peterson, A. C., and Reynolds, B. A. (1996) Multipotent CNS stem cells are present in the adult mammalian spinal cord and ventricular neuroaxis. *J. Neurosci.* **16,** 7599–7609.

94. Palmer, T. D., Ray, J., and Gage, F. H. (1995) FGF-2 responsive neuronal progenitors reside in proliferative and quiescent regions of the adult rodent brain. *Mol. Cell. Neurosci.* **6,** 474–486.

95. Palmer, T. D., Markakis, E. A., Willhoite, A. R., Safar, F., and Gage, F. H. (1999) Fibroblast growth factor-2 activates a latent neurogenic program in neural stem cells from diverse regions of the adult CNS. *J. Neurosci.* **19,** 8487–8497.

96. Holland, E. C., Hively, W. P., DePinho, R., and Varmus, H. E. (1998) Constitutively active epidermal growth factor receptor cooperates with disruption of G1 cell-cycle arrest pathways to induce glioma-like lesions in mice. *Genes Dev.* **12,** 3675–3685.

97. Kleihues, P., Lantos, L., & Magee, P. N. (1976) Chemical carcinogenesis in the nervous system. *Int.Rev.Exp.Pathol.* **15,** 153-232.

98. Kuhn, H. G., Winkler, J., Kempermann, G., Thal, L. J., and Gage, F. H. (1997) Epidermal growth factor and fibroblast growth factor-2 have different effects on neural progenitors in the adult rat brain. *J. Neurosci.* **17,** 5820–5829.

99. Craig, C. G., Tropepe, V., Morshead, C. M., Reynolds, B. A., Weiss, S., and Van der Kooy, D. (1996) *In vivo* growth factor expansion of endogenous subependymal neural precursor cell populations in the adult mouse brain. *J. Neurosci.* **16,** 2649–2658.

5

Neuronal Restricted Precursors

Giri Venkatraman and Marla B. Luskin

INTRODUCTION

The extrinsic and intrinsic factors involved in regulating the developmental progression from totipotent embryonic stem cells to phenotypically restricted neural stem and progenitor cells have been subjects of intense research over the last decade. Two major reasons for this concentrated effort are the gradual acceptance that at least some regions of the brain generate neurons throughout life, and the growing appreciation that these cells could be used to therapeutically treat disorders and injuries of the central nervous system (CNS). Despite early studies by Altman and Das (1) demonstrating ongoing neurogenesis in the adult rodent hippocampus and olfactory bulb, it was widely believed until recently that in mammals the generation of neurons ceases in the late embryonic or early postnatal period. In addition to the persistent neurogenesis in the olfactory bulb and hippocampus, olfactory receptor neurons (ORNs), the first-order neurons in the peripheral olfactory system, were also found to regenerate throughout life in all vertebrates examined (2–4). Postnatal neurogenesis also occurs in the neonatal cerebellum (see ref. 1), although it is more limited in duration than that of the olfactory bulb, the hippocampus, and the ORNs. Populations of neural progenitor cells, which generate the neurons and/or glia in these regions with extended proliferation, have now been identified in the postnatal brain (5–11).

The discovery and characterization of these progenitor cells definitively proved that glia are not the only cells of the nervous system to be generated in the postnatal brain. More recently, brain injury (primarily in the form of seizures) or the administration of neurotrophic factors in the adult rodent brain has also been shown to induce proliferation not only of glia, but also of neurons (12–15). Therefore, the possibility remains that other latent, as yet unidentified, progenitor populations exist in the mature brain.

From: *Stem Cells and CNS Development*
Edited by: M. S. Rao © Humana Press Inc., Totowa, NJ

As the mammalian CNS develops, it is presumed that a gradual restriction occurs in the differentiation potential of stem and progenitor cells. The progeny of these cells generate more restricted progenitors that may be unipotential, i.e., progenitor cells committed to becoming exclusively neurons, astrocytes, or oligodendrocytes (reviewed in ref. *16*). This chapter focuses on two populations of restricted progenitors isolated from the embryonic spinal cord and the neonatal forebrain anterior subventricular zone (SVZa), which are committed to the production of neurons and are therefore called neuronal restricted progenitors (NRPs). To be classified as an NRP, a cell should satisfy the following criteria: (a) an ability to proliferate and (b) expression of a subset of neuronal cell type-specific markers [e.g., type III β-tubulin or microtubule-associated protein (MAP)-2]; NRPs also ordinarily express the embryonic isoform of the neural cell adhesion molecule (E-NCAM). Additionally, we expect the above criteria to be met in vitro as well as in vivo. As a corollary, NRP cells should have an absolute commitment to the neuronal lineage, even in conditions favorable for glial differentiation or in nonnative environments after heterotypic transplantation.

We first review the properties of the mammalian embryonic telencephalic ventricular zone, the prenatal and postnatal telencephalic subventricular zones, and the neuroepithelial cells of the embryonic spinal cord. These are proliferative zones in the developing nervous system that give rise to the majority of neurons and glia in the adult forebrain and spinal cord. We then summarize findings on the postnatal progenitors of the cerebellar, hippocampal, and olfactory receptor neuron populations and contrast the properties of these neural progenitor cell populations to the NRP populations in the embryonic spinal cord and the forebrain anterior subventricular zone. We conclude by discussing the differential effects that neurotrophic factors have on neural progenitor populations in the brain, and how and why the use of NRP cells in transplantation may be therapeutically beneficial. Since these cells are committed to generating neurons, transplantation of NRPs could be useful in CNS diseases with profound neuronal loss.

PROLIFERATIVE ZONES IN THE EMBRYONIC AND POSTNATAL CNS

The diverse array of neurons, astrocytes, and oligodendrocytes present in the mammalian CNS are generated from a single layer of multipotent progenitor cells surrounding the embryonic ventricles and central canal. This layer of multipotent progenitor cells surrounding the ventricle of the embryonic brain is designated the ventricular zone. Precursor cells present in the ventricular zone and, secondarily the subventricular zone (a later appearing layer of cells surrounding the ventricular zone) were found to be the source for most of the neurons and glia comprising the cerebral cortex (reviewed in refs. *17* and *18*). The neurons and glia of the spinal cord arise from analogous stem and progenitor populations surrounding the central canal. By using retroviruses to transduce heritable markers into stem and progenitor cells, it has been shown that by the onset of neurogenesis, in some regions such as the cerebral cortex, there appear to be primarily separate lineages for the neurons, astrocytes, and oligodendrocytes *(19,20)*. In other CNS structures such as the retina, the progenitor cells remain multipotent until their terminal differentiation (reviewed in ref. *21*). Usually, after an appropriate number of mitotic divisions within the proliferative zones, the progenitor cells in the ventricular zone of the telencephalon and embryonic spinal cord exit the cell cycle, and their immature progeny migrate to their permanent position in the CNS. A brief review of the properties of the cells comprising the telencephalic ventricular and subventricular zones and the proliferative layers of embryonic spinal cord follows. Understanding their characteristics will allow us to compare the properties of these multipotent progenitor cells with the developmentally restricted neuronal progenitor cells.

Proliferative Zones in the Embryonic CNS

Telencephalic Ventricular Zone

The earliest layer of dividing cells surrounding the lateral ventricle, which appears during embryogenesis, is the telencephalic ventricular zone (tVZ; see ref. *22*). In vitro, Davis and Temple *(23)* have demonstrated that single tVZ progenitor cells can generate neurons, astrocytes, and oligodendrocytes. Retroviral labeling techniques (see above) have identified late-appearing unipotential cells that generate all pyramidal neurons, all nonpyramidal neurons *(24)*, all astrocytes, or all oligodendrocytes *(25)*. Some studies however, have identified clones containing a mixture of cell types, thus arguing for the presence of bi- or multipotential progenitor cells *(26–28)*. Regardless of whether the neuronal progenitor cells are uni- or multipotential, these cells do not express neuron-specific markers such as type III β-tubulin *(29,30)* or even E-NCAM. The progenitors first exit the cell cycle, and the immature neurons start to migrate, usually along radial glia *(31–33)* to their final destinations in the cerebral cortex. The tVZ progenitor cells, therefore, cannot be considered NRPs since they do not proliferate and express neuron-specific markers concurrently.

Subventricular Zone

The forebrain subventricular zone emerges during embryonic development as a layer of cells surrounding the preexisting tVZ. In the rodent, around E19–20, the tVZ disappears, and cortical neurogenesis ceases. The SVZ, however, persists postnatally after the tVZ has been depleted and is the source of glial progenitors responsible for generating astrocytes and oligodendrocytes *(34)* arising from uni- or, in some cases, bipotential glial precursors *(35–37)*.

Several lines of evidence have indicated, however, that the postnatal SVZ is not exclusively gliogenic. In particular, the SVZ has also been recognized as the region where the neuronal progenitors for the interneurons of the olfactory bulb arise postnatally *(38)*. Additionally, Reynolds and Weiss *(39)* demonstrated the genesis of neurons as well as glia arising from explant cultures from the striatal portion of the postnatal SVZ. Their study suggested that the SVZ might contain multipotent progenitors capable of generating neurons. Furthermore, a lineage analysis using retroviruses capable of expressing bacterial β-galactosidase has revealed that a specialized region of the postnatal subventricular zone surrounding the anterior dorsolateral tip of the lateral ventricle — the SVZa — is composed exclusively of neuronal progenitor cells *(10)*. Other groups have confirmed this finding of neurons originating from the neonatal and adult SVZa (see, for example, refs. *9* and *11*). A more detailed discussion of the SVZa progenitor cells, an NRP population, follows. The portion of the SVZ posterior to the SVZa — henceforth referred to as the SVZp — consists primarily of glial progenitors. The tVZ and the SVZp generate the majority of the neurons and glia in the cerebral cortex; the SVZa generates the interneurons of the olfactory bulb. Together, the progeny of cells in these proliferative regions surrounding the lateral ventricle are responsible for the majority of the neurons and glia in the rodent forebrain, including the striatum, septum, and other structures.

Neuroepithelial Cells of the Embryonic Spinal Cord

In early embryonic development a sheet of morphologically indistinguishable cells termed neuroepithelial (NEP) stem cells folds to form a neural tube at approximately E8.5 in the rat (E7.5 in the mouse). The NEP cells are present in the neural tube as a layer of pseudostratified neuroepithelium encircling the central canal (reviewed in ref. *40*). The progeny of the NEP stem cells migrate away from the embryonic ventricle to their target. Migrating immature neurons in the developing spinal cord express neuron-specific markers, including type III β-tubulin and E-NCAM *(41,42)*, and meet the criteria to be called NRPs. The glial precursors can be identified by their immunoreactivity to specific glial markers such as the antibody A2B5. The progeny of the NRP cells undergo terminal differentiation upon completing their migration.

Rao and his associates *(41,43)* have isolated a pure population of NRP cells from the embryonic spinal cord by plating NEP cells on dishes coated with an antibody to E-NCAM and separating the bound cells — a process known as immunopanning. These experiments have argued that there is an identifiable, phenotypically restricted population of cells distinct from the undifferentiated NEP cells and the differentiated, immature neurons. A similar population may exist in the tVZ but has not been isolated. The spinal cord NRP cells proliferate while expressing neuron-specific markers in vitro, similar to the SVZa progenitors *(44)*. In contrast to the neonatal SVZa, which is present in the postnatal forebrain, this population of cells is apparently present only in the embryonic spinal cord and vanishes by birth.

Proliferative Zones in the Postnatal Brain

Neurogenesis continues in the postnatal rodent in a limited number of regions — the cerebellum, hippocampus, and the olfactory bulb. In addition, the ORNs, the first-order neurons in the olfactory system, are generated throughout life. A summary of the progenitor cells that have been identified in each of these regions follows. The volume of literature characterizing these progenitor populations is immense, and therefore, given the constraints of this chapter, a complete and thorough discussion is not possible. Only the salient points of each study relevant to NRPs will be mentioned.

Hippocampal Progenitors

Neurogenesis in the adult hippocampus has been recognized for several decades *(45–49)*. The progenitor cells are located along a thin strip of cells between the hilar region and the granule cell layer, referred to as the subgranular zone *(5)*. Using the cell proliferation marker bromodeoxyuridine (BrdU) injections (to identify the progeny of the dividing cells), approx 50% of the progeny were found to express a neuronal phenotype and 15% a glial phenotype *(5)*. The phenotype of the other BrdU-positive cells could not be conclusively determined. Similar to stem and multipotent progenitor populations from the SVZ, in the presence of epidermal growth factor (EGF) hippocampal progenitor cells form neurospheres (balls of floating cells) in vitro. Approximately 32% of the cells within a neurosphere differentiate into pyramidal neurons, a percentage that can be increased by the addition of brain-derived neurotrophic factor (BDNF) *(50)*. Thus, one can conclude that the adult hippocampus contains multipotent progenitors whose progeny are both glia and neurons. This multipotent progenitor cell population has also been identified in adult human brains by tagging the dividing progenitor cells with BrdU and determining the phenotype of their progeny *(49)*.

The multipotentiality of hippocampal progenitors has been further substantiated by transplantation experiments. Suhonen et al. *(51)* implanted hippocampal progenitors into the rostral migratory stream (RMS), the pathway traversed by the SVZa-derived cells, to determine whether they adopt the characteristic migratory patterns and phenotype (see below) of the SVZa-derived cells. The transplanted progenitors migrated along the RMS, and a subset of the transplanted cells expressed dopamine — a neurotransmitter never expressed by the endogenous progeny of the hippocampal progenitors, but expressed by the SVZa-derived periglomerular cells in the olfactory bulb. At all the sites examined after migration, however, approx 25% of the progenitors differentiated into glia (assessed by immunoreactivity to glial fibrillary acidic protein [GFAP]). Recently, however, a population of cells has been isolated from the hippocampus that is, apparently, committed to the neuronal lineage *(52)*. More transplantation and in vitro experiments with these cells will shed light on whether these cells meet the above criteria for NRP cells.

Cerebellar Granule Cell Progenitors

Three subsets of precursor cells are present at various stages during cerebellar development. Two of these precursor cell populations are present prenatally (reviewed in refs. *7* and *53*); the third precursor cell population, which is found in the external germinal layer (EGL), comprises the precursors of the granule cells and is present postnatally. The EGL undergoes rapid proliferation at birth until postnatal day 15 (P15) in the rodent. The progenitor cells mature exclusively into granule cell neurons *(54,55)* and maintain this neuronal commitment even after being transplanted into heterotypic sites in the brain. In contrast to the hippocampal progenitors, most of the granule cell progenitors failed to migrate when transplanted into the SVZa *(56)*. Upon transplantation into the hippocampus, however, the cerebellar precursor cells migrated freely and adopted features of host granule or dentate neurons *(57)*, but after transplantation into the striatum, the precursors expressed cerebellum-specific markers *(58)*. At all the sites, after transplantation, the GCL progenitors did not mature into glia, confirming that they are restricted to the neuronal lineage.

Similar to the tVZ progenitors, however, the granule cell progenitors exit the cell cycle before expressing neuronal cell type-specific markers. The progenitor cells in the deeper layers of the eight-cell layer thick EGL first become postmitotic, and the immature neurons migrate along specialized Bergmann glia to their final destinations *(59)* in the internal granule cell layer (IGL, *(60)*. The EGL progenitors express neuron-specific genes such as *Pax-2 (61,62)* and *Math-1 (63)*. Since the EGL progenitors have not been shown to express neuron-specific markers (e.g., type III β-tubulin or MAP-2) while proliferating (Zigova and Luskin, unpublished data), they do not fulfill our criteria for NRPs.

Olfactory Receptor Neuron Progenitors

In addition to the ongoing neurogenesis in the olfactory bulb, which was recognized many years ago, the peripheral ORNs in the nasal epithelium also have a striking ability to sustain neurogenesis routinely as well as to reconstitute themselves after injury *(2,4)*. The average lifespan of an individual ORN in rodents and other vertebrates is estimated to be approx 1 mo. The globose basal cells are the progenitor cell population of the ORNs and reside in the basal layer of the olfactory epithelium *(64)*. As the ORNs mature, there is a gradual migration of the newly generated neurons from the deeper layers of the epithelium to the more superficial layers (toward the nasal cavity). Exposing the olfactory epithelium to detergents such as Triton X-100 or to toxic gases such as methyl bromide destroys the ORNs; however, the basal cells are spared. Following such treatment, the basal cells proliferate and restore the ORN population *(65)*.

The basal cells express some genes and markers associated with neuronal progenitors including the basic helix-loop-helix gene *Mash-1 (66)*, the paired homeobox gene *Pax-6 (67)*, and E-NCAM *(68)*. Their commitment to a neuronal lineage, however, is not absolute. Retroviral labeling studies before and after lesioning with methyl bromide indicate that two types of progenitors are activated after lesioning. The first was a multipotent progenitor giving rise to both neurons and nonneuronal cells, and the second progenitor population generated the ducts, glands, and sustentacular cells, which are the supporting cells in the olfactory epithelium *(69)*. The authors concluded that the progenitors have the capacity to be multipotent and that the globose basal cells were selectively neurogenic in the normal olfactory epithelium, but presumably retained the capacity to be multipotent depending on the cell type(s) that needed to be replenished.

The proliferative zones present in the embryonic and postnatal CNS are ultimately responsible for some of the diverse array of neurons, astrocytes, and oligodendrocytes present in the adult CNS. The genes and factors governing the conversion of a population of multipotent cells into phenotypically distinct cells in the CNS are poorly understood. Again, the above review is not meant to be exhaustive. The reader is referred to other chapters in this volume or to several reviews discussing the characteristics of stem and progenitor cells *(17,40,42,70,71)*. Additionally, despite exhibiting some features of neuronal restricted progenitors, the VZ cells, the NEP stem cells of the spinal cord, and the cerebellar, hippocampal, and ORN progenitors cannot be considered true NRPs because they are multipotent or express neuronal cell type-specific markers only after they cease proliferation. However, the neonatal SVZa and embryonic spinal cord NRPs do fulfill the criteria we have established to be considered true NRPs. Each progenitor population is discussed in turn.

NEURONAL RESTRICTED PROGENITORS

The neonatal SVZa neuronal progenitor cells constitute a distinct population of cells in the neonatal brain separable from the SVZp and the overlying corpus callosum. The true stem cell for the neonatal SVZa progenitors is unknown, but, there is evidence from the adult subventricular zone indicating the presence of an astrocyte-like stem cell within the SVZ *(72)*. However, the neonatal SVZa and the adult SVZ have significant differences with respect to the expression and distribution of astrocyte markers, which is outlined below. Therefore, the identity and characteristics of the multipotent stem cell from which the neonatal SVZa progenitors develop is still an open question.

A convincing candidate for the stem cell for the spinal cord NRP cell, however, has been identified. A multipotent stem cell in the developing spinal cord has been shown to exist *(73,74)*. The NRP cell of the spinal cord is one of three types of progenitor cells that have been isolated from the spinal cord NEP cells. The other progenitor cell populations isolated from the spinal cord neuroepithelial stem cell are the neural crest stem cell (NCSC), which generates peripheral nervous system (PNS) derivatives (see, for example, ref. *75*), and the glial restricted progenitors *(40)*, which generate astrocytes and oligodendrocytes. These findings argue that the spinal cord NRP cells are descended from a pluripotent stem cell and are part of a mosaic of progenitor cells present in the embryonic spinal cord neuroepithelium. We compare and contrast the properties of the SVZa and spinal cord NRP cells to each other and to the other progenitor populations discussed in the above paragraph.

NRPs from the Neonatal Forebrain Anterior Subventricular Zone

As mentioned above in our discussion of the SVZ, retroviruses capable of expressing the bacterial β-galactosidase gene, injected at different points along the entire anteroposterior axis of the neonatal forebrain SVZ, revealed a highly specialized neuronal progenitor population within a discrete region surrounding the dorsolateral tip of the lateral ventricle, corresponding to the anteriormost aspect of the forebrain SVZ (referred to as the SVZa). Without exception, all the cells emanating from the neonatal SVZa were found to generate the interneurons of the olfactory bulb *(10)*. In contrast, virtually all the progeny of the labeled cells situated in the posterior portions of the forebrain SVZ (the aforementioned SVZp) were predominantly fated to become the glia of the cerebral cortex and other forebrain structures. Therefore, the SVZ can be divided into a striatal portion containing multipotent progenitors, a posterior SVZ (SVZp), containing primarily glial progenitors of the cerebral cortex, and the SVZa, which contains exclusively neuronal progenitors. The division between the SVZa and SVZp is further substantiated by our recent studies examining the expression of glial and neuronal markers by polymerase chain reaction, *in situ* hybridization, and immunohistochemistry *(76)*. There are no known anatomic planes or barriers between the neonatal SVZa and SVZp; however, the SVZa is clearly identifiable in the neonatal period for the reasons outlined below.

The mitral cell layer (containing the output neurons of the olfactory bulb) is a discrete layer of postmitotic cells in the olfactory bulb present at birth in the rodent. The progenitors for these cells originate in the VZ *(77,78)*. The interneurons of the olfactory bulb develop postnatally and are generated in large numbers in a short period just after birth *(79,80)*. It stands to reason, therefore, that the progenitors for the interneurons — the SVZa cells — are also present in large numbers during this period. In the early neonatal period, large numbers of SVZa cells and their progeny migrate toward the olfactory bulb, and subsequently the SVZa cell population is substantially reduced in number. The region identifiable in the neonate as the SVZa starts to express glial markers, markedly regresses, and cannot be easily distinguished from the SVZp in juvenile or adult rodents. However, neuronal progenitors still exist in this region and continue to migrate to the olfactory bulb well into adulthood (see, for example, refs. *11, 37,* and *81*). The SVZa-derived cells presumably replace, replenish, or supply new periglomerular and granule cells, thus making the olfactory bulb a highly plastic region in the CNS. However, under particular conditions, the infusion of growth factors like bFGF *(82)* or BDNF *(14)* can expand the SVZa cell population in the adult. Some of these cells express neuronal markers and maintain the ability to divide, and therefore can be considered NRPs.

In vivo, the SVZa progenitors migrate along a well-defined pathway, the RMS *(1)*. The RMS stretches from the SVZa to the subependymal layer in the middle of the olfactory bulb. Most of the migrating SVZa-derived cells exhibit an elongated morphology with a long leading process and a short "tail," typical of migrating neurons. Contrary to most immature neurons arising in the prenatal tVZ, the SVZa progenitors do not migrate along radial glia *(76,85,86)*. The neonatal SVZa-derived cells seem to provide their own substrate for migration and appear to successively "leapfrog" over one another in a process called chain migration *(81,87)* by which one progenitor cell serves as a stepping stone for the adjacent cells. The rate of migration of SVZa-derived cells in the neonatal as well as the adult RMS is significantly faster than that of neurons migrating along radial glia *(88)*. During their migration to the olfactory bulb, the SVZa-derived cells never exit the confines of the RMS. A possible reason for this may be the presence of high levels of proteoglycans and chondroitin sulfate in and/or around the RMS

(89), serving as a barrier, similar to the roles they play in guiding growing axons or growth cones in other regions of the CNS (reviewed in ref. *90*).

A key molecule involved in the migration of the SVZa-derived cells is NCAM. High levels of E-NCAM (also known as polysialated or PSA-NCAM) are expressed in the SVZa and RMS *(81)*. Mice with a genetic deletion of NCAM have significant olfactory bulb defects *(91,92)*. In particular, their olfactory bulbs are smaller in size, and the RMS and SVZa are expanded. Additionally, the SVZa-derived cells are seemingly overly adherent to each other, and migration is curtailed *(92)*. Treloar et al. *(93)* have confirmed and added to these findings and have shown a defect in the glomerulus, a structure where synaptic integration occurs in the olfactory bulb. This discrete genetic deletion points to the importance of NCAM in SVZa cell migration and in the development and organization of the olfactory bulb.

The reason(s) for the highly directional migratory pattern of the SVZa-derived cells is poorly understood. When the olfactory bulb is surgically removed, the SVZa progenitors continue to migrate *(93,94)* (Luskin, unpublished data) toward the ablated bulb, implying that chemorepulsive molecules may influence this process in vivo rather than a chemoattractant molecule from the olfactory bulb. Evidence from in vitro experiments using SVZ neurospheres suggests that the protein Slit may be a candidate repellent molecule *(95)*. Wu et al. have demonstrated that co-culturing SVZ neurospheres with Slit results in the SVZ neuronal progenitors migrating away from the Slit-positive regions; ordinarily, the cells migrate away from the explant in all directions *(11)*. These findings have led to the hypothesis that Slit may direct the neuronal progenitors toward the olfactory bulb. To date, however, this repellent effect induced by Slit has been demonstrated only in vitro. The true role of Slit in this process in vivo is unclear, especially since the RMS is a highly convoluted pathway, and the vertical portion of the RMS actually parallels the septum. Taken together, this evidence indicates that a combination of extracellular matrix molecules, chemorepulsive molecules, and possibly chemoattractant molecules must come into play for the well-orchestrated migration of SVZa progenitor cells.

In contrast to immature neurons arising in the tVZ, migrating neonatal SVZa progenitors express neuronal cell type-specific markers despite being mitotically active. Menezes et al. have shown that virtually all the progenitors in the RMS uniformly express type III β-tubulin *(29)* and MAP-2 (Pencea and Luskin, unpublished data). In addition, the TuJ1-positive cells incorporate the proliferation marker BrdU, in distinct contrast to tVZ cells, which exit the cell cycle prior to expressing cell type-specific markers of differentiating postmitotic neurons. Other markers expressed by the SVZa progenitors include PSA-NCAM (as mentioned earlier), p75 (the low-affinity nerve growth factor receptor), EGF receptors *(96)*, multiple FGF receptor isoforms (Venkatraman and Luskin, unpublished data), and a novel cell surface marker mAb-2F7, which labels neurons and neuronal progenitors exclusively *(97)*. In the neonatal SVZa (P0–7), GFAP expression is essentially absent *(76)*. However, as the animal ages, GFAP immunoreactivity increases along the entire rostrocaudal extent of the migratory pathway and reaches stable expression by P21. Our data therefore demonstrate that the neonatal SVZa progenitors do not require astrocytes or glial processes of any kind for guidance during their peak period of proliferation and migration. The source of the GFAP-positive cells present after P7 is unknown; however, the lack of GFAP protein or mRNA expression further validates our assertion that the neonatal SVZa is a region containing pure neuronal progenitors. A similar claim cannot be made for the adult SVZ, since the region contains a mixture of neuronal progenitors and GFAP-immunoreactive astrocytes. Individual neuronal progenitor cells still exist in the adult SVZ; however, the region itself cannot be considered a pure NRP population.

Luskin and her associates have taken advantage of the clearly identifiable neonatal SVZa to microdissect and culture the SVZa progenitor cells for up to 1 wk *(44)*. They demonstrated that the phenotype of the SVZa cells in vitro is similar to their phenotype in vivo. In particular, while in culture, the SVZa progenitors incorporate BrdU and express neuron-specific type III β-tubulin. Experiments are being conducted in our laboratory to find the optimum conditions for propagating the SVZa cells in vitro for extended periods. In addition, we are investigating the growth factor(s) that promotes neurogenesis.

Homotypic and heterotypic transplantation experiments *(98,99)* using microdissected and dissociated SVZa progenitors further confirm that the SVZa cells are committed to a neuronal phenotype. After homotypic transplantation into the SVZa, the phenotype and proliferative characteristics of the transplanted cells (labeled with PKH26, a lipophilic dye, or with BrdU) were identical to those of endogenous SVZa progenitors. The transplanted cells did not express GFAP, proliferated and expressed neuronal cell type-specific markers in the migratory pathway, and generated granule and periglomerular cells. To investigate whether the cells maintain a neuronal phenotype in a foreign environment, the isolated SVZa cells were transplanted into the neonatal and adult striatum. The cells migrated away from the transplant site, dispersed widely in the striatum, and uniformly continued to express a neuronal phenotype even several weeks after transplantation. Preliminary analysis of their phenotype showed that the transplanted cells morphologically resembled the granule cells in the olfactory bulb *(109)* and some of the cells expressed γ-amino butyric acid

(GABA) (Zigova et al., unpublished data) — the neurotransmitter primarily expressed by the SVZa-derived cells and their progeny in the olfactory bulb. It remains to be determined whether a subset of the transplanted SVZa cells in the striatum expresses dopamine, similar to their endogenous counterparts.

In summary, the neonatal SVZa progenitor cells meet all the criteria for NRPs outlined above. They proliferate while expressing neuronal markers and are uniformly committed to a neuronal fate regardless of the environment. Current evidence indicates that no glial progenitors are present in the neonatal SVZa and that they are a population of pure neuroblasts. Nevertheless, the SVZa in the adult is composed of neuronal progenitors and developing astrocytes, raising the question of whether a β-tubulin III-positive cell generates an astrocyte or, alternatively, a glial precursor invades the SVZa and continues to proliferate. The temporal and phenotypic transformation from the neonatal SVZa to the adult SVZ remains poorly understood and requires investigation.

NRPs from the Embryonic Spinal Cord (Spinal Cord NRPs)

Three groups of progenitor cells can be derived from the neuroepithelial stem cells isolated from the embryonic spinal cord. One group constitutes the NCSCs, whose progeny are the neural cells of the PNS. Much has been learned about the transcription factors involved in neural differentiation (see, for example, refs. *74,101–103*) by studying the NCSCs; however, they will not be considered here. Instead we will focus primarily on two restricted progenitors — the neuronal and glial restricted progenitors (NRPs and GRPs) that mature into neurons and glia, respectively. The restricted progenitors have been isolated in vitro from the spinal cord NEP cells by immunopanning, which takes advantage of the differential expression of cell surface markers to separate the two populations of restricted progenitors from each other.

Immunopanning has been described in detail *(40,41)*; briefly, NEP cells are harvested from the embryonic spinal cord and plated on dishes coated with an antibody against E-NCAM. The bound cells are considered to be an enriched population of NRP cells, since neuronal progenitors but not GRPs in the spinal cord uniformly express E-NCAM. In the same vein, by using the antibody A2B5 instead of E-NCAM, GRP cells can be isolated. The spinal cord NRP cells have been successfully cultured for long periods (weeks to months), allowing them to be extensively studied.

The phenotype of the spinal cord NRP cells has been well-characterized in vitro. For details, the reader is referred to ref. *43*. The results and the interpretations of the authors' experiments reported by Kalyani and Rao *(43)* are summarized below. The NRP cells express the neuron-specific markers (e.g., E-NCAM, type III β-tubulin, and MAP-2), proliferate (assayed by BrdU incorporation), and never express GFAP. The removal of FGF and addition of retinoic acid leads to withdrawal from the cell cycle, increase in soma size, and process formation, similar to the in vitro responses of P19 cells or dorsal root ganglion cells to retinoic acid in vitro *(104,105)*. Additionally, markers indicative of ongoing differentiation are also expressed after the retinoic acid exposure, including NCAM, the midmolecular weight isoform of neurofilament (NF-M), and synaptophysin — none of which are expressed by acutely dissociated spinal cord NEP cells.

In addition to expressing a number of neuron-specific markers, the spinal cord NRP cells also respond to the addition of bone morphogenic protein (BMP)-2 in vitro by decreasing cell division and expressing one or more neurotransmitters. They synthesize multiple neurotransmitters including a combination of GABA, glutamate, acetylcholine, or dopamine, a transmitter not expressed by spinal cord neurons. Undoubtedly, the ultimate phenotype acquired by the NRP cells in vivo reflects integrated responses to multiple intrinsic and extracellular signals.

To investigate the responses of the spinal cord NRP cells to a foreign environment, we transplanted cultured spinal cord NRP cells into the neonatal SVZa — a region containing NRPs and where the microenvironment is conducive to migration and proliferation. A subset of the transplanted spinal cord NRP cells migrated along the RMS to the olfactory bulb, similar to the neonatal SVZa progenitors. However, in contrast to the neonatal SVZa progenitors, the transplanted spinal cord NRP cells migrated to discrete areas in the frontal cortex, occipital cortex, and cerebellum and assumed multiple morphologies akin to host neurons present at their final destination. This indicates that the spinal cord NRP cells are capable of migrating along radial glia-independent pathways like the RMS. Similar to heterotypically and homotypically transplanted SVZa cells, the spinal cord NRP cells retain their neuronal identity following transplantation. Even in glia-rich regions like the corpus callosum, the transplanted spinal cord NRP cells expressed neuronal cell type-specific markers (e.g., MAP-2 and type III β-tubulin) and not glial markers.

For a limited time after transplantation into the SVZa, the spinal cord NRP cells also continue to proliferate, suggesting that they are still transiently responsive to mitogenic cues in the host brain. All the NRP cells expressed neurotransmitters (e.g., GABA, and glutamate) or their synthetic enzymes (e.g., choline acetyltransferase [ChAT] — the synthetic enzyme for acetylcholine), similar to the NRP cells in vitro. The expression of dopamine by the transplanted NRP cells was notably absent, unlike the NRP cells in vitro. Interestingly, GABA and glutamate were expressed only in the olfactory bulb and anterior olfactory nucleus (AON) surrounding the

RMS, whereas ChAT expression was evident in nearly all the transplanted NRP cells. The transplanted cells also expressed markers reflecting ongoing differentiation, such as NF-M and synaptophysin. Similar to their properties in vitro, spinal cord NRP cells in vivo are able to generate heterogeneous progeny, and their ultimate phenotype is dictated by their microenvironment.

Therefore, both the spinal cord NRP cells and the neonatal SVZa progenitors are true neuroblasts and fulfill the criteria we have established above to be classified as NRPs. To date they remain the only true NRP populations isolated from the CNS. A review of the similarities and differences among the neonatal SVZa NRPs, the spinal cord NRPs, and telencephalic VZ progenitors are outlined in Table 1.

Again, although the VZ progenitors and cerebellar progenitors give rise to neurons, they do not proliferate while expressing neuronal cell type-specific markers and therefore cannot be considered NRPs. The olfactory receptor neuron and hippocampal progenitors can adopt nonneuronal fates and therefore cannot be classified as NRPs. It is worth reinforcing that during nervous system development, it is likely that multipotent progenitors differentiate into an NRP or a glial-restricted progenitor prior to terminal differentiation by a process of asymmetric division and restriction of phenotype *(16)*. Therefore, other NRP populations probably exist in the developing CNS. Also, our definition of an NRP is based on currently available neuron-specific markers such as MAP-2, E-NCAM, and type III β-tubulin. Isolation and characterization of other such markers (such as the recently described U-β-6, which recognizes unphosphorylated forms of type III β-tubulin) *(106)* or neuronal progenitor-specific markers, will help in future identification of other NRP populations, or possibly reclassify the neuronally committed unipotent tVZ cells or cerebellar progenitor cells as NRPs.

SOURCES OF NEW NEURONS IN THE ADULT BRAIN

Neural stem cells can be divided, albeit somewhat arbitrarily, into EGF-dependent neurospheres containing stem cells or FGF-dependent stem cells *(40)*. Hippocampal progenitors and striatal SVZ progenitors usually form neurospheres; stem cells from the neural crest, however, usually grow as FGF-dependent monolayers. EGF and FGF are required for the progenitor populations to survive and/or proliferate in vitro. The neurospheres and FGF-dependent stem cell cultures are a mixture of stem cells, neuronal progenitors, and glial progenitors. Adding or deleting a specific trophic factor can affect a particular fate. For example, addition of BDNF to hippocampal neurospheres leads to an increased number of neurons within the neurospheres in culture *(50)*, whereas withdrawal of FGF from neural crest cultures leads to gliogenesis *(107)*.

Table 1

Comparison of Three Progenitor Populations in the Rodent CNS

Progenitor	Expression of neuron-specific markers	Divide and express neuronal cell type-specific markers	Migration patterns of progeny in vivo	Phenotype of by progeny	NT expressed
Neonatal SVZa NRPs	Yes	Yes	Radial glia independent migration. Dispersed in striatum after transplantation.	Olfactory bulb interneurons	GABA, DA (in vivo)
Spinal cord NRPs	Yes	Yes	Migration primarily along radial glia in the spinal cord. Radial glia independent migration in host RMS after transplantation.	Primarily cholinergic motor neurons	ACh in vivo ACh, GABA, Glu, DA in vitro. ACh, GABA, glutamate after transplantation
Telencephalic VZ	No	No	Migration along radial glia. Unable to traverse RMS after transplantation.	Pyramidal and nonpyramidal neurons, glia	and glutamate (in vivo)

Based on these in vitro findings, several investigators have induced neural proliferation in the postnatal brain by the intraventricular administration of these neurotrophic factors. Infusion of BDNF into the lateral ventricle leads to an increased number of granule cell neurons in the olfactory bulb *(14)*. Although this study did not address whether the increased number of cells is due to increased proliferation and/or survival of the progenitors, a study by Kirschenbaum and Goldman *(108)* showed increased neuronal survival of SVZ cells exposed to BDNF, which suggests that a similar phenomenon may occur in vivo. Infusion of bFGF or EGF into the lateral ventricles *(6,12,82,84,109,110)* or even peripherally *(12,84)* in rodents leads to proliferation in the striatal SVZ that surrounds the lateral ventricle; however, astroglia seem to be preferentially generated in the presence of these trophic factors *(12,84)*. BDNF infusion into the lateral ventricle of adult rodents, however, leads to a higher percentage of neurons being generated. It has not been established, whether the growth factor-derived neurons are functionally active. In particular, their ability to form synapses, their response to neurotransmitters, and their electrical activity all remain to be determined. Studies are being conducted in our lab and others to examine some of these issues and also to try and increase the ratio of neurons to glia from the administration of exogenous neurotrophic factors.

The induction of seizures also leads to neurogenesis, especially in the hippocampus and temporal lobe *(15,111,112)*. Aberrant neural activity can be brought about by drugs (e.g., kainic acid) or by a process known as kindling. Regardless of the methods used to induce seizures, the dentate granule cell progenitors in the hippocampus proliferate. The vast majority of the newly generated cells are neurons, *(15,113)*, which were found in ectopic locations and projected inappropriately. Electrical kindling of the amygdala also resulted in hippocampal progenitor proliferation, but only after multiple seizure episodes *(114)*. Studies by Gould and her colleagues *(115–117)* have shown that estrogen, thyroid hormone, and a stress-free environment can induce proliferation in the hippocampus; conversely, stressors and testosterone lead to decreased proliferation.

The above evidence, a small part of a large body of work, suggests that neurogenesis in the postnatal brain can be regulated. The neurons generated following seizure in the hippocampus seem to interact functionally with existing neurons, although the connections are aberrant *(15)*.

SYNERGISTIC EFFECTS OF EXTRINSIC AND INTRINSIC FACTORS DICTATE DIFFERENT YET EXCLUSIVELY NEURONAL PROPERTIES IN SVZa AND SPINAL CORD NRP CELLS

Although the NRP cells of the neonatal SVZa and embryonic spinal cord generate exclusively neurons, we have found significant differences in their differentiation potential and their ability to respond to extracellular cues. First, they generate phenotypically different neurons in vivo. Second, although the progeny of the two NRP populations are neuronal, they demonstrate significant differences in their migration patterns and ultimate phenotypes after transplantation, suggesting either an intrinsic bias and/or a differential response to common cues.

As discussed in our reviews of the individual NRP populations, the homotypically transplanted SVZa cells proliferated and matured into periglomerular and granule cells expressing both GABA and dopamine, similar to their endogenous counterparts. Spinal cord NRP cells transplanted into the SVZa also traversed the RMS to the granule cell layer of the olfactory bulb, similar to SVZa progenitor cells, but a contingent of the transplanted NRP cells exited the RMS and, at all their destinations, primarily expressed ChAT. However, GABA and glutamate production (in addition to ChAT expression) was evident in the olfactory bulb and anterior olfactory nucleus. This leads us to hypothesize that ChAT expression may be a default differentiation pathway adopted by the spinal cord NRP cells when extrinsic cues necessary to induce GABA or glutamate expression (which are presumably present in the olfactory bulb and anterior olfactory nucleus) are not present. The cues from the olfactory bulb, however, that induce dopamine expression in the SVZa cells did not exert the same influence on the spinal cord NRP cells, although these cells can express dopamine in vitro *(43)*. Therefore, common extrinsic cues in the RMS and the olfactory bulb, in concert with intrinsic properties of the different NRP populations, induced completely different yet exclusively neuronal phenotypes. This further reinforces the notion that the ultimate phenotype of an individual neuron involves the integrated response to multiple extracellular and intracellular signals during development.

The cues within the RMS also affected the migration of the two NRP populations differently. Signals from the RMS and olfactory bulb confined the homotypically transplanted SVZa progenitor cells to the RMS; however, a cohort of the spinal cord NRP cells were immune to these signals and exited the RMS. Analysis of the migration patterns of the transplanted NRP cells indicates that this was not a random diffusion of cells; the spinal cord NRP cells en route to the cerebellum and occipital cortex preferentially bypassed regions closer to the SVZa such as the striatum and hippocampus. Neither the intrinsic properties of the spinal cord NRP cells enabling them to exit the RMS nor the cues from the neonatal cerebral cortex and cerebellum that influence spinal cord NRP cell migration are not known. The SVZa progenitor cells, however, seem to register another set of cues in a foreign environment. After heterotypic transplantation into the neonatal or adult striatum, the SVZa cells dispersed from the site of implantation but were virtually confined to the striatum. The transplanted cells were not detected in the adjacent cortical regions. Thus, the SVZa cells show different migratory patterns in various regions. The underlying molecular basis of this complex set of choices has not been revealed.

Dissimilarities in migratory ability between the NRP populations from the neonatal SVZa compared with the embryonic spinal cord have been noted; such differences between the NRP populations and other neural progenitor populations exist as well. Experiments using cerebellar and hippocampal progenitors *(51,56)* transplanted into the RMS also reveal differences in the ability of these progenitor cell populations to interpret and respond to the migratory cues from the RMS. When the cerebellar precursors were implanted into the SVZa, they were largely restricted to the implantation site, whereas the hippocampal progenitors migrated extensively to the olfactory bulb. A similar set of extracellular cues present in the RMS and olfactory bulb therefore induces different migratory patterns, reflecting the varied intrinsic properties of the different sets of progenitor populations.

Several studies have demonstrated that these differences in migratory ability and developmental potential are evident very early in development. For example, when transplanted into the adult SVZ or the embryonic brain, the precursors from the medial ganglionic eminence (MGE) were able to migrate extensively; however, precursors from the lateral ganglionic eminence (LGE) were not. The LGE precursors, however, migrated efficiently to the olfactory bulb when transplanted into the SVZ *(118)*. Similar differentiation biases and regionalization can also be seen in the developing forebrain *(119)*, where distinct domains are formed in early embryogenesis.

Our transplantation studies with spinal cord and SVZa NRP cells reinforce the notion of intrinsic programmed biases in progenitor cells from various regions of the CNS. Recently, Mujtaba et al. *(120)* successfully isolated neural cells from cultured embryonic stem (ES) cells, which meet the criteria required to be considered NRPs. However, since they were isolated from cells present much earlier in embryogenesis, and not from specific regions in the developing or neonatal CNS, intrinsic biases such as those present in the neonatal SVZa, the spinal cord NRP, or the LGE or MGE precursor cells may not exist in the ES-NRP cells. It is altogether possible, however, that their developmental potential could be more restricted, since in vitro they may not be exposed to factors required for proper maturation and could be in a permanent NRP-like state. This issue could be resolved by transplantation experiments comparing ES-NRP cells with NRP cells from the embryonic spinal cord and the neonatal SVZa. If, for example, the

ES-NRP cells differentiated into cells that morphologically resemble the host region in which they were transplanted, one could contend that the ES-NRP cells have no intrinsic differentiation biases. These types of transplantation experiments using ES-NRP cells will, therefore, enable us to better understand the role of extrinsic cues in directing differentiation.

THERAPEUTIC USES OF NRPS

Our studies with spinal cord NRP cells and neonatal SVZa cells suggest that they could be used for therapeutic transplantation and that these restricted neuronal progenitor cells have a number of properties that would offer distinct advantages over other cell types. The transplanted NRP cells are generally not restricted to the site of implantation in the host brain and undergo extensive migration after transplantation, indicating that they are able to penetrate the extracellular matrix in the adult brain. In contrast to NRP cells, when multipotent cells are transplanted into the intact or diseased brain, most of the cells remain at the site of transplantation, thereby limiting their usefulness *(121–123)*. The reasons why transplanted stem and multipotent progenitor cells do not disperse are unknown, but one possibility is that the cells are unable to penetrate the extracellular matrix of the adult brain. Another reason may be related to the sequence of events that occur during normal neural development. Most of the multipotent progenitors are restricted to the neuroepithelium surrounding the ventricular zone; their progeny migrate extensively. NRP cells, therefore, may respond to the migratory signals better than stem cells. Immortalized stem cell populations have been shown to migrate extensively *(124,125)*; however, using these cells for therapeutic replacement is fraught with risk. This ability of NRP cells to migrate, therefore, is a considerable advantage where cells may have to function after traveling to reach distant and appropriate locations.

Unlike the multipotent stem cells, the NRP cells themselves do not differentiate into glia. This is important for two reasons. First, diseases and injuries in which neuronal losses predominate can be better treated with NRP cells. Additionally, neurons unlike glia, do not express class II antigens and do not upregulate HLA antigens as a response to inflammatory cytokines. Thus neuronal replacements theoretically would be less immunogenic than stem cell or multipotent progenitor cell replacements.

Although NRP cells are a restricted progenitor population, they still retain their ability to proliferate. Spinal cord NRP cells, for example, continue to divide for 2 d after transplantation into the neonatal SVZa *(126)*. Since the progeny of NRP cells are all neurons, it may be necessary to transplant only a limited number of NRP cells. Although stem cells do proliferate, a significant portion of their progeny seems to be glial, which would be of limited value in most disease states or injury, in which neuronal losses usually predominate.

Our data, indicating that the SVZa and spinal cord NRP cells have intrinsic developmental biases that determine or limit their differentiation pathway, suggest that it will be important to isolate precursors from appropriate regions to optimize transplantation results. In the case of spinal cord injuries, for example, neuronal precursors isolated from the fetal spinal cord may be the best precursor population for neuronal replacement strategies. For Parkinson's disease, the optimal strategy could be to use the SVZa precursors, a subset of which express dopamine. Certainly, finding conditions and factors that promote expression of dopamine by the SVZa progenitors would be of considerable value in this scenario. We are beginning to realize the potential therapeutic benefits of using NRP cells for transplantation; future experiments will only make these roles clearer.

FUTURE DIRECTIONS

Since the recent identification and characterization of the neonatal SVZa and embryonic spinal cord NRP populations, much has already been learned by studying these cells in vitro, in vivo, and after transplantation. The behavior of these cells after heterotypic transplantation suggests that these cells hold immense promise in neuronal replacement therapy, given their seemingly unwavering commitment to the neuronal lineage. NRP cells may replace missing neurons or their projections and may serve to bridge connections between cells upstream of and downstream of an injury. NRP cells could also form novel connections to circumvent the function of missing or damaged neurons. In addition to neuronal replacement, NRP cells could provide trophic support for injured neurons, promote remyelination, and serve to regulate or minimize the glial scarring that accompanies CNS injury (127,128). Further experiments using these cells to deliver not only the NRPs themselves, but also NRPs transfected with selected genes or factors, are being planned, which will potentially augment their therapeutic usefulness. In some instances, intrinsic biases or developmental restrictions present in individual NRP populations could be useful. For example, in Alzheimer's disease, in which cholinergic losses predominate, spinal cord NRP cells could be extremely beneficial.

To realize fully the potential of NRP cells in increasing our understanding of neural development, further transplantation experiments are needed. The development of the CNS is a complex interplay between intrinsic factors present within the cell and the extracellular milieu. The different, yet equally important roles of these factors in determining a particular neuronal phenotype can be studied independently using NRP cells. Similar to our experiments, the same cell population can be transplanted in different locations to learn more about the role of extracellular factors. Conversely, transplanting different NRP populations into the same location will increase our understanding of intrinsic factors. For example, in our transplantation experiments, we can infer that additional extrinsic factors must be present in the olfactory bulb to induce GABA and glutamate expression in addition to ChAT and to affect the intrinsic cholinergic bias in the transplanted spinal cord NRP cells. Additionally, extrinsic factors from the olfactory bulb dictate a dopaminergic phenotype in a subset of endogenous and homotopically transplanted SVZa-derived cells. The spinal cord NRP cells, however, did not respond to the same "dopaminergic" cues of the olfactory bulb, although they migrated to the bulb along the RMS. Therefore, transcription factors or regulatory elements (intrinsic factors), which are presumably present in the SVZa cells but missing in the spinal cord NRP cells, direct the dopaminergic phenotype. We are actively investigating the role of such transcription factors in the SVZa cells. Similar experiments, using SVZa or spinal cord NRP cells (or other such cells identified in the future) transplanted in multiple locations (e.g., the hippocampus or cerebellum) will enable us to identify both the transcription factors and the extracellular elements required to generate a unique neuronal phenotype.

Homologous cell populations in the developing human brain and spinal cord remain to be identified, although recently Quinn et al. *(129)* described a multipotent progenitor population from the embryonic human spinal cord. Certainly, isolation and characterization of human NRP populations will soon follow. Fully understanding the properties of rodent NRPs, and their usefulness and role in transplantation therapy, as our current studies have started to do, will enable us to use human NRP cells to their maximum potential in the future.

ACKNOWLEDGMENTS

The authors thank Drs. Mahendra Rao and Kelly Ciombor in particular for their suggestions as well as criticisms of the manuscript. We also thank the members of the Luskin lab, past and present. This work was supported in part by an R01 grant to M. B. L. from the NIDCD.

REFERENCES

1. Altman, J. and Das, G. D. (1966) Autoradiographic and histological studies of postnatal neurogenesis. I. A longitudinal investigation of the kinetics, migration and transformation of cells incorporating tritiated thymidine in neonate rats, with special reference to postnatal neurogenesis in some brain regions. *J. Comp. Neurol.* **126,** 337–389.

2. Graziadei, P. P., Levine, R. R., and Monti Graziadei, G. A. (1979) Plasticity of connections of the olfactory sensory neuron: regeneration into the fore-brain following bulbectomy in the neonatal mouse. *Neuroscience* **4**, 713–727.

3. Graziadei, P. P. and Monti Graziadei, A. G. (1983) Regeneration in the olfactory system of vertebrates. *Am. J. Otolaryngol.* **4**, 228–233.

4. Graziadei, P. P. and Monti Graziadei, G. A. (1980) Neurogenesis and neuron regeneration in the olfactory system of mammals. III. Deafferentation and reinnervation of the olfactory bulb following section of the fila olfactoria in rat. *J. Neurocytol.* **9**, 145–162.

5. Gage, F. H., Kempermann, G., Palmer, T. D., Peterson, D. A., and Ray, J. (1998) Multipotent progenitor cells in the adult dentate gyrus. *J. Neurobiol.* **36**, 249–266.

6. Gage, F. H., Coates, P. W., Palmer, T. D., Kuhn, H. G., Fisher, L. J., Suhonen, J. O., et al. (1995) Survival and differentiation of adult neuronal progenitor cells transplanted to the adult brain. *Proc. Natl. Acad. Sci. USA* **92**, 11,879–11,883.

7. Hatten, M. E. and Heintz, N. (1995) Mechanisms of neural patterning and specification in the developing cerebellum. *Ann. Rev. Neurosci.* **18**, 385–408.

8. Graziadei, P. P. and Monti Graziadei, G. A. (1985) Neurogenesis and plasticity of the olfactory sensory neurons. *Ann. NY Acad. Sci.* **457**, 127–142.

9. Reynolds, B. A., Tetzlaff, W., and Weiss, S. (1992) A multipotent EGF-responsive striatal embryonic progenitor cell produces neurons and astrocytes. *J. Neurosci.* **12**, 4565–4574.

10. Luskin, M. B. (1993) Restricted proliferation and migration of postnatally generated neurons derived from the forebrain subventricular zone. *Neuron* **11**, 173–189.

11. Lois, C. and Alvarez-Buylla, A. (1993) Proliferating subventricular zone cells in the adult mammalian forebrain can differentiate into neurons and glia. *Proc. Natl. Acad. Sci. USA* **90**, 2074–2077.

12. Kuhn, H. G., Winkler, J., Kempermann, G., Thal, L. J., and Gage, F. H. (1997) Epidermal growth factor and fibroblast growth factor-2 have different effects on neural progenitors in the adult rat brain. *J. Neurosci.* **17**, 5820–5829.

13. Tao, Y., Black, I. B., and DiCicco-Bloom, E. (1997) In vivo neurogenesis is inhibited by neutralizing antibodies to basic fibroblast growth factor. *J. Neurobiol.* **33**, 289–296.

14. Zigova, T., Pencea, V., Wiegand, S. J., and Luskin, M. B. (1998) Intraventricular administration of BDNF increases the number of newly generated neurons in the adult olfactory bulb. *Mol. Cell Neurosci.* **11**, 234–245.

15. Parent, J. M., Yu, T. W., Leibowitz, R. T., Geschwind, D. H., Sloviter, R. S., and Lowenstein, D. H. (1997) Dentate granule cell neurogenesis is increased by seizures and contributes to aberrant network reorganization in the adult rat hippocampus. *J. Neurosci.* **17**, 3727–3738.

16. Rao, M. S. (1999) Multipotent and restricted precursors in the central nervous system. *Anat. Rec.* **257**, 137–148.

17. Luskin, M. B. (1994) Neuronal cell lineage in the vertebrate central nervous system. *FASEB J.* **8**, 722–730.

18. Price, J., Williams, B. P., and Gotz, M. (1995) The generation of cellular diversity in the cerebral cortex. *Ciba Found. Symp.* **193,** 71–84; discussion 117–126.
19. Luskin, M. B., Pearlman, A. L., and Sanes, J. R. (1988) Cell lineage in the cerebral cortex of the mouse studied in vivo and in vitro with a recombinant retrovirus. *Neuron* **1,** 635–647.
20. Luskin, M. B., Parnavelas, J. G., and Barfield, J. A. (1993) Neurons, astrocytes, and oligodendrocytes of the rat cerebral cortex originate from separate progenitor cells: an ultrastructural analysis of clonally related cells. *J. Neurosci.* **13,** 1730–1750.
21. Cepko, C. L., Austin, C. P., Yang, X., Alexiades, M., and Ezzeddine, D. (1996) Cell fate determination in the vertebrate retina. *Proc. Natl. Acad. Sci.USA* **93,** 589–595.
22. Boulder Committee. (1970) Embryonic vertebrate central nervous system: revised terminology. *Anat. Rec.* **166,** 257–261.
23. Davis, A. A. and Temple, S. (1994) A self-renewing multipotential stem cell in embryonic rat cerebral cortex. *Nature* **372,** 263–266.
24. Parnavelas, J. G., Barfield, J. A., Franke, E., and Luskin, M. B. (1991) Separate progenitor cells give rise to pyramidal and nonpyramidal neurons in the rat telencephalon. *Cereb. Cortex* **1,** 463–468.
25. Grove, E. A., Williams, B. P., Li, D. Q., Hajihosseini, M., Friedrich, A., and Price, J. (1993) Multiple restricted lineages in the embryonic rat cerebral cortex. *Development* **117,** 553–561.
26. Carnow, T. B., Barbarese, E., and Carson, J. H. (1991) Diversification of glial lineages: a novel method to clone brain cells in vitro on nitrocellulose substratum. *Glia* **4,** 256–268.
27. Kilpatrick, T. J. and Bartlett, P. F. (1995) Cloned multipotential precursors from the mouse cerebrum require FGF-2, whereas glial restricted precursors are stimulated with either FGF-2 or EGF. *J. Neurosci.* **15,** 3653–3661.
28. Birling, M. C. and Price, J. (1998) A study of the potential of the embryonic rat telencephalon to generate oligodendrocytes. *Dev. Biol.* **193,** 100–113.
29. Menezes, J. R. and Luskin, M. B. (1994) Expression of neuron-specific tubulin defines a novel population in the proliferative layers of the developing telencephalon. *J. Neurosci.* **14,** 5399–5416.
30. Valverde, F., De Carlos, J. A., and Lopez-Mascaraque, L. (1995) Time of origin and early fate of preplate cells in the cerebral cortex of the rat. *Cereb. Cortex* **5,** 483–493.
31. Rakic, P. (1971) Guidance of neurons migrating to the fetal monkey neocortex. *Brain Res.* **33,** 471–476.
32. Rakic, P. (1971) Neuron-glia relationship during granule cell migration in developing cerebellar cortex. A Golgi and electronmicroscopic study in Macacus Rhesus. *J. Comp. Neurol.* **141,** 283–312.
33. Walsh, C. and Cepko, C. L. (1990) Cell lineage and cell migration in the developing cerebral cortex. *Experientia* **46,** 940–947.
34. Privat, A. (1975) Postnatal gliogenesis in the mammalian brain. *Int. Rev. Cytol.* **40,** 281–323.

35. Luskin, M. B. and McDermott, K. (1994) Divergent lineages for oligodendrocytes and astrocytes originating in the neonatal forebrain subventricular zone. *Glia* **11,** 211–226.

36. Levison, S. W. and Goldman, J. E. (1993) Both oligodendrocytes and astrocytes develop from progenitors in the subventricular zone of postnatal rat forebrain. *Neuron* **10,** 201–212.

37. Levison, S. W. and Goldman, J. E. (1997) Multipotential and lineage restricted precursors coexist in the mammalian perinatal subventricular zone. *J. Neurosci. Res.* **48,** 83–94.

38. Farbman, A. I. (1991) Developmental neurobiology of the olfactory system, in *Smell and Taste in Health and Disease* (Getchel, T. V., ed.), Raven, New York, pp. 19–33.

39. Reynolds, B. A. and Weiss, S. (1992) Generation of neurons and astrocytes from isolated cells of the adult mammalian central nervous system [see comments]. *Science* **255,** 1707–1710.

40. Kalyani, A. J. and Rao, M. S. (1998) Cell lineage in the developing neural tube. *Biochem. Cell. Biol.* **76,** 1051–1068.

41. Mayer-Proschel, M., Kalyani, A. J., Mujtaba, T., and Rao, M. S. (1997) Isolation of lineage-restricted neuronal precursors from multipotent neuroepithelial stem cells. *Neuron* **19,** 773–785.

42. Mayer-Proschel, M. (1999) Cell differentiation in the embryonic mammalian spinal cord. *J. Neural. Transm. Suppl.* **55,** 1–8.

43. Kalyani, A. J., Piper, D., Mujtaba, T., Lucero, M. T., and Rao, M. S. (1998) Spinal cord neuronal precursors generate multiple neuronal phenotypes in culture. *J. Neurosci.* **18,** 7856–7868.

44. Luskin, M. B., Zigova, T., Soteres, B. J., and Stewart, R. R. (1997) Neuronal progenitor cells derived from the anterior subventricular zone of the neonatal rat forebrain continue to proliferate in vitro and express a neuronal phenotype. *Mol. Cell. Neurosci.* **8,** 351–366.

45. Altman, J. and Das, G. D. (1965) Autoradiographic and histological evidence of postnatal hippocampal neurogenesis in rats. *J. Comp. Neurol.* **124,** 319–335.

46. Reznikov, K. Y. (1991) Cell proliferation and cytogenesis in the mouse hippocampus. *Adv. Anat. Embryol. Cell Biol.* **122,** 1–74.

47. Kuhn, H. G., Dickinson-Anson, H., and Gage, F. H. (1996) Neurogenesis in the dentate gyrus of the adult rat: age-related decrease of neuronal progenitor proliferation. *J. Neurosci.* **16,** 2027–2033.

48. Slomianka, L. and Geneser, F. A. (1997) Postnatal development of zinc-containing cells and neuropil in the hippocampal region of the mouse. *Hippocampus* **7,** 321–340.

49. Eriksson, P. S., Perfilieva, E., Bjork-Eriksson, T., Alborn, A. M., Nordborg, C., Peterson, D. A., and Gage, F. H. (1998) Neurogenesis in the adult human hippocampus [see comments]. *Nature Med.* **4,** 1313–1317.

50. Shetty, A. K. and Turner, D. A. (1998) In vitro survival and differentiation of neurons derived from epidermal growth factor-responsive postnatal hippocampal stem cells: inducing effects of brain-derived neurotrophic factor. *J. Neurobiol.* **35,** 395–425.

51. Suhonen, J. O., Peterson, D. A., Ray, J., and Gage, F. H. (1996) Differentiation of adult hippocampus-derived progenitors into olfactory neurons in vivo. *Nature* **383,** 624–627.
52. Roy, N. S., Wang, S., Li, J., Benraiss, A., Harrison-Restelli, C., Fraser, R. A., et al. (2000) In vitro neurogenesis by progenitor cells isolated from the adult human hippocampus. *Nature Med.* **6,** 271–277.
53. Hatten, M. E., Alder, J., Zimmerman, K., and Heintz, N. (1997) Genes involved in cerebellar cell specification and differentiation.*Curr. Opin. Neurobiol.* **7,** 40–47.
54. Gao, W. Q. and Hatten, M. E. (1994) Immortalizing oncogenes subvert the establishment of granule cell identity in developing cerebellum. *Development* **120,** 1059–1070.
55. Jankovski, A., Rossi, F., and Sotelo, C. (1996) Neuronal precursors in the postnatal mouse cerebellum are fully committed cells: evidence from heterochronic transplantations. *Eur. J. Neurosci.* **8,** 2308–2319.
56. Jankovski, A. and Sotelo, C. (1996) Subventricular zone-olfactory bulb migratory pathway in the adult mouse: cellular composition and specificity as determined by heterochronic and heterotopic transplantation. *J. Comp. Neurol.* **371,** 376–396.
57. Vicario-Abejon, C., Cunningham, M. G., and McKay, R. D. (1995) Cerebellar precursors transplanted to the neonatal dentate gyrus express features characteristic of hippocampal neurons. *J. Neurosci.* **15,** 6351–6363.
58. Sommer, C., Bele, S., and Kiessling, M. (1997) Expression of cerebellar specific glutamate and $GABA_A$ receptor subunits in heterotopic cerebellar grafts. *Dev. Brain Res.* **102,** 225–230.
59. Feng, L., Hatten, M. E., and Heintz, N. (1994) Brain lipid-binding protein (BLBP): a novel signaling system in the developing mammalian CNS. *Neuron* **12,** 895–908.
60. Kuhar, S. G., Feng, L., Vidan, S., Ross, M. E., Hatten, M. E., and Heintz, N. (1993) Changing patterns of gene expression define four stages of cerebellar granule neuron differentiation. *Development* **117,** 97–104.
61. Urbanek, P., Fetka, I., Meisler, M. H., and Busslinger, M. (1997) Cooperation of Pax2 and Pax5 in midbrain and cerebellum development. *Proc. Natl. Acad. Sci. USA* **94,** 5703–5708.
62. Rowitch, D. H., Kispert, A., and McMahon, A. P. (1999) Pax-2 regulatory sequences that direct transgene expression in the developing neural plate and external granule cell layer of the cerebellum. *Dev. Brain Res.* **117,** 99–108.
63. Ben-Arie, N., Bellen, H. J., Armstrong, D. L., McCall, A. E., Gordadze, P. R., Guo, Q., Matzuk, M. M., and Zoghbi, H. Y. (1997) Math1 is essential for genesis of cerebellar granule neurons. *Nature* **390,** 169–172.
64. Caggiano, M., Kauer, J. S., and Hunter, D. D. (1994) Globose basal cells are neuronal progenitors in the olfactory epithelium: a lineage analysis using a replication-incompetent retrovirus. *Neuron* **13,** 339–352.
65. Walters, E., Grillo, M., Oestreicher, A. B., and Margolis, F. L. (1996) LacZ and OMP are co-expressed during ontogeny and regeneration in olfactory

receptor neurons of OMP promoter-lacZ transgenic mice. *Int. J. Dev. Neurosci.* **14,** 813–822.

66. Guillemot, F., Lo, L. C., Johnson, J. E., Auerbach, A., Anderson, D. J., and Joyner, A. L. (1993) Mammalian achaete-scute homolog 1 is required for the early development of olfactory and autonomic neurons. *Cell* **75,** 463–476.

67. Davis, J. A. and Reed, R. R. (1996) Role of Olf-1 and Pax-6 transcription factors in neurodevelopment. *J. Neurosci.* **16,** 5082–5094.

68. Whitesides, J. G., 3rd and LaMantia, A. S. (1996) Differential adhesion and the initial assembly of the mammalian olfactory nerve. *J. Comp. Neurol.* **373,** 240–254.

69. Huard, J. M., Youngentob, S. L., Goldstein, B. J., Luskin, M. B., and Schwob, J. E. (1998) Adult olfactory epithelium contains multipotent progenitors that give rise to neurons and non-neural cells. *J. Com. Neurol.* **400,** 469–486.

70. McConnell, S. K. (1995) Constructing the cerebral cortex: neurogenesis and fate determination. *Neuron* **15,** 761–768.

71. Stemple, D. L. and Mahanthappa, N. K. (1997) Neural stem cells are blasting off. *Neuron* **18,** 1–4.

72. Doetsch, F., Caille, I., Lim, D. A., Garcia-Verdugo, J. M., and Alvarez-Buylla, A. (1999) Subventricular zone astrocytes are neural stem cells in the adult mammalian brain. *Cell* **97,** 703–716.

73. Morrison, S. J., White, P. M., Zock, C., and Anderson, D. J. (1999) Prospective identification, isolation by flow cytometry, and in vivo self-renewal of multipotent mammalian neural crest stem cells. *Cell* **96,** 737–749.

74. Shah, N. M., Groves, A. K., and Anderson, D. J. (1996) Alternative neural crest cell fates are instructively promoted by TGFbeta superfamily members. *Cell* **85,** 331–343.

75. Rao, M. S. and Anderson, D. J. (1997) Immortalization and controlled in vitro differentiation of murine multipotent neural crest stem cells. *J. Neurobiol.* **32,** 722–746.

76. Law, A. K., Pencea, V., Buck, C. R., and Luskin, M. B. (1999) Neurogenesis and neuronal migration in the neonatal rat forebrain anterior subventricular zone do not require GFAP-positive astrocytes. *Dev. Biol.* **216,** 622–634.

77. Santacana, M., Heredia, M., Valverde, F. (1992) Transient pattern of exuberant projections of olfactory axons during development in the rat. *Dev. Brain Res.* **70,** 213–222.

78. Hinds, J. W. and Ruffett, T. L. (1973) Mitral cell development in the mouse olfactory bulb: reorientation of the perikaryon and maturation of the axon initial segment. *J. Comp. Neurol.* **151,** 281–306.

79. Skeen, L. C., Due, B. R., and Douglas, F. E. (1985) Effects of early anosmia on two classes of granule cells in developing mouse olfactory bulbs. *Neurosci Lett.* **54,** 301–306.

80. Cummings, D. M. and Brunjes, P. C. (1997) The effects of variable periods of functional deprivation on olfactory bulb development in rats. *Exp. Neurol.* **148,** 360–366.

81. Lois, C. and Alvarez-Buylla, A. (1994) Long-distance neuronal migration in the adult mammalian brain. *Science* **264,** 1145–1148.

82. Wagner, J. P., Black, I. B., and DiCicco-Bloom, E. (1999) Stimulation of neonatal and adult brain neurogenesis by subcutaneous injection of basic fibroblast growth factor. *J. Neurosci.* **19,** 6006–6016.

83. Tropepe, V., Sibilia, M., Ciruna, B. G., Rossant, J., Wagner, E. F., and van der Kooy, D. (1999) Distinct neural stem cells proliferate in response to EGF and FGF in the developing mouse telencephalon. *Dev. Biol.* **208,** 166–188.

84. Craig, C. G., Tropepe, V., Morshead, C. M., Reynolds, B. A., Weiss, S., and van der Kooy, D. (1996) In vivo growth factor expansion of endogenous subependymal neural precursor cell populations in the adult mouse brain. *J. Neurosci.* **16,** 2649–2658.

85. Miragall, F., Kadmon, G., Faissner, A., Antonicek, H., and Schachner, M. (1990) Retention of J1/tenascin and the polysialylated form of the neural cell adhesion molecule (N-CAM) in the adult olfactory bulb. *J. Neurocytol.* **19,** 899–914.

86. Kishi, K. (1987) Golgi studies on the development of granule cells of the rat olfactory bulb with reference to migration in the subependymal layer. *J. C. Neurol.* **258,** 112–124.

87. Rousselot, P., Lois, C., and Alvarez-Buylla, A. (1995) Embryonic (PSA) N-CAM reveals chains of migrating neuroblasts between the lateral ventricle and the olfactory bulb of adult mice. *J. Comp. Neurol.* **351,** 51–61.

88. Smith, C. M. and Luskin, M. B. (1998) Cell cycle length of olfactory bulb neuronal progenitors in the rostral migratory stream. *Dev. Dyn.* **213,** 220–227.

89. Thomas, L. B., Gates, M. A., and Steindler, D. A. (1996) Young neurons from the adult subependymal zone proliferate and migrate along an astrocyte, extracellular matrix-rich pathway. *Glia* **17,** 1–14.

90. Silver, J. (1994) Inhibitory molecules in development and regeneration. *J. Neurol.* **242,** S22–S24.

91. Cremer, H., Lange, R., Christoph, A., Plomann, M., Vopper, G., Roes, J., Brown, R., Baldwin, S., Kraemer, P., and Scheff, S. (1994) Inactivation of the N-CAM gene in mice results in size reduction of the olfactory bulb and deficits in spatial learning. *Nature* **367,** 455–459.

92. Hu, H., Tomasiewicz, H., Magnuson, T., and Rutishauser, U. (1996) The role of polysialic acid in migration of olfactory bulb interneuron precursors in the subventricular zone. *Neuron* **16,** 735–743.

93. Kirschenbaum, B., Doetsch, F., Lois, C., and Alvarez-Buylla, A. (1999) Adult subventricular zone neuronal precursors continue to proliferate and migrate in the absence of the olfactory bulb. *J. Neurosci.* **19,** 2171–2180.

94. Jankovski, A., Garcia, C., Soriano, E., and Sotelo, C. (1998) Proliferation, migration and differentiation of neuronal progenitor cells in the adult mouse subventricular zone surgically separated from its olfactory bulb. *Eur. J. Neurosci.* **10,** 3853–3868.

95. Wu, W., Wong, K., Chen, J., Jiang, Z., Dupuis, S., Wu, J. Y., and Rao, Y. (1999) Directional guidance of neuronal migration in the olfactory system by the protein Slit [sce comments]. *Nature* **400,** 331–336.

96. Okano, H. J., Pfaff, D. W., and Gibbs, R. B. (1996) Expression of EGFR-, p75NGFR-, and PSTAIR (cdc-2)-like immunoreactivity by proliferating cells in the adult rat hippocampal formation and forebrain. *Dev. Neurosci.* **18,** 199–209.

97. Schubert, W., Coskun, V., Tahmina, M., Rao, M. S., Luskin, M. B., and Kaprielian, Z. (2000) Characterization and distribution of a new cell surface marker of neuronal precursors. *Dev. Neurosci.* **22,** 154–166.

98. Zigova, T., Pencea, V., Betarbet, R., Wiegand, S. J., Alexander, C., Bakay, R. A., and Luskin, M. B. (1998) Neuronal progenitor cells of the neonatal subventricular zone differentiate and disperse following transplantation into the adult rat striatum. *Cell Transplant.* **7,** 137–156.

99. Zigova, T., Betarbet, R., Soteres, B. J., Brock, S., Bakay, R. A., and Luskin, M. B. (1996) A comparison of the patterns of migration and the destinations of homotopically transplanted neonatal subventricular zone cells and heterotopically transplanted telencephalic ventricular zone cells. *Dev. Biol.* **173,** 459–474.

100. Sommer, L., Shah, N., Rao, M., and Anderson, D. J. (1995) The cellular function of MASH1 in autonomic neurogenesis. *Neuron* **15,** 1245–1258.

101. Groves, A. K. and Anderson, D. J. (1996) Role of environmental signals and transcriptional regulators in neural crest development. *Dev. Genet.* **18,** 64–72.

102. Anderson, D. J., Groves, A., Lo, L., Ma, Q., Rao, M., Shah, N. M., and Sommer, L. (1997) Cell lineage determination and the control of neuronal identity in the neural crest. *Cold Spring Harbor Symp. Quant. Biol.* **62,** 493–504.

103. Anderson, D. J. (1997) Cellular and molecular biology of neural crest cell lineage determination.*Trends Genet.* **13,** 276–280.

104. Quinn, S. D. and De Boni, U. (1991. Enhanced neuronal regeneration by retinoic acid of murine dorsal root ganglia and of fetal murine and human spinal cord in vitro. *In Vitro Cell. Dev. Biol.* **27,** 55–62.

105. Jones-Villeneuve, E. M., Rudnicki, M. A., Harris, J. F., and McBurney, M. W. (1983) Retinoic acid-induced neural differentiation of embryonal carcinoma cells. *Mol. Cell. Biol.* **3,** 2271–2279.

106. Fanarraga, M. L., Avila, J., and Zabala, J. C. (1999) Expression of unphosphorylated class III beta-tubulin isotype in neuroepithelial cells demonstrates neuroblast commitment and differentiation. *Eur. J. Neurosci.* **11,** 517–527.

107. Rao, M. S. and Mayer-Proschel, M. (1997) Glial-restricted precursors are derived from multipotent neuroepithelial stem cells. *Dev. Biol.* **188,** 48–63.

108. Kirschenbaum, B. and Goldman, S. A. (1995) Brain-derived neurotrophic factor promotes the survival of neurons arising from the adult rat forebrain subependymal zone. *Proc. Natl. Acad. Sci. USA* **92,** 210–214.

109. Gritti, A., Parati, E. A., Cova, L., Frolichsthal, P., Galli, R., Wanke, E., et al. (1996) Multipotential stem cells from the adult mouse brain proliferate and self-renew in response to basic fibroblast growth factor. *J. Neurosci.* **16,** 1091–1100.

110. Palmer, T. D., Ray, J., and Gage, F. H. (1995) FGF-2-responsive neuronal progenitors reside in proliferative and quiescent regions of the adult rodent brain. *Mol. Cell. Neurosci.* **6,** 474–486.

111. Holmes, G. L., Gairsa, J. L., Chevassus-Au-Louis, N., and Ben-Ari, Y. (1998) Consequences of neonatal seizures in the rat: morphological and behavioral effects. *Ann. Neurol.* **44,** 845–857.

112. Holmes, G. L., Sarkisian, M., Ben-Ari, Y., and Chevassus-Au-Louis, N. (1999) Mossy fiber sprouting after recurrent seizures during early development in rats. *J. Comp. Neurol.* **404,** 537–553.

113. Parent, J. M. and Lowenstein, D. H. (1997) Mossy fiber reorganization in the epileptic hippocampus. *Curr. Opin. Neurol* **10,** 103–109.
114. Parent, J. M., Janumpalli, S., McNamara, J. O., and Lowenstein, D. H. (1998) Increased dentate granule cell neurogenesis following amygdala kindling in the adult rat. *Neurosci. Lett.* **247,** 9–12.
115. Gould, E. and Tanapat, P. (1999) Stress and hippocampal neurogenesis. *Biol. Psychiatry* **46,** 1472–1479.
116. Gould, E., Reeves, A. J., Fallah, M., Tanapat, P., Gross, C. G., and Fuchs, E. (1999) Hippocampal neurogenesis in adult Old World primates. *Proc. Natl. Acad. Sci. USA* **96,** 5263–5267.
117. Gould, E., Beylin, A., Tanapat, P., Reeves, A., and Shors, T. J. (1999) Learning enhances adult neurogenesis in the hippocampal formation [see comments]. *Nat. Neurosci.* **2,** 260–265.
118. Wichterle, H., Garcia-Verdugo, J. M., Herrera, D. G., and Alvarez-Buylla, A. (1999) Young neurons from medial ganglionic eminence disperse in adult and embryonic brain. *Nat. Neurosci.* **2,** 461–466.
119. Shimamura, K. and Rubenstein, J. L. (1997) Inductive interactions direct early regionalization of the mouse forebrain. *Development* **124,** 2709–2718.
120. Mujtaba, T., Piper, D. R., Kalyani, A., Groves, A. K., Lucero, M. T., and Rao, M. S. (1999) Lineage-restricted neural precursors can be isolated from both the mouse neural tube and cultured ES cells. *Dev. Biol.* **214,** 113–127.
121. Kukekov, V. G., Laywell, E. D., Thomas, L. B., and Steindler, D. A. (1997) A nestin-negative precursor cell from the adult mouse brain gives rise to neurons and glia. *Glia* **21,** 399–407.
122. Brustle, O. and McKay, R. D. (1996) Neuronal progenitors as tools for cell replacement in the nervous system. *Curr. Opin. Neurobiol.* **6,** 688–695.
123. Lundberg, C. and Bjorklund, A. (1996) Host regulation of glial markers in intrastriatal grafts of conditionally immortalized neural stem cell lines. *Neuroreport* **7,** 847–852.
124. Flax, J. D., Aurora, S., Yang, C., Simonin, C., Wills, A. M., Billinghurst, L. L., et al. (1998) Engraftable human neural stem cells respond to developmental cues, replace neurons, and express foreign genes. *Nat. Biotechnol.* **16,** 1033–1039.
125. Lundberg, C., Martinez-Serrano, A., Cattaneo, E., McKay, R. D., and Bjorklund, A. (1997) Survival, integration, and differentiation of neural stem cell lines after transplantation to the adult rat striatum. *Exp. Neurol.* **145,** 342–360.
126. Yang, H., Mujtaba, T., Venkatraman, G., Wu, Y., Rao, M. S., and Luskin, M. B. (2000) Region-specific differentiation of neuronal restricted progenitor cells from the embryonic spinal cord after heterotopic transplatation into the neonatal rat forebrain. *Proc. Natl. Acad. Sci. USA, in press.*
127. Tuszynski, M. H. and Gage, F. H. (1996) Somatic gene therapy for nervous system disease. *Ciba Found. Symp.* **196,** 85–94.
128. Martinez-Serrano, A. and Bjorklund, A. (1995-96) Gene transfer to the mammalian brain using neural stem cells: a focus on trophic factors, neuroregeneration, and cholingergic neuron systems. *Clin. Neurosci.* **3,** 301–309.
129. Quinn, S. M., Walters, W. M., Vescovi, A. L., and Whittemore, S. R. (1999) Lineage restriction of neuroepithelial precursor cells from fetal human spinal cord. *J. Neurosci. Res.* **57,** 590–602.

6

Glial Restricted Precursors

Mark Noble and Margot Mayer-Pröschel

The most numerous of the cells of the central nervous system (CNS) are the glia. These cells, which include the myelin-forming oligodendrocytes of the white matter and the ubiquitous astrocytes, play many roles in normal development and in disease. A subject of study since the time of del Rio de Hortega, a great deal of knowledge has been obtained regarding the development and function of these cells. Nonetheless, it must be recognized that we are still far from having a comprehensive understanding of the origins and biology of the glia. The extent to which our knowledge is still in its early stages is reflected in the often inadequate nomenclature with which to discuss the complexity already believed to exist. As this can make the literature confusing, Table 1 provides a list of terms used in this field and the cells to which these terms correspond.

By far the better understood of the glia is the oligodendrocyte, which is generally thought of as having the sole function of myelinating CNS axons. Such a view of this cell is almost certainly an oversimplification, as indicated by observations that oligodendrocytes may play a role in the production of such important bioactive proteins as nerve growth factor (1,2) and transferrin (3,4). They also play a role in the inhibition of axonal growth following injury in white matter tracts (5–7). As with all cell types, it is likely that such hints of other functions represent only a superficial view of the full range of regulatory networks in which these cells are engaged.

Development of the oligodendrocyte is understood in a detailed manner that far exceeds that for neurons and astrocytes (see Fig. 1 for summary). Indeed, studies on oligodendrocyte development are sophisticated enough to have allowed certain probing questions in developmental biology that have not been possible with any other cell type. As a number of aspects of oligodendrocyte development have been the subject of relatively recent

From: *Stem Cells and CNS Development*
Edited by: M. S. Rao © Humana Press Inc., Totowa, NJ

Table 1
Neural Cell types of the CNS

Type	Characteristics
Neural stem cell (NSC) Neuroepithelial stem cell (NEP cell)	Multipotent neuroepithelial stem cells (nestin positive; lineage-specific antigen negative)
Neuron-restricted precursor (NRP) cell	Cells with the ability to generate neurons but not glia (E-NCAM$^+$)
Glial-restricted precursor (GRP) cell	Tripotential glial-restricted precursor cells, which can generate oligodendrocytes, type 1 astrocytes and type 2 astrocytes (A2B5$^+$)
Oligodendrocyte type 2 astrocyte (O-2A) progenitor cell; Oligodendrocyte precursor cell (OPC)	Bipotential glial-restricted precursor cells, which can generate oligodendrocytes and type 2 astrocytes. (A2B5$^+$; also 04$^+$/GalC$^-$)
Astrocyte precursor cell (APC)	precursor cells that can generate astrocytes but not oligodendrocytes (A2B5$^{+/-}$/Ran-2$^+$/GalC$^-$)
Oligodendrocyte	Cells with the antigenic phenotype of myelin- producing cells (GalC$^+$/MPB$^+$/O1$^+$ but A2B5$^-$)
Type 1 astrocyte	Cells with the antigenic phenotype of the type 1 astrocytes of the rat optic nerve, (Ran-2$^+$/GFAP$^+$/FGFR3$^+$ and A2B5$^-$)
Type 2 astrocyte	Cells with the antigenic phenotype of the type 2 astrocytes derived from O-2A progenitor cells (A2B5$^+$/GFAP$^+$ but FGFR3$^-$/Ran-2$^-$)

reviews, attention is focused here on issues that have not recently been widely addressed. We draw on published data and work in progress in an attempt to provide an overview of oligodendrocyte generation indicative of the currently perceived level of complexity of this process. We also consider data indicating that understanding such details may be critical for understanding the biologic basis of a wide range of instances of aberrant neurologic development.

OLIGODENDROCYTE GENERATION REQUIRES MULTIPLE STAGES OF SEQUENTIAL LINEAGE RESTRICTION WITHIN DEVELOPMENTALLY NESTED PRECURSOR CELL POPULATIONS

As with all cells of the CNS, the development of the oligodendrocyte begins with the totipotent neuroepithelial stem cell (NSC). Studies on NSCs derived from spinal cords of embryonic rats have indicated that in order for

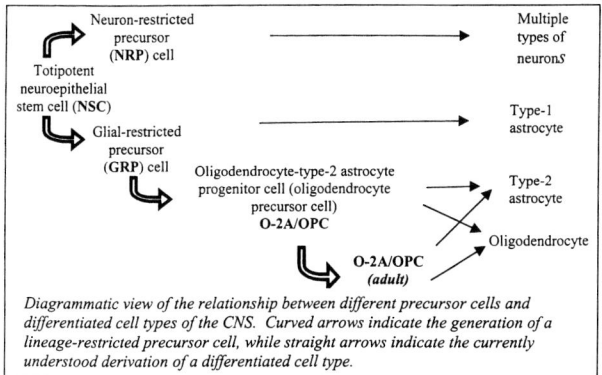

Diagrammatic view of the relationship between different precursor cells and differentiated cell types of the CNS. Curved arrows indicate the generation of a lineage-restricted precursor cell, while straight arrows indicate the currently understood derivation of a differentiated cell type.

Fig. 1. Precursor cells of the central nervous system. Diagrammatic view of the relationship between different precursor cells and differentiated cell types of the CNS. Curved arrows indicate the generation of a lineage-restricted precursor cell, and straight arrows indicate the currently understood derivation of a differentiated cell type.

these cells to generate glia they must first pass through a stage of lineage restriction. This lineage restriction leads to the generation of two distinct populations of precursor cells. On one path is the neuron-restricted precursor (NRP) cell, which gives rise solely to neurons and not to glial cells *(8)*. Complementary to this option is the glial path, which is defined by a glial-restricted precursor (GRP) cell *(9,10)*. This precursor cell can give rise to at least three distinct glial cell types (oligodendrocytes and two types of astrocytes) but not to neurons. GRP and NRP cells first appear at embryonic day (E)12-E13 of spinal cord development *(8,10)*.

Distinct from the GRP cell is a bipotential glial precursor cell that goes by the alternative names of the oligodendrocyte type 2 astrocyte (O-2A) progenitor cell *(11)* and the oligodendrocyte precursor cell (OPC; *12*). Thus far, it does not appear that this glial precursor cell population gives rise to type 2 astrocytes during development *(12,13)*, at least in the optic nerve, although these precursors can give rise to astrocytes following transplantation into lesion sites *(14)*. There is strong disagreement over which terminology is more appropriate, but the terms O-2A progenitor cell and OPC both refer to the same cell population. This review refers to these cells as O-2A/OPCs.

Numerous features distinguish O-2A/OPCs and GRP cells from each other (summarized in Table 2). For example, GRP cells exposed to fetal calf scrum give rise to both type 2- and type 1-like astrocytes *(10)*, whereas O-2A/OPCs in these conditions give rise exclusively to type 2 astrocytes *(11, 15–17)*. GRP cells can, however, be induced to differentiate exclusively into type 2-like astrocytes by growing them in the presence of basic

Table 2
Characteristics of O-2A and GRP cells[a]

Characteristics	O-2A cells (p7 optic nerve)	GRP cells (E13.5 spinal cord)
Antigenic phenotype		
A2B5	+	+
PDGFR-α	+	–
GD3	+	–
O4	–/+	–
nestin	+	+
Growth factor response		
PDGF/T3	Oligodendrocytes	Cell death!
10% FCS	Type 2 astrocytes	Type 1 and type 2 astrocytes
bFGF/CNTF	Progenitor cells	Type 2 astrocytes
bFGF/BMP4	Type 2 astrocytes	Type 1 astrocytes

[a]O-2A cells isolated from the postnatal optic nerve display different and similar features in comparison with embryonic derived GRP cells. The growth factor response is based on pure cultures grown in the presence of the indicated cytokines (10 ng/mL) for 3 d followed by immunocytochemistry to determine the predominant cell type that is generated (bFGF/CNTF and bFGF/BMP4 response, unpublished data, M.M.P.).

fibroblast growth factor (bFGF) together with ciliary neurotrophic factor (CNTF; *10*), a condition in which O-2A progenitor cells instead undergo self-renewal or differentiate into oligodendrocytes *(18,19)*. GRP cells and O-2A/OPCs also differ in their response to mitogens, their survival factor requirements, and their behavior in mitogen-free conditions. O-2A/OPCs divide extensively when exposed to platelet-derived growth factor (PDGF), an O-2A/OPC mitogen produced by type 1 astrocytes, (e.g., refs. *20–27*), and respond to PDGF, CNTF, insulin-like growth factor-I (IGF-I), and insulin as survival factors *(28–33)*. When grown in chemically defined medium lacking mitogens, O-2A/OPCs undergo default differentiation into oligodendrocytes *(11,22,34,35)*. Freshly isolated GRP cells from E13.5 spinal cord, in contrast, do not respond to PDGF and do not express PDGF receptors, although they will acquire these receptors when grown in vitro for several days *(10)*. They do not respond to PDGF, CNTF, insulin, or IGF-1 as survival factors. When grown in mitogen-free chemically defined medium, GRP cells do not differentiate but instead undergo cell death. GRP cells and O-2A/OPCs also differ in their substrate requirements. GRP cells exhibit optimal survival on a combination of fibronectin and laminin *(10)*, whereas O-2A/OPCs can be readily grown on poly cationic surfaces (such as poly-L-lysine *(11,35,36)*.

As GRP cells are the earliest appearing precursor cell restricted to glial development thus far found in the CNS, we have suggested that these cells might be the ancestor of all CNS glia *(10,37)*. If this suggestion is correct, then it follows that these cells should not only be able to give rise to all the glial cell types, but also that they should be able to give rise to other glial precursor cells that arise in the CNS at later developmental stages, such as the O-2A/OPC. Such a result would be supportive of a developmental model in which the generation of a differentiated cell type involves progressive lineage restriction through multiple layers of lineage-restricted precursor cell populations. Alternatively, tripotential GRP cells and bipotential O-2A/OPCs might represent two different developmental paths that lead to the same differentiated cell type. The derivation of a particular cell type from more than one precursor cell lineage would not be unprecedented, as exemplified by the generation of, e.g., cartilage from both mesenchymal and neural crest precursor cells *(38–40)*.

GRP cells can generate oligodendrocytes in a variety of in vitro conditions *(41)*. All these oligodendrocyte-containing cultures also contained cells that expressed an antigenic phenotype, consistent with the hypothesis that they were O-2A/OPCs. To define oligodendrocytes, cultures were labeled with antibodies against galactocerebroside (GalC), a myelin-specific glycolipid *(42)*. To define precursor cells that can give rise to oligodendrocytes, cultures also were labeled with the 04 antibody *(43)*. This antibody labels oligodendrocytes at two developmental stages, and reacts with at least two distinct antigens *(44)*. One of the antigens recognized by the 04 antibody is sulfatide, which appears on oligodendrocytes subsequent to the appearance of GalC. A variety of careful studies have demonstrated that a different and still unknown antigen is expressed in this lineage prior to the expression of GalC *(44)*, and that cells expressing an 04^+GalC^- antigenic phenotype that are isolated from the CNS of postnatal rats express many of the major characteristics of O-2A/OPCs *(45–47)*.

Confirmation that 04^+GalC^- cells derived from GRP cells were O-2A/OPCs was obtained by purifying these cells and analyzing their differentiation potential in clonal growth conditions. This clonal analysis confirmed that 04^+GalC^- cells derived from GRP cells generate type 2 astrocytes in response to exposure to serum, whereas GRP cells make both type 1 and type 2 astrocytes in these conditions. These 04^+GalC^- cells also differentiated into oligodendrocytes when grown in chemically defined medium and expressed other characteristics of O-2As/OPCs.

In analogy with observations that GRP cells may be a necessary intermediate in the generation of glia from the totipotent neuroepithelial (NEP) stem cells of the spinal cord, it now appears that the bipotential O-2A/OPC may be a necessary intermediate in the generation of oligodendrocytes from GRP cells.

It is intriguing to speculate whether the knowledge obtained thus far allows a description of the entire pathway of lineage restriction from the embryonic stem cell to the oligodendrocyte (as summarized in Fig. 2). Studies by Okabe et al. *(48)* and Fraichard et al. *(49)* have demonstrated that ES cells can give rise to totipotent neurospheres, which share the ability of the NSCs of the embryonic spinal cord to give rise to all the differentiated cell types of the CNS *(50)*. Although the oligodendrocyte precursor cells generated from neurospheres (e.g., as in ref. *51*) have not yet been characterized sufficiently to determine whether these cells are GRP cells, O-2A/OPCs, or still another population of glial precursor cells, it thus far appears that embryonic NSCs give rise directly to GRP cells *(9)*. As O-2A/OPCs give rise directly to oligodendrocytes *(11,17,25,34,35)*, the generation of O-2A/OPCs from GRP cells raises the possibility that the developmental lineage depicted in Fig. 2 might represent a complete description of one path to the creation of a differentiated oligodendrocyte. As more markers become available to allow the detailed study of other stages of gliogenesis, it will be possible to test this possibility more stringently.

REGULATION OF OLIGODENDROCYTE GENERATION FROM DIVIDING O-2A/OPCS INVOLVES AT LEAST THREE DISTINCT PROCESSES

The process by which O-2A/OPCs generate oligodendrocytes has been the subject of extensive analysis. Of particular importance in such research has been the fact that O-2A/OPCs can be readily grown in vitro as a purified cell population and in individual clones *(19,22,52)*. Studies on this lineage also benefit from the fact that the bipolar morphology of O-2A/OPCs and the multipolar morphology of oligodendrocytes is sufficiently distinct that it is possible to identify each of these cell types by visual inspection, thus allowing detailed clonal analysis of differentiation to be readily carried out.

Initial studies on the generation of oligodendrocytes from dividing O-2A/OPCs indicated that this differentiation process was controlled by a cell-intrinsic program that causes all clonally related cells to undergo differentiation in a synchronous and symmetric manner *(53)*. As discussed below, it now seems likely that this symmetric clonal differentiation was the outcome of conducting experiments in the presence of both type 1 astrocytes and thyroid hormone (TH) *(22,52,54)*, and did not reflect the O-2A/OPCs most fundamental differentiation characteristics. We now know that these conditions are highly potent inducers of oligodendrocyte generation and

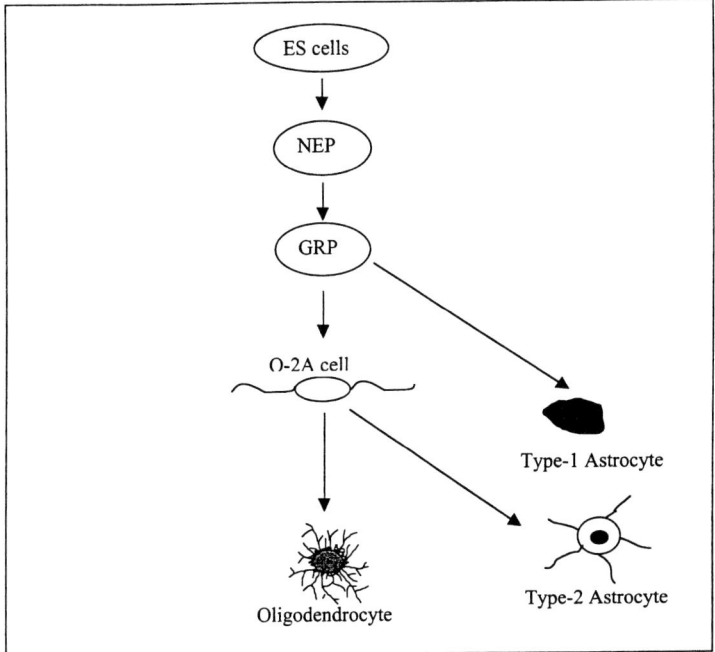

Fig. 2. Theoretical summary of the complete lineage history of the oligodendrocyte. Pluripotent embryonic stem cells (ES cells) can generate multipotent neuroepithelial (NEP) cells. NEP cells give rise to glial restricted precursor (GRP). GRP cells are tripotential, giving rise to oligodendrocytes and two distinct types of astrocytes: type 1 and type 2 astrocytes. The generation of oligodendrocytes and type 2 astrocytes seems to involve another glial restricted precursor cell, the oligodendrocyte type 2 astrocyte progenitor cell (O-2A cell).

cause what is basically an asymmetric system to read out as though the differentiation pattern were symmetric.

It currently appears that three distinct stages of oligodendrocyte generation can be recognized, as summarized in Table 3. The initial generation of oligodendrocytes seems to be controlled by a cell-intrinsic clock that, in a still unknown manner, causes dividing progenitor cells to begin to generate oligodendrocytes with a highly stereotyped timing. For example, in cultures derived from brain or optic nerve of embryonic rats of a variety of ages, the first oligodendrocytes appear at a time equivalent to the date of birth of the rat, the same time at which oligodendrocytes first appear in vivo in these tissues (55,56). This appropriately timed initial generation of oligodendrocytes is a robust phenomenon that has been observed in a variety of conditions, including in high-density cultures from brain or optic nerve

Table 3
The Three Stages of Oligodendrocyte Generation from Dividing O-2A/OPCs

Stage	
1	Oligodendrocyte generation is initiated by cell-intrinsic mechanisms
2	The probability of self-renewal vs differentiation is regulated by cell-extrinsic signals
3	Cell-intrinsic mechanisms limit the period of division as an O-2A/OPC with perinatal characteristics, unless overridden by environmental signals

grown in medium supplemented with fetal calf serum (FCS) *(55)*, and in high- or low-density cultures supplemented with medium conditioned by type 1 astrocytes or with PDGF *(22,25,57)*.

One of the most remarkable features of the initial generation of oligodendrocytes is that it is a highly asymmetric process *(22)*. Clones of O-2A/OPCs derived from the brains of E15 rats induced to divide by exposure to PDGF will generate their first oligodendrocytes after 6 d of in vitro expansion. The percentage of oligodendrocytes generated per clone varies over a broad range, but in the great majority of clones represents <30% of the total cells and in many clones represents <10% of the total cells. There is little correlation between the size of the clone and the extent of oligodendrocyte generation that occurs, with some clones at this time containing less than 15 cells and others more than 300 cells. We interpret such an outcome to mean two things.

First, the timer that controls initial oligodendrocyte generation does not count cell divisions but instead actually measures time in some manner. This interpretation is based on the fact that clones of 15 cells or less would have undergone four divisions and those with 300 cells or more would have undergone nine or more divisions, yet both types of clone, and all other sizes of clone, exhibit a high probability of generating their first oligodendrocytes with a highly stereotyped timing. For example, in our experiments *(22)*, 60% of clones generated their first oligodendrocytes at a time equivalent to the date of birth of the rat, with virtually no clones generating any oligodendrocytes even 1 d earlier. The second conclusion we draw is that the mechanism that causes oligodendrocyte generation is hard-wired to the extent that it can be "activated" nine divisions in advance (for instance, in those clones containing 300 or more cells) but is encoded in such a manner that the actual penetrance (i.e., the extent of differentiation) is asymmetric over a broad range. At present, no known molecular mechanism would account for a process that is temporally regulated with such precision yet allows for such a range of asymmetry and variability at the level of individual clones.

Detailed analysis of oligodendrocyte generation in individual clones of O-2A/OPCs isolated from embryonic rat brain or postnatal optic nerve indicates that once oligodendrocyte generation is initiated within a clone, a second process is initiated in which the balance between self-renewal and differentiation is modulated by environmental signals *(22)*. During this second period, O-2A/OPCs (isolated from optic nerves of postnatal rats and induced to divide by PDGF) exposed to certain signaling molecules [e.g., neurotrophin-3 (NT-3) or bFGF] tend to undergo self-renewing divisions in which progenitor cell expansion is dominant over differentiation. In contrast, exposure to other signaling molecules (e.g., TH or CNTF) promotes the generation of oligodendrocytes and reduces self-renewal *(19,22,52)*. The extent of plasticity that it is possible to achieve is impressive and can range from induction of differentiation of most cells within a clonal family within only a small number of divisions *(22,52,53)* to the almost complete suppression of differentiation and continuous promotion of self-renewal for many weeks *(58)*.

In the third stage of oligodendrocyte generation, cell-intrinsic mechanisms appear to cause dividing progenitor O-2A/OPCs to differentiate either into oligodendrocytes or into a second generation of O-2A/OPCs progenitor cells with properties more appropriate for the physiologic requirements of the adult CNS *(59)*; as discussed later in this review. The cell-intrinsic control of differentiation in this third stage is not absolute, however, and can be environmentally overridden, for example, by continuous exposure to PDGF + bFGF *(58)*.

THE ADULT-SPECIFIC O-2A/OPC AS THE FIRST DEMONSTRATION THAT A SINGLE LINEAGE MAY CONTAIN MULTIPLE PRECURSOR CELL POPULATIONS WITH DIFFERENT PROPERTIES

Complex as the picture of oligodendrocyte generation might seem thus far, it is already clear that this picture needs to be expanded to include consideration of multiple O-2A/OPC populations. At the time of this writing, we know of four distinct groups of O-2A/OPCs with different biologic properties.

The idea that a single precursor cell population, as defined by its lineage restriction, may actually contain multiple precursor cells with differing phenotypes was first indicated by findings that O-2A/OPCs isolated from the optic nerves of developing and adult rats expressed strikingly different properties *(60)*. We interpret the data discussed below in the context of observations that the physiologic demands made on precursor cell populations during different developmental epochs can be remarkably divergent, particularly in tissues that do not undergo constant turnover in the adult animal. In such tissues, with the CNS being a prime example, the rapid increase in numbers of precursor cells and their differentiated progeny becomes turned off over a relatively short period. Once this period of rapid cell generation ends, precursor cell populations may be maintained in the adult tissue, to participate in homeostatic cell replacement during normal function and to repair damaged tissue following injury. The regulation of the precursor cells resident in adult tissue must, however, be different in some manner from that which characterizes the explosive growth of early development.

Initial studies on O-2A/OPCs of the adult CNS revealed that these cells differ from their perinatal counterparts in a variety of properties *(60,61)*. For example, O-2A/OPCs[perinatal] divide rapidly (18 h), migrate rapidly (21 μm/h), and differentiate rapidly into oligodendrocytes when grown in serum-free medium in the absence of mitogens. O-2A/OPCs[adult], in contrast, divide, differentiate, and migrate much more slowly *(60;* Table 4). In addition, O-2A/OPCs[perinatal] can be induced by external signaling molecules to undergo a pattern of differentiation in which all members of a single clone turn into oligodendrocytes in a relatively synchronous and symmetric manner *(22,25,52,53)*. In contrast, in these same conditions, O-2A/OPCs[adult] predominantly generate oligodendrocytes in association with asymmetric division and differentiation *(59)*. These several differences appear to be cell intrinsic, in that *perinatal* and *adult* O-2A/OPCs seem to co-exist in the optic nerve in vivo between 2 and 4 wk after birth and express their characteristic phenotypes even when they are grown together in the same tissue culture dish *(62)*.

The existence of two O-2A/OPC populations with properties so distinct as those of the perinatal and adult CNS raises the question of what the developmental relationship between these two populations might be. Time-lapse microcinematography studies and serial passaging experiments revealed that O-2A/OPCs[perinatal] isolated from optic nerves of 7-d-old rat pups can give rise directly to cells with the characteristics of O-2A/OPCs[adult], thus demonstrating a direct lineage relationship between these populations *(59)*. Such observations have recently been repeated and extended to show that the transition from *perinatal* to *adult* phenotype can occur independently of interactions with cells outside of the O-2A/OPC lineage and also can occur in the absence of such differentiation-inducing signals as TH *(63)*.

Studies on the maturation in the O-2A/OPC lineage provide a very different view of cellular aging than is derived from studies on cellular senescence. As the O-2A/OPC[perinatal] goes through its continuing cell cycles, it does not appear to reach a stage where division is no longer possible. Instead, as shown previously *(64)*, rapidly dividing cells from late embryos intrinsically mature into the less rapidly dividing cells of the postnatal

Table 4
Key Distinguishing Characteristics of Perinatal and Adult O-2A/OPCs

	O-2A/OPC^{perinatal}	O-2A/OPC^{adult}
Cell cycle length	18 ± 4 h	65 ± 18 h
Migration rate	21.4 ± 1.6__m h^{-1}	4.3 ± 0.7 _m h^{-1}
Time to differentiation of 50% of cells into oligodendrocytes in mitogen-free conditions	48 h	72–120 h

animal, which themselves give rise to slowly dividing *adult* cells *(59,63)*. Thus, the transition that occurs with continued division of O-2A/OPCs in vitro is very different from that which occurs in, e.g., dividing fibroblasts, which reach a point (termed senescence) at which they no longer divide at all. Aging in the O-2A/OPC lineage is associated instead with the emergence of a new population of precursor cells with fundamentally different properties, including a much slower cell cycle length. Such results give strong support to the view that maturational processes associated with aging of precursor cells include biologic alterations that may prove far more subtle than, e.g., telomere shortening.

THE DEVELOPING CNS CONTAINS MULTIPLE O-2A/OPC POPULATIONS WITH DIFFERENT PROPERTIES

The initial proposal that the differences between *perinatal* and *adult* O-2A/OPCs indicates that the specific physiologic requirements of a particular tissue are reflected in the intrinsic properties of the precursor cells resident in that tissue has recently been extended to consider different CNS regions of animals of the same age. In these studies, striking differences between the properties of O-2A/OPCs isolated from different regions of the CNS of 7-d-old rat pups have been found, differences that once again appear to be highly relevant to the understanding of development.

One of the striking aspects of CNS development is that different regions of this tissue develop according to different schedules. For example, neuron production in the rat spinal cord is largely complete by the time of birth, is still ongoing in the cerebellum for at least several days after birth, and continues in the olfactory system and in some regions of the hippocampus throughout life. Great variation is also seen in the time course of gliogenesis. For example, myelination has long been observed to progress in a rostral-caudal direction and begins in the spinal cord significantly earlier than in the brain *(65–67)*. Even within a single CNS region, myelination is not synchronous. For example, in the optic nerve myelinogenesis occurs with a retinal-to-chiasmal gradient, with regions of the nerve nearest the retina becoming myelinated first *(65,68)*. The cortex itself shows the widest range of timings of myelination, both initiating later than many other CNS regions *(65–67)* and exhibiting an ongoing myelinogenesis that can extend over long periods of time. This latter characteristic is seen perhaps most dramatically in the human brain, for which it has been suggested that myelination may not be complete until after several decades of life *(69,70)*.

What might account for the dissimilar rates of development in different CNS regions? One possibility is that the plasticity exhibited by precursor cells is modulated in a specific manner by local microenvironments, such that variant periods of production of modulators of differentiation cause differing behaviors in precursor cell populations that are widely distributed in the CNS but are fundamentally equivalent to each other. In such a situation, variations in developmental timing might be explained entirely by environmental modulation of precursor cell fate. At the other extreme, precursor cell populations from diverse CNS regions might exhibit biologically distinct properties that are related to the time course of differentiation. It is even conceivable that precursor cells from different CNS regions might be identical in respect to lineage restriction but might diverge widely in other biologic properties that are relevant to specific aspects of development in the CNS region of residence.

The hypothesis that environmental modulation of precursor cell plasticity might account for differing time courses of differentiation is consistent with previous analyses of the relative contributions of cell-intrinsic and cell-extrinsic mechanisms to the generation of oligodendrocytes *(22)*. The plasticity that exists in the second stage of oligodendrocyte generation from dividing O-2A/OPCs allows for a great range in the timing over which large numbers of these cells are produced, as discussed earlier. Exposure to NT-3 or to bFGF enables more extended division of O-2A/OPCs (at least in vitro) and could spread the generation of oligodendrocytes over a longer period. In contrast, exposure to TH or to CNTF (which promote oligodendrocyte generation) could cause an earlier differentiation. Even though the extent to which such plasticity is utilized in vivo is not known, it appears from in vitro studies that the basic biology of O-2A/OPCs is such that myelination could occur over a range of time courses if local microenvironments were sufficiently different from each other.

Despite the expression of great plasticity in the timing of oligodendrocyte generation in vitro, this does not rule out the existence of multiple oligodendrocyte precursor cell populations that differ in their biologic properties in manners that are related to development of the CNS regions from which they are isolated, much in the manner that *perinatal* and *adult* O-2A/OPCs differ from each other.

We have recently made the surprising discovery that the CNS of early postnatal (P7) rats contains multiple O-2A/OPC populations that exhibit markedly different properties, both in respect to their intrinsic tendency to undergo self-renewing divisions and in their response to inducers of oligodendrocyte generation (J. Power, J. Smith, M. Mayer-Pröschel, and M. Noble, manuscript in preparation). Specifically, characterization of O-2A/OPCs isolated from optic nerve, optic chiasm, and cortex of P7 rats has revealed that each of these populations expresses distinct biologic properties. To discuss these different populations efficiently, we introduce the terminology O-2A/OPC(ON), O-2A/OPC(OC), and O-2A/OPC(CX) to refer to O-2A/OPCs isolated from optic nerve, optic chiasm, and cortex, respectively.

One of the surprising findings to emerge from our recent studies was the extent to which self-renewal characteristics differed in progenitor cells that exhibit an apparently identical lineage restriction and are isolated from postnatal animals of a single age. For example, O-2A/OPC populations isolated even from adjoining CNS regions, the optic nerve, and the optic chiasm exhibit marked differences in their probability of undergoing differentiation when grown in the presence of PDGF sans inducers of differentiation (e.g., TH, CNTF), such that cells derived from the chiasm are intrinsically far more capable of undergoing extended self-renewal than cells isolated from the nerve. Still more different from optic nerve-derived cells are O-2A/OPC(CX) progenitor cells, which are able to undergo continuous self-renewal for many days when grown in basal division conditions, generating large clones of up to 300 cells that consist predominantly of progenitors and containing very few oligodendrocytes. Even after 10 d of in vitro growth in these conditions, the average clonal composition in O-2A/OPC(CX) progenitor cell cultures consisted of 99.5% progenitors and 0.5% oligodendrocytes. The average clonal composition of O-2A/OPC(ON) progenitor cells at this time point, in contrast, was 37% progenitors and 63% oligodendrocytes. Addition of TH to the basal division medium (which contained PDGF alone) was able to promote oligodendrocyte generation in all of the different O-2A/OPC populations, but even here marked differences were seen. In particular, even when exposed to TH, the great majority of O-2A/OPCs derived from the cortex continued to undergo extensive self-renewal.

In analogy with the proposal that the differences between O-2A/OPCs[adult] and O-2A/OPCs[perinatal] of the rat optic nerve are reflective of the differing physiologic requirements of tissue development and tissue homeostasis, we propose that the differences that distinguish O-2A/OPC(ON), O-2A/OPC(OC), and O-2A/OPC(CX) populations are reflective of differing physiologic requirements of the tissues to which these cell contribute. For example, a variety of experiments have indicated that the O-2A/OPC population of the optic nerve arises from a germinal zone located in or near the optic chiasm and enters the nerve by migration *(71,72)*. Thus, it would not be surprising if progenitor cells of the optic chiasm expressed properties expected of cells at a potentially earlier developmental stage than those cells that are isolated from optic nerve of the same physiologic age. Such properties would be expected to include the capacity to undergo a greater extent of self-renewal, much as has been seen when the properties of O-2A/OPCs from optic nerves of embryonic rats and postnatal rats have been compared *(64)*.

In respect to the properties of cortical progenitor cells, physiologic considerations also appear to be consistent with our observations. The cortex is one of the last regions of the CNS in which myelination is initiated, and the process of myelination also can continue for extended periods in this region *(65–67)*. If the biology of a precursor cell population is reflective of the developmental characteristics of the tissue in which it resides, then one might expect that O-2A/OPC isolated from this tissue would not initiate oligodendrocyte generation until a later time than occurs with progenitor cells isolated from structures in which myelination occurs earlier. In addition, cortical progenitor cells might be physiologically required to make oligodendrocytes for a longer time owing to the long period of continued development in this tissue, at least as this has been defined in the human CNS *(69,70)*.

Is it possible that the properties of O-2A/OPC(ON), O-2A/OPC(OC), and O-2A/OPC(CX) populations represent a developmental progression within a single population of precursor cells? Such a possibility is not unreasonable, particularly in light of our previous demonstrations that O-2A^{adult} progenitor cells can be directly generated from O-2A^{perinatal} progenitor cells of the optic nerve *(59)*. Similarly, others have shown that O-2A/OPCs of the embryonic optic nerve give rise in vitro to populations with the properties of cells derived from the postnatal optic nerve *(64)*. Proposing such a developmental relationship between O-2A/OPCs(OC) and O-2A/OPCs(ON) is also not unreasonable in light of indications that the O-2A/OPCs of the optic nerve have their origin in or near the optic chiasm *(71,72)*. Despite the above considerations, however, it also is possible that the different O-2A/OPC populations we have isolated from P7 rats represent distinct developmental lineages that are all able to give rise to the same differentiated cell types. The ability of a single type of differentiated cell to arise from more than a single precursor cell population is well-known, with the generation of cartilage and skeletal muscle from both mesenchymal and neural crest lineages being just one well-documented example of such an occurrence *(38,40)*. Future research will hopefully reveal which of these alternatives is correct.

Regardless of whether or not O-2A(CX), O-2A(OC), and O-2A(ON) progenitor cells represent a developmental progression within a single lineage or multiple precursor cell populations that share an identical lineage restriction, it is striking that the self-renewal properties of each population appear to be cell intrinsic. At least in this regard, the differences between these populations do not seem to be owing to production by the progenitor cells of diffusible factors that cause otherwise similar cells to exhibit dissimilar behaviors.

OTHER OLIGODENDROCYTE PRECURSOR CELLS

One of the important aspects of oligodendrocyte development that has not yet been integrated into the information provided thus far is that there may be still other oligodendrocyte precursor cells whose biologic properties are currently less well-understood. In particular, the precise relationship that the tripotential GRP cell or the O-2A/OPC has to other glial restricted progenitors, such as the polysialated neural cell adhesion molecule (PSA-NCAM)-expressing pre-progenitor cell *(73–77)*, or the cells generated by exposure of totipotent neuroepithelial stem cells to B104 conditioned medium *(78)*, is unclear at present. Moreover, it is clear that the totipotent NSCs isolated from a variety of regions of the CNS, and even from the adult CNS, are also able to generate oligodendrocytes in vitro and following transplantation *(79–85)*. It is not known whether the generation of oligodendrocytes from all these sources follows the same developmental sequence. It is certainly possible, for example, that the GRP cell is a specialized cell of the embryonic animal and that the NSCs of the adult animal generate glia without first progressing through a GRP cell stage. Unraveling these relationships remains an important challenge.

DEVELOPMENTAL TRANSITIONS AND DEVELOPMENTAL MALADIES

There are multiple reasons for attempting to understand the complexities of glial cell development. Deciphering the mysteries of differentiation is certainly a worthy challenge. Moreover, the use of precursor cells for CNS repair may allow treatment of medical disorders that presently have no treatment, and understanding the biology of these cells is important in such regard. Indeed, the possibility of using precursor cell transplantation to repair demyelinating damage has long been demonstrated and is the subject of active research in multiple locations. Still another reason for being vitally interested in precursor biology, however, and one that has perhaps not yet received the attention that it deserves is the opportunity to understand the cellular basis for defective development of the CNS.

Aberrant neurologic development is associated with a wide range of physiologic insults. The diverse causes of such problems include various nutritional deficiency disorders, hypothyroidism, fetal alcohol syndrome, treatment of CNS cancers of childhood by radiation, and treatment of even some non-CNS cancers of childhood by chemotherapy. Common to most — and perhaps even to all — of the physiologic challenges associated with neurologic impairment are problems related to formation or maintenance of myelin.

A number of the physiologic insults that result in neurologic impairment and defective myelination, such as TH, iron, and nutritional deficiencies (i.e., the syndromes on which this application is focused), are well known to have their most severe outcomes if they occur during critical developmental periods. In 1921 the suggestion was first made that perturbations to development have their maximal effect *(86)* if they occur during critical developmental periods, but it was not until 45 yr later that it was specifically hypothesized that the period of myelination represents a particularly important critical period in brain development *(87)*. This hypothesis is well-supported by a variety of observations (as discussed below). Still, relatively little is known about the actual biological processes involved in defining these critical periods, at either a biochemical or developmental level.

Several different kinds of developmental processes could be relevant to the existence of critical periods, as represented in Fig. 3. These include cell survival, cell division, and expression of differentiated functions. Moreover, these processes could be altered in differentiated cells or in the precursor cells that give rise to the differentiated cells. For example, reductions in the number of a particular differentiated cell type could be as a result of a failure of that cell to survive after it is generated. Reductions in cell number could equally result, however, from death or reduced division of ancestral precursor cells, from a failure of the precursor cells to undergo normal patterns of differentiation, or from other perturbations discussed in later portions of this application.

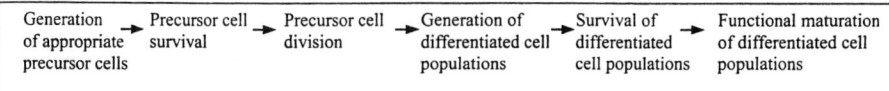

Fig. 3. Summary of the points in the history of a cell population at which normal development can be disrupted by physiologic insult.

Thyroid Hormone and Oligodendrocyte Development

Currently, the best available data suggesting that developmental transitions in precursor cell populations might be intimately involved in the biology of critical periods come from studies on the action of TH in glial development. Hypothyroidism, usually associated with iodine deficiency, is a major cause of mental retardation and other developmental disorders *(88,89)*. It is well-established in animal models that perinatal hypothyroidism is associated with defects in myelination and a reduced production of myelin-specific gene products and that these defects can be at least partially ameliorated if thyroid hormone therapy is initiated early enough in postnatal life *(90–94)*.

We have earlier mentioned that TH is an effective promoter of oligodendrocyte generation from dividing O-2A/OPCs, and both in vitro and in vivo experimentation has made it clear that a likely effect of hypothyroidism is a reduction in the number of oligodendrocytes generated *(22,95)*. As we found that oligodendrocytes are readily generated in the absence of TH, but that the proportion of oligodendrocytes per clone is reduced if this hormone is not present, we suggested that TH has its primary effect on regulating the proportion of dividing progenitor cells that undergo differentiation *(22)*. Data in vivo are consistent with this analysis. Hypothyroid animals are known to generate reduced amounts of myelin, indicating that some oligodendrocytes are generated in these animals *(90–92,96)*. Although we did not examine oligodendrocyte representation in the optic nerve at the date of birth in our own experiments *(22)*, we did find that oligodendrocytes were readily detected at 2 d after birth, but at only 20% of the prevalence seen in normal animals. Intriguingly, this was the same degree of difference in oligodendrocyte prevalence observed in embryonic brain cultures grown in the presence or absence of TH.

We have also considered the possibility that a failure to generate appropriate numbers of oligodendrocytes could be owing to a failure to generate sufficient numbers of O-2A/OPCs, the ancestors of oligodendrocytes. To examine the possibility that this stage of oligodendrocyte development might also be a possible point of developmental disruption in hypothryoidism, we determined whether TH plays any role in modulating the generation of O-2A/OPCs from their ancestral GRP cells. As mentioned earlier, we found that GRP cells can generate oligodendrocytes and cells with the antigenic phenotype of O-2A/OPCs in a variety of in vitro conditions, but the most potent are those involving exposure to TH *(41)*. Thus, it is possible that regulation of this transition could also be aberrant in deficiency syndromes and other developmental syndromes associated with a failure of myelination.

Nutritional Deficiency

Critical periods for the generation of defects in myelination have been most clearly established in the case of nutritional deficiency, with highly focused studies being carried out in rodent models. Lasting and significant myelin deficits are associated with undernourishment that is established at birth (or earlier, through maternal deprivation) and maintained through weaning *(97–99)*. Strikingly, the deficit in myelin can be quite specific, in that this parameter is affected even in cases of undernourishment in which brain weight remains normal. Moreover, nutritional rehabilitation post weaning may be associated with a restoration of brain weight to normal or near-normal levels, even in the absence of a restoration of myelin content to normal levels. That critical developmental periods are important in repair of this damage is indicated by observations that restoration of normal levels of myelin requires that normal nutrition be itself restored well before weaning. In addition, severe starvation imposed from 14 to 30 d after birth (i.e., after the critical period ends) produces no lasting deficit in myelin accumulation *(98)*, thus stressing the importance of critical periods in the establishment of this deficit.

We are not yet aware of any studies that have examined nutritional deficiency in a manner directly analogous to our studies on TH deficiency. Nonetheless, both in vitro and in vivo data raise the possibility that oligodendrocyte generation is impaired in at least some models of undernourishment. It has been reported that there is a relative reduction in the numbers of oligodendrocytes that are generated in glucose-deprived cultures *(100)*. In addition, severe malnutrition regimes are associated with a clear reduction in glial cell numbers in vivo *(101)*, although cell type-specific markers were not utilized to determine whether this reduction preferentially affected oligodendrocytes rather than astrocytes.

Still another point in oligodendrocyte development at which a failure of normal regulatory mechanisms would be associated with disruptions in the myelination process involves oligodendrocyte differentiation itself. Here too there is evidence that TH and nutritional status could play important roles in allowing normal development to occur. The ability of TH to modulate expression of myelin-specific genes has been extensively documented *(90–92,96,101,102)*. Regulation of myelin-specific genes by nutritional status has also been reported both in vivo and in vitro. In vivo, it is well established that the myelin deficits associated with undernutrition are even observed in animals in which oligodendrocyte number appears to be normal *(103)*. In such animals, however, it has been reported *(104)* that the mRNAs for three important myelin proteins [myelin-associated glycoprotein (MAG), proteolipid protein (PLP), and myelin basic protein (MBP)] do not undergo the normal increases seen in brains of well-nourished animals. Increases are delayed for several days beyond the normal time (i.e., days 7–9) at which they are observed, and the increases are lower in extent.

In vitro studies using glucose deprivation as a model for caloric undernutrition *(105)* have raised the surprising possibility that transient caloric restriction at critical periods may lead to long-term effects on differentiated function. In these experiments, mixed cultures were generated from newborn rat brain and exposed to different glucose concentrations, ranging from 0.55 to10 mg/mL; the lower doses are within the range that occurs in clinical hypoglycemia. Low glucose concentrations were associated with markedly lower levels of increases in MAG, PLP, and MBP mRNA levels and with a subsequent and abnormal downregulation in these mRNA levels. These effects were specific, in that total mRNA levels in the cultures were normal. Most importantly, these effects appeared to be irreversible if the glucose deprivation was applied over a time period that mirrored the critical period for nutritional deprivation in vivo. Deprivation coincident with the normal time of myelin gene activation and the period of rapid upregulation (6–14 d in vitro [DIV]) was irreversible. Deprivation at a later stage was instead associated with only transient depressing effects. These results are very important in raising the possibility that critical developmental periods may in part reflect the timing of nuclear events related to temporal and tissue-specific patterns of gene expression. It is also intriguing that recent studies have suggested that these effects are not seen if glucose deprivation is applied to cultures of purified oligodendrocytes, an observation the authors have suggested may indicate the involvement of an intermediary cell type to achieve suppression of myelin gene expression *(106)*. It is important to note, however, that the presence of other cells would promote division of O-2A/OPCs and generation of immature oligodendrocytes *(35)*, thus raising the possibility that the lack of any affect also may reflect effects at particular developmental stages.

ASTROCYTE

We have focused our attention in this review on the development of the oligodendrocyte because of the wealth of information that has been generated on this topic. Even though we think it likely that the oligodendrocyte is functionally more complex than is thought, and also believe that oligodendrocytes may themselves comprise a more heterogeneous collection of cells than is currently thought, it seems unlikely that this cell type is as complex as the astrocyte.

The range of functions that have been attributed to astrocytes is broader than for any other cell type of which we know. Astrocytes provide substrates for axonal growth, secrete mitogens and survival factors for neurons and for O-2A/OPCs *(24,35,107–111)*, induce endothelial cells to differentiate to form the blood-brain barrier *(112,113)*, can function as antigen-presenting cells *(114–116)*, have been found to produce complement and inflammatory cytokines *(117,118)*, inactivate neurotransmitters and neuropeptides through both uptake and enzymatic inactivation *(119,120)*, modulate ion fluxes in the brain, may modulate neuronal activity through their own calcium fluxes *(121,122)*, can produce and secrete compounds with neurotransmitter activity *(123,124)*, play a central role in the generation of glial scar tissue *(125–128)*, and express a host of still other functions *(129–131)*.

The correlation between diversity of function in astrocytes and development of these cells, at this stage, represents wholly unknown territory. A number of astrocyte precursor cells have been described, including the A2B5+ astrocyte precursor cell present in E17 spinal cord and originally described by Miller and colleagues *(132,133)*, the putative astrocyte precursor cells from the embryonic mouse cerebellum described by Seidman et al. *(134)*, and the astrocyte precursor cell described by Mi and Barres *(135)*. In addition, Miller and colleagues *(136)* have described five distinct astrocyte populations in the spinal cord based on heritable morphologic differences. At this stage, little or nothing is known about how these various populations of astrocytes are developmentally related to each other, or what induces the appearance of specialized astrocyte functions. Moreover, little is known about the extent to which physiologic insults disrupt normal astrocytic development and function. Considering the multiple roles played by these cells, it would be very surprising if disruption of their function did not have profound neurologic consequences.

One of the greatest problems hindering the study of astrocyte diversity is the lack of markers. The extensive characterization of O-2A/OPCs and the existence of markers (the O4 and anti-GalC antibodies) that enable purification of a specific subset of progeny cells derived from GRP cells allowed the demonstration of derivation of a bipotential O-2A/OPC-like cell from a tripotential GRP cell. We do not yet have sufficient markers and/or knowledge of growth conditions to determine whether GRP cells also can give rise to such other glial precursor cell populations as the astrocyte precursor cells that have been described by others *(132–135)* or perhaps even to the pre-O-2A/OPC described by Grinspan and colleagues *(136)*.

The extent of the need for new markers in the study of astrocytes is particularly well-illustrated by the recent studies of Alvarez-Buylla and colleagues, who have described in the subventricular zone of the adult mammalian brain a precursor cell that expresses glial fibrillary acidic protein (GFAP), the supposedly "definitive" marker of an astrocyte; *(137)* yet is able to give rise to neuroblasts that migrate into the olfactory stream and differentiate into neurons when they reach the olfactory bulb *(138)*. That we currently have no suitable markers for differentiating even these GFAP+ cells from astrocytes committed to the range of CNS functions mentioned in the beginning of this closing section is a stark indication of how far we must progress to analyze the development of what is the major cell type in the CNS. Considering the complexity of astrocytes, it seems most likely that their developmental history will be at least as filled with novel discoveries as has been the case for their companion in the glial kingdom, the oligodendrocyte.

REFERENCES

1. Gonzalez, D., Dees, W. L., Hiney, J. K., Ojeda, S. R., and Saneto, R. P. (1990) Expression of beta-nerve growth factor in cultured cells derived from the hypothalamus and cerebral cortex. *Brain Res.* **511,** 249–258.
2. Micera, A., Vigneti, E., and Aloe, L. (1998) Changes of NGF presence in nonneuronal cells in response to experimental allergic encephalomyelitis in Lewis rats. *Exp. Neurol.* **154,** 41–46.
3. Bloch, B., Popovici, T., Levin, M. J., Tuil, D., and Kahn, A. (1985) Transferrin gene expression visualized in oligodendrocytes of the rat brain by using in situ hybridization and immunohistochemistry. *Proc. Natl. Acad. Sci. USA* **82,** 6706–6710.
4. Espinosa de los Monteros, A., Kumar, S., Scully, S., Cole, R., and de Vellis, J. (1990) Transferrin gene expression and secretion by rat brain cells in vitro. *J. Neurosci. Res.* **25,** 576–580.
5. Chen, M. S., Huber, A. B., van der Haar, M. E., Frank, M., Schnell, L., Spillmann, A. A., et al. (2000) Nogo-A is a myelin-associated neurite outgrowth inhibitor and an antigen for monoclonal antibody IN-1. *Nature* **403,** 434–439.
6. GrandPre, T., Nakamura, F., Vartanian, T., and Strittmatter, S. M. (2000) Identification of the Nogo inhibitor of axon regeneration as a Reticulon protein. *Nature* **403,** 439–444.
7. Huang, D. W., McKerracher, L., Braun, P. E., and David, S. (1999) A therapeutic vaccine approach to stimulate axon regeneration in the adult mammalian spinal cord. *Neuron* **24,** 639–647.
8. Mayer-Pröschel, M., Kalyani, A., Mujtaba, T., and Rao, M. S. (1997) Isolation of lineage-restricted neuronal precursors from multipotent neuroepithelial stem cells. *Neuron* **19,** 773–785.
9. Rao, M. and Mayer-Pröschel, M. (1997) Glial restricted precursors are derived from multipotent neuroepithelial stem cells. *Dev. Biol.* **188,** 48–63.
10. Rao, M., Noble, M., and Mayer-Pröschel, M. (1998) A tripotential glial precursor cell is present in the developing spinal cord. *Proc. Natl. Acad. Sci. USA* **95,** 3996–4001.
11. Raff, M. C., Miller, R. H., and Noble, M. (1983) A glial progenitor cell that develops in vitro into an astrocyte or an oligodendrocyte depending on the culture medium. *Nature* **303,** 390–396.

12. Skoff, R. P. (1996) The lineages of neuroglial cells. *Neuroscientist* **2**, 335–344.
13. Fulton, B. P., Burne, J. F., and Raff, M. C. (1991) Glial cells in the rat optic nerve. The search for the type-2 astrocyte. *Ann. N.Y. Acad. Sci.* **633**, 27–34.
14. Franklin, R. J. and Blakemore, W. F. (1995) Glial-cell transplantation and plasticity in the O-2A lineage—implications for CNS repair. *Trends Neurosci.* **18**, 151–156.
15. Lillien, L. E. and Raff, M. C. (1990) Analysis of the cell-cell interactions that control type-2 astrocyte development in vitro. *Neuron* **4**, 525–534.
16. Lillien, L. E. and Raff, M. C. (1990) Differentiation signals in the CNS: type-2 astrocyte development in vitro as a model system. *Neuron* **5**, 5896–6273.
17. Raff, M. C., Williams, B. P., and Miller, R. H. (1984) The in vitro differentiation of a bipotential glial progenitor cell. *EMBO J.* **3**, 1857–1864.
18. Barres, B., Burne, J., Holtmann, B., Thoenen, H., Sendtner, M., and Raff, M. (1996) Ciliary neurotrophic factor enhances the rate of oligodendrocyte generation. *Mol. Cell. Neurosci.* **8**, 146–156.
19. Mayer, M., Bhakoo, K., and Noble, M. (1994) Ciliary neurotrophic factor and leukemia inhibitory factor promote the generation, maturation and survival of oligodendrocytes in vitro. *Development* **120**, 142–153.
20. Gard, A. L. and Pfeiffer, S. E. (1993) Glial cell mitogens bFGF and PDGF differentially regulate development of O4 + GalC - oligodendrocyte progenitors. *Dev. Biol.* **159**, 618–630.
21. Grinspan, J. B., Reddy, U. R., Stern, J. L., Hardy, M., Williams, M., Baird, L., and Pleasure, D. (1990) Oligodendroglia express PDGF beta-receptor protein and are stimulated to proliferate by PDGF. *Ann. Y. Acad. Sci.* **605**, 71–80.
22. Ibarrola, N., Mayer-Proschel, M., Rodriguez-Pena, A., and Noble, M. (1996) Evidence for the existence of at least two timing mechanisms that contribute to oligodendrocyte generation in vitro. *Dev. Biol.* **180**, 1–21.
23. McKinnon, R. D., Matsui, T., Dubois-Dalcq, M., and Aaronson, S. A. (1990) FGF modulates the PDGF-driven pathway of oligodendrocytic development. *Neuron* **5**, 603–614.
24. Noble, M., Murray, K., Stroobant, P., Waterfield, M. D., and Riddle, P. (1988) Platelet-derived growth factor promotes division and motility and inhibits premature differentiation of the oligodendrocyte/type-2 astrocyte progenitor cell. *Nature* **333**, 560–562.
25. Raff, M. C., Lillien, L. E., Richardson, W. D., Burne, J. F., and Noble, M. D. (1988) Platelet-derived growth factor from astrocytes drives the clock that times oligodendrocyte development in culture. *Nature* **333**, 562–565.
26. Richardson, W. D., Pringle, N., Mosley, M., Westermark, B., and Dubois-Dalcq, M. (1988) A role for platelet-derived growth factor in normal gliogenesis in the central nervous system. *Cell* **53**, 309–319.
27. Satoh, J. and Kim, S. U. (1994) Proliferation and differentiation of fetal human oligodendrocytes in culture. *J. Neurosci. Res.* **39**, 260–272.
28. Barres, B. A. and Raff, M. C. (1994) Control of oligodendrocyte number in the developing rat optic nerve. *Neuron* **12**, 935–942.
29. Barres, B. A., Raff, M. C., Gaese, F., Bartke, I., Dechant, G., and Barde, Y. A. (1994) A crucial role for neurotrophin-3 in oligodendrocyte development. *Nature* **367**, 371–375.

30. Barres, B. A., Schmidt, R., Sendnter, M., and Raff, M. C. (1993) Multiple extracellular signals are required for long-term oligodendrocyte survival. *Development* **118**, 283–295.
31. Gard, A. L., Burrell, M. R., Pfeiffer, S. E., Rudge, J. S., and Williams, W. C. N. (1995) Astroglial control of oligodendrocyte survival mediated by PDGF and leukemia inhibitory factor-like protein. *Development* **121**, 2187–2197.
32. Louis, J. C., Magal, E., Takayama, S., and Varon, S. (1993) CNTF protection of oligodendrocytes against natural and tumor necrosis factor-induced death. *Science* **259**, 689–692.
33. Mayer, M. and Noble, M. (1994) N-acetyl-L-cysteine is a pluripotent protector against cell death and enhancer of trophic factor-mediated cell survival in vitro. *Proc. Natl. Acad. Sci. USA* **91**, 7496–7500.
34. Noble, M., Barnett, S. C., Bogler, O., Land, H., Wolswijk, G., and Wren, D. (1990) Control of division and differentiation in oligodendrocyte-type-2 astrocyte progenitor cells. *Ciba Found Symp.* **150**, 227–243.
35. Noble, M. and Murray, K. (1984) Purified astrocytes promote the in vitro division of a bipotential glial progenitor cell. *EMBO J.* **3**, 2243–2247.
36. Agresti, C., Aloisi, F., and Levi, G. (1991) Heterotypic and homotypic cellular interactions influencing the growth and differentiation of bipotential oligodendrocyte-type-2 astrocyte progenitors in culture. *Dev Biol.* **144**, 16–29.
37. Mayer-Pröschel, M., Rao, M. S., and Noble, M. (1997) Progenitor cells of the central nervous system: a boon for clinical neuroscience. *J. NIH Res.* **9**, 31–37.
38. Baroffio, A., Dupin, E., and Le Dourin, N. M. (1991) Common precursors for neural and mesestodermal derivatives in the cephalic neural crest. *Development* **112**, 301–305.
39. Cassiede, P., Dennis, J. E., Ma, F., and Caplan, A. I. (1996) Osteochondrogenic potential of marrow mesenchymal progenitor cells exposed to TGF-beta 1 or PDGF-BB as assayed in vivo and in vitro. *J. Bone Miner. Res.* **11**, 1264–1273.
40. Selleck, M. A., Scherson, T. Y., and Bronner-Fraser, M. (1993) Origins of neural crest cell diversity. *Dev. Biol.* **159**, 1–11.
41. Gregori, N., Bernard, S., Noble, M., and Mayer-Pröschel, M. (2000) Tripotential glial-restricted precursor (GRP) cells can give rise to bipotential oligodendrocyte-type-2 astrocyte (O-2A) progenitor cells: evidence for a role of multiple layers of lineage restriction in oligodendrocyte development, submitted.
42. Ranscht, B., Clapshaw, P. A., Price, J., Noble, M., and Seifert, W. (1982) Development of oligodendrocytes and Schwann cells studied with a monoclonal antibody against galactocerebroside. *Proc. Natl. Acad. Sci. USA* **79**, 2709–2713.
43. Sommer, I. and Schachner, M. (1981) Monoclonal Antibody (O1-O4) to oligodendrocyte cell surfaces: An immunocytological study in the central nervous system. *Dev. Biol.* **83**, 311–327.
44. Bansal, R., Washington, A. E., Gard, A. L., and Pfeiffer, S. E. (1989) Multiple and novel specificities of monoclonal antibodies O1, O4 and R-mAb used in the analysis of oligodendrocyte development. *J. Neurosci. Res.* **24**, 548–557.
45. Barnett, S. C., Hutchins, A. M., and Noble, M. (1993) Purification of olfactory nerve ensheathing cells from the olfactory bulb. *Dev. Biol.* **155**, 337–350.

46. Gard, A. L., Williams, W. C. N., and Burrell, M. R. (1995) Oligodendroblasts distinguished from O-2A glial progenitors by surface phenotype (O4 + GalC-) and response to cytokines using signal transducer LIFR beta. *Dev. Biol.* **167,** 596–608.

47. Trotter, J. and Schachner, M. (1989) Cells positive for the O4 surface antigen isolated by cell sorting are able to differentiate into astrocytes or oligodendrocytes. *Brain Res. Dev. Brain Res.* **46,** 115–122.

48. Okabe, S., Forsberg-Nilsson, K., Spiro, A. C., Segal, M., and McKay, R. D. (1996) Development of neuronal precursor cells and functional postmitotic neurons from embryonic stem cells in vitro. *Mech. Dev.* **59,** 89–102.

49. Fraichard, A., Chassande, O., Bilbaut, G., Dehay, C., Savatier, P., and Samarut, J. (1995) In vitro differentiation of embryonic stem cells into glial cells and functional neurons. *J. Cell Sci.* **108,** 3181–3188.

50. Kalyani, A., Hobson, K., and Rao, M. S. (1997) Neuroepithelial stem cells: isolation, characterization and clonal analysis. *Dev. Biol.* **187,** 203–226.

51. Laywell, E. D., Kukekov, V. G., and Steindler, D. A. (1999) Multipotent neurospheres can be derived from forebrain subependymal zone and spinal cord of adult mice after protracted postmortem intervals. *Exp. Neurol.* **156,** 430–433.

52. Barres, B. A., Lazar, M. A., and Raff, M. C. (1994) A novel role for thyroid hormone, glucocorticoids and retinoic acid in timing oligodendrocyte development. *Development* **120,** 1097–1108.

53. Temple, S. and Raff, M. C. (1986) Clonal analysis of oligodendrocyte development in culture: evidence for a developmental clock that counts cell division. *Cell* **44,** 773–779.

54. Mayer, M., Bogler, O., and Noble, M. (1993) The inhibition of oligodendrocytic differentiation of O-2A progenitors caused by basic fibroblast growth factor is overridden by astrocytes. *Glia* **8,** 12–19.

55. Abney, E., Bartlett, P., and Raff, M. (1981) Astrocytes, ependymal cells, an oligodendrocytes develop on schedule in dissociated cell cultures of embryonic rat brain. *Dev. Biol.* **83,** 301–310.

56. Miller, R. H., Ffrench Constant, C., and Raff, M. C. (1989) The macroglial cells of the rat optic nerve. *Annu. Rev. Neurosci.* **12,** 517–534.

57. Raff, M. C., Abney, E. R., and Fok-Seang, J. (1985) Reconstitution of a developmental clock in vitro: a critical role for astrocytes in the timing of oligodendrocyte differentiation. *Cell* **42,** 61–69.

58. Bögler, O., Wren, D., Barnett, S. C., Land, H., and Noble, M. (1990) Cooperation between two growth factors promotes extended self-renewal and inhibits differentiation of oligodendrocyte-type-2 astrocytes (O-2A) progenitor cells. *Proc. Natl. Acad. Sci. USA* **87,** 6368–6372.

59. Wren, D., Wolswijk, G., and Noble, M. (1992) In vitro analysis of the origin and maintenance of O-2A[adult] progenitor cells. *J. Cell. Biol.* **116,** 167–176.

60. Wolswijk, G. and Noble, M. (1989) Identification of an adult-specific glial progenitor cell. *Development* **105,** 387–400.

61. Ffrench-Constant, C. and Raff, M. C. (1986) Proliferating bipotential glial progenitor cells in adult rat optic nerve. *Nature* **319,** 499–502.

62. Wolswijk, G., Riddle, P. N., and Noble, M. (1990) Coexistence of perinatal and adult forms of a glial progenitor cell during development of the rat optic nerve. *Development* **109,** 691–698.

63. Tang, D. G., Tokumoto, Y. M., and Raff, M. C. (2000) Long-term culture of purified oligodendrocyte precursor cells: Evidence for an intrinsic maturation program that plays out over months. *J. Cell Biol.* **148,** 971–984.

64. Gao, F. and Raff, M. (1997) Cell size control and a cell-intrinsic maturation program in proliferating oligodendrocyte precursor cells. *J. Cell Biol.* **138,** 1367–1377.

65. Foran, D. R. and Peterson, A. C. (1992) Myelin acquisition in the central nervous system of the mouse revealed by an MBP-LacZ transgene. *J. Neurosci.* **12,** 4890–4897.

66. Kinney, H. C., Brody, B. A., Kloman, A. S., and Gilles, F. H. (1988) Sequence of central nervous sytem myelination in human infancy. II. Patterns of myelination in autopsied infants. *J. Neuropath. Exp. Neurol.* **47,** 217–234.

67. Macklin, W. B. and Weill, C. L. (1985) Appearance of myelin proteins during development in the chick central nervous system. *Dev. Neurosci.* **7,** 170–178.

68. Skoff, R. P., Toland, D., and Nast, E. (1980) Pattern of myelination and distribution of neuroglial cells along the developing optic system of the rat and rabbit. J. *Comp. Neurol.* **191,** 237–253.

69. Benes, F. M., Turtle, M., Khan, Y., and Farol, P. (1994) Myelination of a key relay zone in the hippocampal formation occurs in the human brain during childhood, adolescence and adulthood. *Arch. Gen. Psychiatry* **51,** 477–484.

70. Yakovlev, P. L. and Lecours, A. R. (1967) The myelogenetic cycles of regional maturation of the brain, in *Regional Development of the Brain in Early Life.* (Minkowski, A., et al., eds.), Blackwell, Oxford. pp. 3–70.

71. Ono, K., Bansal, R., Payne, J., Rutishauser, U., and Miller, R. H. (1995) Early development and dispersal of oligodendrocyte precursors in the embryonic chick spinal cord. *Development* **121,** 1743–1754.

72. Small, R. K., Riddle, P., and Noble, M. (1987) Evidence for migration of oligodendrocyte-type-2 astrocyte progenitor cells into the developing rat optic nerve. *Nature* **328,** 155–157.

73. Ben-Hur, T., Rogister, B., Murray, K., Rougon, G., and Dubois-Dalcq, M. (1998) Growth and fate of PSA-NCAM + precursors of the postnatal brain. *J. Neurosci.* **18,** 5777–5788.

74. Grinspan, J. B., Stern, J. L., Pustilnik, S. M., and Pleasure, D. (1990) Cerebral white matter contains PDGF-responsive precursors to O2A cells. *J. Neurosci.* **10,** 1866–1873.

75. Keirstead, H., Ben-Hur, T., Rogister, B., O'Leary, M., Dubois-Dalcq, M., and Blakemore, W. (1999) Polysialylated neural cell adhesion molecule-positive CNS precursors generate both oligodendrocytes and Schwann cells to remyelinate the CNS after transplantation. *J. Neurosci.* **19,** 7529–7536.

76. Nait-Oumesmar, B., Decker, L., Lachapelle, F., Avellana-Adalid, V., Bachelin, C., and Van Evercooren, A. B. (1999) Progenitor cells of the adult mouse subventricular zone proliferate, migrate and differentiate into oligodendrocytes after demyelination. *Eur. J. Neurosci.* **11,** 4357–4366.

77. Vitry, S., Avellana-Adalid, V., Hardy, R., Lachapelle, F., and Baron-Van Evercooren, A. (1999) Mouse oligospheres: from pre-progenitors to functional oligodendrocytes. *J. Neurosci. Res.* **58,** 735–751.
78. Zhang, S. C., Lipsitz, D., and Duncan, I. D. (1998) Self-renewing canine oligodendroglial progenitor expanded as oligospheres. *J. Neurosci. Res.* **54,** 181–190.
79. Carpenter, M., Cui, X., Hu, Z., Jackson, J., Sherman, S., Seiger, A., and Wahlberg, L. (1999) In vitro expansion of a multipotent population of human neural progenitor cells. *Exp. Neurol.* **158,** 265–278.
80. Hammang, J., Archer, D., and Duncan, I. (1997) Myelination following transplantation of EGF-responsive neural stem cells into a myelin-deficient environment. *Exp. Neurol.* **147,** 84–95.
81. Milward, E., Lundberg, C., Ge, B., Lipsitz, D., Zhao, M., and Duncan, I. (1997) Isolation and transplantation of multipotential populations of epidermal growth factor-responsive, neural progenitor cells from the canine brain. *J. Neurosci. Res.* **50,** 862–871.
82. Svendsen, C., Caldwell, M., and Ostenfeld, T. (1999) Human neural stem cells: isolation, expansion and transplantation. *Brain Pathol.* **9,** 499–513.
83. Vescovi, A., Gritti, A., Galli, R., and Parati, E. (1999) Isolation and intracerebral grafting of nontransformed multipotential embryonic human CNS stem cells. *J. Neurotrauma.* **16,** 689–693.
84. Vescovi, A., Parati, E., Gritti, A., Poulin, P., Ferrario, M., Wanke, E., et al. (1999) Isolation and cloning of multipotential stem cells from the embryonic human CNS and establishment of transplantable human neural stem cell lines by epigenetic stimulation. *Exp. Neurol.* **156,** 71–83.
85. Vescovi, A. and Snyder, E. (1999) Establishment and properties of neural stem cell clones: plasticity in vitro and in vivo. *Brain Pathol.* **9,** 569–598.
86. Stockard, C. R. (1921) Developmental reate and structural expression in experimental study of twins, 'double monsters' and single deformities, and the interaction among embryonic organs during their origin and development. *Am. J. Anat.* **28,** 115–277.
87. Davison, A. N. and Dobbings, J. (1966) Myelination as a vulnerable period in brain development. *Br. Med. Bull.* **22,** 40–44.
88. Delange, F. (1994) The disorders induced by iodine deficiency. *Thyroid* **4,** 107–128.
89. Lazarus, J. H. (1999) Thyroid hormone and intellectual development: a clinician's view. *Thyroid* **9,** 659–660.
90. Bernal, J. and Nunez, J. (1995) Thyroid hormones and brain development. *Eur. J. Endocrinol.* **133,** 390–398.
91. Ibarrola, N. and Rodriguez-Pena, A. (1997) Hypothyroidism coordinately and transiently affects myelin protein gene expression in most rat brain regions during postnatal development. *Brain Res.* **752,** 285–293.
92. Marta, C. B., Adamo, A. M., Soto, E. F., and Pasquini, J. M. (1998) Sustained neonatal hyperthyroidism in the rat affects myelination in the central nervous system. *J. Neurosci. Res.* **53,** 251–259.
93. Munoz, A., Rodriguez-Pena, A., Perez-Castillo, A., Ferreiro, B., Sutcliffe, J. G., and Bernal, J. (1991) Effects of neonatal hypothyroidism on rat brain gene expression. *Mol. Endocrinol.* **5,** 273–280.

94. Noguchi, T., Sugisaki, T., Satoh, I., and Kudo, M. (1985) Partial restoration of cerebral myelination of the congenitally hypothyroid mouse by parenteral or breast milk administration of thyroxine. *J. Neurochem.* **45**, 1419–1426.

95. Ahlgren, S., Wallace, H., Bishop, J., Neophytou, C., and Raff, M. (1997) Effects of thyroid hormone on embryonic oligodendrocyte precursor cell development in vivo and in vitro. *Mol. Cell Neurosci.* **9**, 420–432.

96. Rodriguez-Pena, A., Ibarrola, N., Iniguez, M., Munoz, A., and Bernal, J. (1993) Neonatal hypothyroidism affects the timely expression of myelin-associated glycoprotein in the rat brain. *J. Clin. Invest.* **91**, 812–818.

97. Wiggins, R. C., Benjamins, J. A., Krigman, M. R., and Morell, P. (1974) Synthesis of myelin proteins during starvation. *Brain Res.* **80**, 345–349.

98. Wiggins, R. C. and Fuller, G. N. (1978) Early postnatal starvation causes lasting brain hypomyelination. *J. Neurochem.* **30**, 1231–1237.

99. Wiggins, R. C., Miller, S. L., Benjamins, J. A., Krigman, M. R., and Morell, P. (1976) Myelin synthesis during postnatal nutritional deprivation and subsequent rehabilitation. *Brain Res.* **107**, 257–273.

100. Zuppinger, K., Wiesmann, U., Siegrist, H. P., Schafer, T., Sandru, L., Schwarz, H. P., and Herschkowitz, N. (1981) Effect of glucose deprivation on sulfatide synthesis and oligodendrocytes in cultured brain cells of newborn mice. *Pediatr. Res.* **15**, 319–325.

101. Pasquini, J. M. and Adamo, A. M. (1994) Thyroid hormones and the central nervous system. *Dev. Neurosci.* **16**, 1–8.

101. Krigman, M. R. and Hogan, E. L. (1976) Undernutrition in the developing rat: effect upon myelination. *Brain Res.* **107**, 239–255.

102. Rodriguez-Pena, A. (1999) Oligodendrocyte development and thyroid hormone. *J. Neurobiol.* **40**, 497–512.

103. Sikes, R. W., Fuller, G. N., Colbert, C., Chronister, R. B., DeFrance, J., and Wiggins, R. C. (1981) The relative numbers of oligodendroglia in different brain regions of normal and postnatally undernourished rats. *Brain Res. Bull.* **6**, 385–391.

104. Royland, J. E., Konat, G., and Wiggins, R. C. (1993) Abnormal upregulation of myelin genes underlies the critical period of myelination in undernourished developing rat brain. *Brain Res.* **607**, 113–116.

105. Royland, J. E., Konat, G. W., and Wiggins, R. C. (1993) Myelin gene activation: a glucose sensitive critical period in development. *J. Neurosci. Res.* **36**, 399–404.

106. Royland, J. E., Konat, G. W., and Wiggins, R. C. (1999) Differentiation dependent activation of the myelin genes in purified oligodendrocytes is highly resistant to hypoglycemia. *Metab. Brain Dis.* **14**, 189–195.

107. Carroll, P., Sendtner, M., Meyer, M., and Thoenen, H. (1993) Rat ciliary neurotrophic factor (CNTF): gene structure and regulation of mRNA levels in glial cell cultures. *Glia* **9**, 176–187.

108. Chernausek, S. D. (1993) Insulin-like growth factor-I (IGF-I) production by astroglial cells: regulation and importance for epidermal growth factor-induced cell replication. *J. Neurosci. Res.* **34**, 189–197.

109. Hammarberg, H., Risling, M., Hokfelt, T., Cullheim, S., and Piehl, F. (1998) Expression of insulin-like growth factors and corresponding binding proteins

(IGFBP 1-6) in rat spinal cord and peripheral nerve after axonal injuries. *J. Comp. Neurol.* **400,** 57–72.

110. Moretto, G., Xu, R. Y., Walker, D. G., and Kim, S. U. (1994) Co-expression of mRNA for neurotrophic factors in human neurons and glial cells in culture. *J. Neuropathol. Exp. Neurol.* **53,** 78–85.

111. Richardson, W. D., Pringle, N., Mosley, M., Westermark, B., and Dubois-Dalcq, M. (1988) A role for platelet-derived growth factor in normal gliogenesis in the central nervous system. *Cell* **53,** 309–319.

112. Janzer, R. C. and Raff, M. C. (1987) Astrocytes induce blood-brain barrier properties in endothelial cells. *Nature* **325,** 253–257.

113. Rubin, L. L., Hall, D. E., Porter, S., Barbu, K., Cannon, C., Horner, H. C. et al. (1991) A cell culture model of the blood-brain barrier. *J. Cell. Biol.* **115,** 1725–1735.

114. Fierz, W., Endler, B., Reske, K., Wekerle, H., and Fontana, A. (1985) Astrocytes as antigen-presenting cells. I. Induction of Ia antigen expression on astrocytes by T cells via immune interferon and its effect on antigen presentation. *J. Immunol.* **134,** 3785–3793.

115. Fontana, A., Erb, P., Pircher, H., Zinkernagel, R., Weber, E., and Fierz, W. (1986) Astrocytes as antigen-presenting cells. Part II: Unlike H-2K-dependent cytotoxic T cells, H-2Ia-restricted T cells are only stimulated in the presence of interferon-gamma. *J. Neuroimmunol.* **12,** 15–28.

116. Frei, K. and Fontana, A. (1997) Antigen presentation in the CNS. *Mol. Psychiatry* **2,** 96–98.

117. Oh, J. W., Schwiebert, L. M., and Benveniste, E. N. (1999) Cytokine regulation of CC and CXC chemokine expression by human astrocytes. *J. Neurovirol.* **5,** 82–94.

118. Van Wagoner, N. J., Oh, J. W., Repovic, P., and Benveniste, E. N. (1999) Interleukin-6 (IL-6) production by astrocytes: autocrine regulation by IL-6 and the soluble IL-6 receptor. *J. Neurosci.* **19,** 5236–5244.

119. Hansson, E. and Ronnback, L. (1992) Adrenergic receptor regulation of amino acid neurotransmitter uptake in astrocytes. *Brain Res. Bull.* **29,** 297–301.

120. Mentlein, R., and Dahms, P. (1994) Endopeptidases 24.16 and 24.15 are responsible for the degradation of somatostatin, neurotensin, and other neuropeptides by cultivated rat cortical astrocytes. *J. Neurochem.* **62,** 27–36.

121. Araque, A., Sanzgiri, R. P., Parpura, V., and Haydon, P. G. (1999) Astrocyte-induced modulation of synaptic transmission. *Can. J. Physiol. Pharmacol.* **77,** 699–706.

122. Vesce, S., Bezzi, P., and Volterra, A. (1999) The highly integrated dialogue between neurons and astrocytes in brain function. *Sci. Prog.* **82,** 251–270.

123. Brand, A., Leibfritz, D., Hamprecht, B., and Dringen, R. (1998) Metabolism of cysteine in astroglial cells: synthesis of hypotaurine and taurine. *J. Neurochem.* **71,** 827–832.

124. Hertz, L., Dringen, R., Schousboe, A., and Robinson, S. R. (1999) Astrocytes: glutamate producers for neurons. *J. Neurosci. Res.* **57,** 417–428.

125. Bush, T. G., Puvanachandra, N., Horner, C. H., Polito, A., Ostenfeld, T., Svendsen, C. N., et al. (1999) Leukocyte infiltration, neuronal degeneration,

and neurite outgrowth after ablation of scar-forming, reactive astrocytes in adult transgenic mice. *Neuron* **23,** 297–308.

126. Fitch, M. T., Doller, C., Combs, C. K., Landreth, G. E., and Silver, J. (1999) Cellular and molecular mechanisms of glial scarring and progressive cavitation: in vivo and in vitro analysis of inflammation-induced secondary injury after CNS trauma. *J. Neurosci.* **19,** 8182–8198.

127. McKeon, R. J., Jurynec, M. J., and Buck, C. R. (1999) The chondroitin sulfate proteoglycans neurocan and phosphacan are expressed by reactive astrocytes in the chronic CNS glial scar. *J. Neurosci.* **19,** 10,778–10,788.

128. Ridet, J. L., Malhotra, S. K., Privat, A., and Gage, F. H. (1997) Reactive astrocytes: cellular and molecular cues to biological function. *Trends Neurosci.* **20,** 570–577.

129. Drukarch, B., Schepens, E., Stoof, J. C., Langeveld, C. H., and Van Muiswinkel, F. L. (1998) Astrocyte-enhanced neuronal survival is mediated by scavenging of extracellular reactive oxygen species. *Free. Radic. Biol. Med.* **25,** 217–220.

130. Fillenz, M., Lowry, J. P., Boutelle, M. G., and Fray, A. E. (1999) The role of astrocytes and noradrenaline in neuronal glucose metabolism. *Acta Physiol. Scand.* **167,** 275–284.

131. Porter, J. T. and McCarthy, K. D. (1997) Astrocytic neurotransmitter receptors in situ and in vivo. *Prog. Neurobiol.* **51,** 439–455.

132. Fok-Seang, J. and Miller, H. R. (1992) Astrocyte precursors in neonatal rat spinal cord cultures. *J. Neurosci.* **12,** 2751–2764.

133. Fok-Seang, J. and Miller, R. H. (1994) Distribution and differentiation of A2B5 + glial precursors in the developing rat spinal cord. *J. Neurosci. Res.* **37,** 219–235.

134. Seidman, K., Teng, A., Rosenkopf, R., Spilotro, P., and Weyhenmeyer, J. (1997) Isolation, cloning and characterization of a putative type-1 astrocyte cell line. *Brain Res.* **753,** 18–26.

135. Mi, H. and Barres, B. A. (1999) Purification and characterization of astrocyte precursor cells in the developing rat optic nerve. *J. Neurosci.* **19,** 1049–1061.

136. Miller, R. H., Zhang, H., and Fok-Seang, J. (1994) Glial cell heterogeneity in the mammalian spinal cord. *Perspect. Dev. Neurobiol.* **2,** 225–231.

136. Grinspan, J. B., Stern, J. L., Pustilnik, S. M., and Pleasure, D. (1990) Cerebral white matter contains PDGF-responsive precursors to O2A cells. *J. Neurosci.* **10,** 1866–1873.

137. Bignami, A., Eng, L. F., Dahl, D., and Uyeda, C. T. (1972) Localization of the glial fibrillary acidic protein in astrocytes by immunofluorescence. *Brain Res.* **43,** 429–435.

138. Doetsch, F., Caille, I., Lim, D. A., Garcia-Verdugo, J. M., and Alvarez-Buylla, A. (1999) Subventricular zone astrocytes are neural stem cells in the adult mammalian brain. *Cell* **97,** 703–716.

7

PNS Precursor Cells

Tanya A. Moreno and Marianne Bronner-Fraser

INTRODUCTION

The peripheral nervous system (PNS) is comprised of groups of neurons and support cells whose cell bodies lie outside the spinal cord and brain. These peripheral ganglia relay sensory input back to the central nervous system (CNS), where the information is processed and physical responses are generated. The PNS is primarily derived from a population of precursor cells called neural crest cells that arise within the developing CNS but subsequently migrate to the periphery and are highly versatile with respect to the types of derivatives that they form.

The neural crest is one of the defining features of vertebrates. Neural crest cells originate in the ectoderm of the early embryo and develop as a ridge of cells flanking the rostrocaudal length of the open neural tube, resembling a "crest." Initially, these cells appear to be multipotent and subsequently give rise to both neuronal and nonneuronal derivatives, including neurons and support cells of the PNS, pigment cells, smooth muscle cells, and cartilage and bone of the face and skull *(1,2)*. More recently, it has been shown that some neural crest cells are stem cells that self-renew in vivo and can contribute to at least some of the derivatives generated by the neural crest *(3)*.

Interest in the mechanisms of induction, migration, and differentiation of neural crest cells has occupied developmental biologists for more than 130 years *(4–7* and *8*, reprinted in *2)*. Much is known about the later steps of neural crest development such as migration pathways and cell fate decisions *(1, 9–12)*. However, molecular aspects of these processes have only begun to be uncovered within the last two decades. This review summarizes recent findings regarding neural crest induction and the isolation and characterization of neural crest stem cells.

From: *Stem Cells and CNS Development*
Edited by: M. S. Rao © Humana Press Inc., Totowa, NJ

ORIGIN AND INDUCTION OF THE NEURAL CREST

Neural Crest Origin

The ectoderm is the source of the tissues that eventually form the epidermis, CNS, and PNS of all vertebrates. It is initially patterned into neural and nonneural ectoderm by signals emanating from a mesodermal organizing center during gastrulation, i.e., the dorsal lip of the blastopore (Spemann's organizer) in amphibians, Hensen's node in avians, the node in the mouse, and the embryonic shield in zebrafish. This process is called neural induction *(13–16)*. Later, the underlying mesoderm also plays a role in supplying rostrocaudal positional information to the neural ectoderm. At the start of neural induction, a broad domain of ectoderm adjacent to the midline thickens to form a columnar epithelium called the neural plate. The ectoderm outside of the neural plate will give rise to the epidermis and, in the head region, placodes. Placodes are regional thickenings of the ectoderm that will contribute to the cranial sensory ganglia and the sense organs of the head such as the eyes, ears, and nose *(17,18)*. They form the remainder of the PNS that is not generated by the neural crest.

Induction of the neural crest occurs at the border region between the future epidermis and the neural plate (reviewed in refs. *19* and *20*). As development proceeds, the neural plate begins to roll into a tube, causing its lateral edges to form folds that eventually approximate at the dorsal midline of the embryo. The neural folds typically contain the premigratory neural crest cells, although there are some exceptions. For example, in the frog, *Xenopus*, the cranial neural crest is not incorporated into the neural tube, but remains as a separate condensed mass of cells in the border region. Thus, neural crest cells delaminate from the neuroepithelium and begin to migrate before neural tube closure in some species (e.g., mouse, *Xenopus*) *(2,21–25)*, whereas in other species (e.g., chicken), they migrate only after apposition of the neural folds *(26)*. Thus, the CNS is formed from the rolled-up neural plate, and the PNS is formed from the ectodermal placodes and the neural crest cells residing in and around the dorsal neural tube, which delaminate from the neural epithelium and migrate throughout the embryo (Fig. 1).

It was originally thought that the neural crest was a segregated population of cells, largely based on the fact that these cells appear morphologically distinct from neural tube cells in some species (e.g., axolotl and zebrafish). In other species, however, presumptive neural crest cells are not readily distinguishable from dorsal neural tube cells. Moreover, single-cell lineage analyses of the dorsal neural tube have shown that individual precursors in the neural tube can form both neural crest and neural tube derivatives in chick *(27,28)*, frog *(29)*, and mouse *(30,31)*. Even more strikingly, prior to

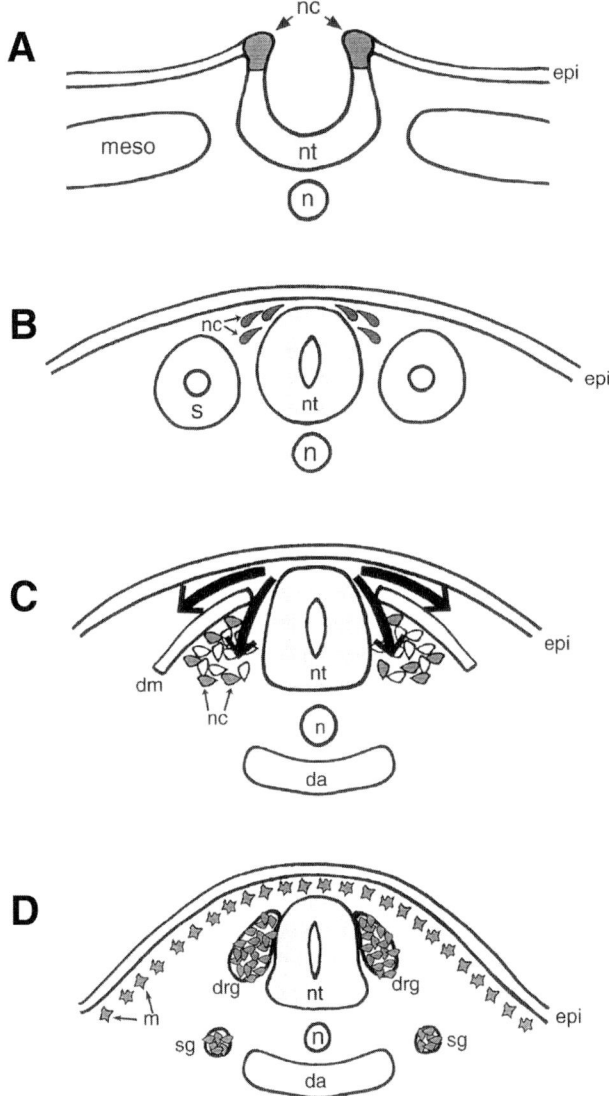

Fig. 1. Neural crest-forming regions and migration pathways in avians: cross-sectional view. (**A**) E1.5–2. Thickened epithelium at the midline begins to fold into a tube. The border of the neural and nonneural ectoderm is the site of neural crest formation. (**B**) E2–2.5 Neural crest cells delaminate from the the dorsal neural tube and begin to migrate. (**C**) E3. Two migration pathways are shown in the trunk: the dorsolateral pathway passes between the dermomyotome and epidermis, and the ventral pathway passes through the sclerotome of the somites. (**D**) E4. Neural crest cells in the trunk populate the dorsal root ganglia and sympathetic ganglia and form melanocytes in the skin. da, dorsal aorta; dm, dermomyotome; drg, dorsal root ganglion; epi, epidermis; m, melanocyte; meso, nonaxial mesoderm; nc, neural crest; nt, neural tube; s, somite; sg, sympathetic ganglion.

neural tube closure, the neural folds can give rise to all three ectodermal derviatives: epidermis, neural tube, and neural crest *(12)*. Recently, genetic screens in zebrafish have identified a mutation called *narrowminded*, which supports a shared lineage between CNS and PNS cells. This mutant lacks both early neural crest cells (PNS) and sensory neurons in the neural tube (CNS) *(32)*. Further evidence for a common neural progenitor comes from isolation of stem cells from the spinal cord neuroepithelium (NEP) cells that can form both CNS and PNS derivatives *(33)*.

Not only has it been shown the neural tube/neural crest lineage is shared, but it has also been demonstrated that these cells are not irreversibly committed to either fate until relatively late in development. The ability of the neural tube to produce neural crest cells may persist for long periods. Sharma et al. *(34)* identified a late-emigrating population of neural tube cells that form neural crest-like derivatives. When transplanted into neural crest migratory pathways of younger embryos, these cells can migrate and differentiate into neural crest derivatives *(35)*. Conversely, it has been shown that early-migrating neural crest cells can reincorporate into the ventral neural tube and express markers characteristic of floor plate cells when challenged by transplantation *(36)*.

Neural Crest Induction

Cell–Cell Interaction at the Neural Plate Border

Several theories of neural crest induction exist (reviewed in ref. *19*). Both the mesoderm and the epidermal ectoderm have been shown to have the ability to induce neural crest. (This section discusses the evidence for ectodermal interactions; see below for a discussion of mesoderm.) The best supported model for neural crest induction is one in which cell–cell interaction at the border between neural and nonneural ectoderm is responsible for inducing the neural crest. In vivo grafting experiments suggest that interactions between presumptive epidermis and neural plate can form neural crest cells. In amphibians, epidermis grafted into the neural plate generates neural crest cells *(37,38)*. In avians and frogs, neural plate tissue grafted into the epidermal ectoderm results in the production of migratory cells expressing neural crest cell markers *(12,39,40)*. In vitro co-culture experiments have similarly provided evidence for the sufficiency of the neural plate/epidermal ectoderm interaction to generate neural crest cells *(12,39,40)*. Interestingly, both the epidermis and the neural plate cells contributed to the neural crest cell population *(12,38)*.

The potential for more ventral neural tube cells to generate neural crest was examined in ablation experiments in which the dorsal region of the neural folds containing the presumptive neural crest cells was removed, thus bringing more ventral regions of the tube into contact with epidermal ecto-derm. In this situation, neural crest cells were regenerated at the zone of con-tact *(41–44;* ref. *45),* for a limited period. These data show that a very important mechanism of neural crest induction is mediated through cell-cell interactions at the border between the epidermal ectoderm and the neural plate.

The Role of BMPs in Neural Crest Induction: Setting up the Neural Plate Border Region

There is growing evidence, particularly from the *Xenopus* system, that members of the tumor growth factor-β (TGF-β) superfamily of signaling molecules play an integral role in setting up the border between neural and nonneural ectoderm. Given that neural crest cells arise at this border, it is likely that these cells are an important target of this signaling process.

Several lines of evidence support the idea that bone morphogenic protein (BMP) molecules play a role in neural induction (for review, see refs. *14* and *15). Xenopus* BMP-4 is expressed throughout the ectoderm prior to neural induction and then is lost from regions fated to become the neural plate *(46–48).* The secreted BMP antagonists noggin *(49,50),* chordin *(51,52),* and follistatin *(53,54)* all are expressed in Spemann's organizer, the tissue responsible for patterning the ectoderm. Thus, the neural plate forms adjacent to the organizer, the source of BMP inhibition, whereas the nonneural ectoderm lies distal to the organizer (Fig. 2).

One possibility is that inhibition of BMP signaling is sufficient to gener-ate both the neural plate and the neural crest, with high levels of inhibition yielding neural tissue and intermediate levels yielding neural crest. The idea that a diffusible morphogen could act to instruct the ectoderm to assume the various available fates was first proposed by Raven and Kloos *(55),* who hypothesized that an "evocator" present in a graded fashion could generate neural crest at low levels and neural plate and neural crest at high levels (reviewed in ref. *19).* In *Xenopus* ectodermal explants (animal caps), vary-ing the level of BMP activity leads to varying fates of ectoderm *(56,57).* Overexpression of a dominant-negative BMP receptor *(58)* or the BMP an-tagonist chordin in *Xenopus* ectodermal explants causes neural crest marker expression and in whole embryos enhances the neural crest domain in a dose-dependent fashion *(59).* In contrast, the reciprocal experiment of overexpressing BMP-4 itself in intact embryos does not influence the size of the neural crest domain. Instead, the size of the neural plate decreases in a dose-dependent fashion, thus moving the location, but not the extent, of the presumptive neural crest *(59).* Furthermore, chordin by itself cannot induce robust expression of neural crest markers in *Xenopus* animal caps *(59).* Taken together, these results indicate that inhibition of BMP signaling alone is not sufficient to induce neural crest formation.

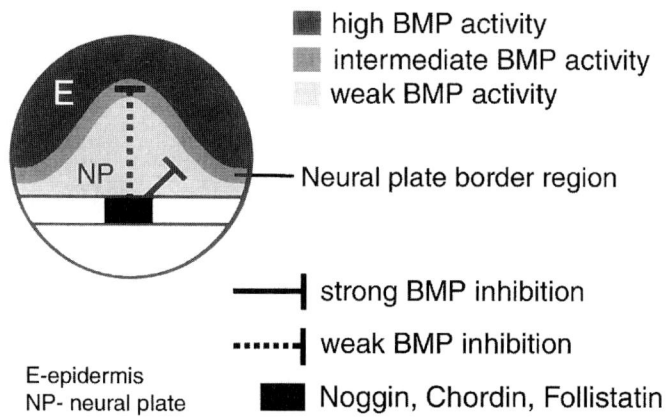

Fig. 2. Schematic diagram of the *Xenopus* model of neural induction. The BMP antagonists noggin, chordin, and follistatin are secreted from Spemann's organizer (black box) to modulate BMP activity in the ectoderm. The activity of BMP molecules establishes three fates of ectoderm: lowest activity = neural plate; intermediate activity = neural crest; highest activity = epidermis. This simplistic model does not include the evidence for the involvement of other molecules in neural and neural crest induction, but is intended as a simplified model of neural induction. (Modified from ref. *59*.)

Genetic Evidence for the Involvement of TGF-β Family Members in Neural Crest Induction

Genetic evidence in the zebrafish supports a role for TGF-β family molecules influencing the fate of the ectoderm. Nguyen et al. *(60)* have investigated *swirl (61,62)*, a mutation in the zebrafish BMP-2 gene. *swirl* mutants display a loss of neural crest progenitors, whereas mutations in genes downstream of BMP-2b such as *somitabun* (mutation in Smad5, a BMP signal transducer) expand the neural crest domain *(60)*. The zebrafish mutant *radar*, which affects a dpp-Vg1-related molecule distinct from the BMP-2/4 and BMP-5/8 subgroups (*63*; and see ref. *64* for a review of TGF-β relationships), results in the loss of the neural crest marker msxC and selected neural crest derivatives. Conversely, overexpression of the *radar* gene causes upregulation of msxC expression, but only in areas contiguous with the endogenous msxC domain *(63)*. In these mutants, however, the mesoderm underlying the neural crest is also affected, allowing for the possibility that the strength of the phenotype is not solely due to changes in BMP signaling in the ectoderm. This suggests that the activity of TGF-β family members contributes to the patterning of the ectoderm. However, only certain regions are competent to respond to these molecules, suggesting that other gene activities may be required for the establishment of the neural crest.

Transgenic mice bearing null mutations in BMP-4 *(65)*, follistatin *(66)*, or noggin *(67)* do not display the neural defects that would be expected by extrapolation from the experiments in *Xenopus* described above. It has been suggested that redundancy between different BMP family members, the antagonizing molecules, or other developmental defects may obscure the phenotype (reviewed in ref. *68*). Alternatively, there may be interesting species differences in the process of neural induction and neural crest formation. Indeed, many studies in the chick embryo have added to the interspecific discrepancies that are found upon investigation of the role of TGF-β signaling as a mechanism for neural induction and neural crest formation.

BMPs Can Induce Neural Crest in Culture

In the chick embryo, BMP-4 and BMP-7 are expressed in the epidermal ectoderm that contacts the neural tube *(69,70)*. As development proceeds, however, expression is lost in the epidermal ectoderm but BMP-4 is expressed in the neural folds and dorsal neural tube *(71)*, along with another TGF-β family member, dorsalin-1, which is upregulated after neural tube closure *(69,72)*. When added to isolated intermediate neural plates in tissue culture, both BMP-4 and -7 have been shown to induce neural crest markers and migratory cells *(69)*. This seemingly contrasts with the results in *Xenopus*, where inhibition of BMP signaling yields neural fates. However, the paradigm for neural induction by BMP repression in the neural plate does not appear to function in the chick embryo at the time of neural crest/neural plate border formation. Chordin, which inhibits BMP activity, is expressed in the avian organizer (Hensen's node) but alone cannot neuralize ectoderm *(73)*. Additionally, neither BMP-4 nor -7 is sufficient to repress neural induction in the neural plate when ectopically expressed *(73)*.

Furthermore, by implanting noggin-producing cells into the neural tube or under the neural fold regions, it has been shown that BMP signaling is required in the chick neural tube for expression of neural crest markers, but not at the stage at which BMP is expressed in the ectoderm *(74)*. Pera et al. *(75)* found that ectopic expression of BMP-2 or -4 under the neural/nonneural border region distorts the neural plate and causes epidermal ectoderm marker expression in areas that would normally give rise to neural plate. Taken together, these results seem to indicate that BMP signaling plays several important roles in neural crest development, beginning with the positioning of the neural plate border and continuing with the maintenance of neural crest induction. Importantly, it is likely that other molecules are involved in the initiation of neural crest induction. Later, BMPs in the dorsal neural tube induce roof plate cells and sensory neurons *(76)*. Still later, BMPs are involved in the differentiation of sympathoadrenal precursors from neural crest cells *(77–79)*.

There is no direct evidence that either BMP-4 or -7 is the molecule that diffuses from the epidermal ectoderm to induce crest cells *(69)*. Indeed, it was shown that BMP-4 induces epidermis at the expense of neural tissue *(80)*. The ability of BMP-4 and -7 to induce neural crest from neural plate cultures *(69,76)* may be a reflection of the molecule having first induced epidermis, which in turn interacted with the neural plate to induce neural crest. Another possibility is that exogenous BMP bypasses an epidermal signaling event and mimics a later action of endogenous BMP signaling in the dorsal neural tube that is sufficient to generate neural crest cells. This possibility is supported by the later neural tube requirement for BMP signaling to produce neural crest cells, as demonstrated by Selleck et al. *(74)*. Thus, the action of BMPs may be required within the responding tissues to maintain crest production, rather than being a property of the initial induction (reviewed in ref. *20*).

It is important to bear in mind that although many experimental differences between species are reported in the literature, these are most likely to be a result of the rather striking differences among the organisms that are used for study. Differences in morphology and timing of development must require differences in gene expression to achieve the overall goal of properly forming the animal. For example, the frog embryo begins as a hollow ball of cells, whereas the chick embryo begins as a flat sheet of cells. In the frog embryo, development relies for a period on maternal stores of messenger RNAs, which contrasts with the chick embryo. Moreover, the distances between signaling centers and their responding tissues may require different mechanisms in order to effect induction of neural tissue and other developmental events. Although there are many apparent species differences, these may reflect variations in the finer details that accommodate spatial and temporal variations among organisms; the general mechanisms are likely to be common for all vertebrates *(20)*.

Other Sources of Neural Crest-Inducing Signals:

The Mesoderm

It would be overly simplistic to assume that a single signaling event within the ectoderm is sufficient to account for induction of the neural crest. Many lines of evidence suggest that the nonaxial mesoderm is also involved in inducing the neural crest. Although conjugating epidermis and neural plate in vitro is sufficient to induce neural crest markers in the absence of mesoderm *(12,38–40)*, mesoderm could represent an important modifier. Mesoderm/neural plate conjugates do not induce early neural crest markers *(12,81)*. However, it was demonstrated that paraxial mesoderm conjugated with neural plate could induce the formation of melanocytes, a neural crest derivative *(12)*. Similarly, non-axial mesoderm from both chick and frog can induce neural crest markers in neural plate co-culture experiments *(58,82,83)*, and removal of the nonaxial mesoderm before neural induction is complete results in a failure of the ectoderm to express neural crest markers *(58,83)*. The evidence that mesoderm can influence neural crest formation suggests that there may be other molecules involved in the early steps of neural crest induction.

Wnt Family Members

As discussed above, it seems likely that inhibition of BMP alone cannot account for neural crest induction, making it probable that other signaling systems are involved. Possible candidates for involvement in this process are secreted molecules expressed in both mesoderm and ectoderm, that have been implicated in patterning the neural tube. These include members of the wingless/int family known in vertebrates as Wnts *(84)* and the fibroblast growth factor (FGF) family *(85,86)*. In *Xenopus* ectodermal explants (animal caps), Wnt1 and Wnt3a *(87)*, Wnt7b *(88)*, and Wnt8 *(59)*, in conjunction with inhibition of BMP signaling (i.e., neural induction) can induce the expression of neural crest markers. Furthermore, overexpression of β-catenin (a downstream component of the Wnt signaling pathway) expands the neural crest domain; expression of a dominant-negative Wnt ligand eliminates the neural crest domain in *Xenopus* embryos *(59)*.

One of the earliest neural crest markers in *Xenopus* is the zinc finger transcription factor, Slug *(89)* (Fig. 3). When animal caps overexpressing Slug are juxtaposed to Wnt8-expressing explants, neural crest markers are induced, thus bypassing the requirement for inhibition of BMP signaling *(59)*. In contrast, Slug alone cannot induce neural crest *(59)*. Slug, in turn, can expand its own expression domain when overexpressed in the whole embryo *(59)*. These results suggest that a two-signal model may account for the events underlying neural crest formation, such that Wnt signaling together with inhibition of BMP signaling induces the neural crest marker, Slug, with Slug expression abrogating further need for BMP inhibition *(59)*.

Many Wnt molecules are expressed in spatiotemporal patterns appropriate for involvement in various aspects of neural crest development. *Xenopus* Wnt8 is expressed in the ventrolateral mesoderm *(90)*, a tissue that has been shown to be a neural crest inducer when conjugated with neural plate in vitro *(58,82,83)*, and avian Wnt-8C is similarly expressed in the nonaxial mesoderm *(91)*. *Xenopus* Wnt7b is expressed throughout the ectoderm at gastrulation *(88)*, and other Wnts may well be expressed in the ectoderm.

In chick *(39,92)*, frog *(93,94)*, and mouse embryos *(95,96)*, Wnt1 and Wnt3a are expressed in the dorsal neural tube well after the initial expression

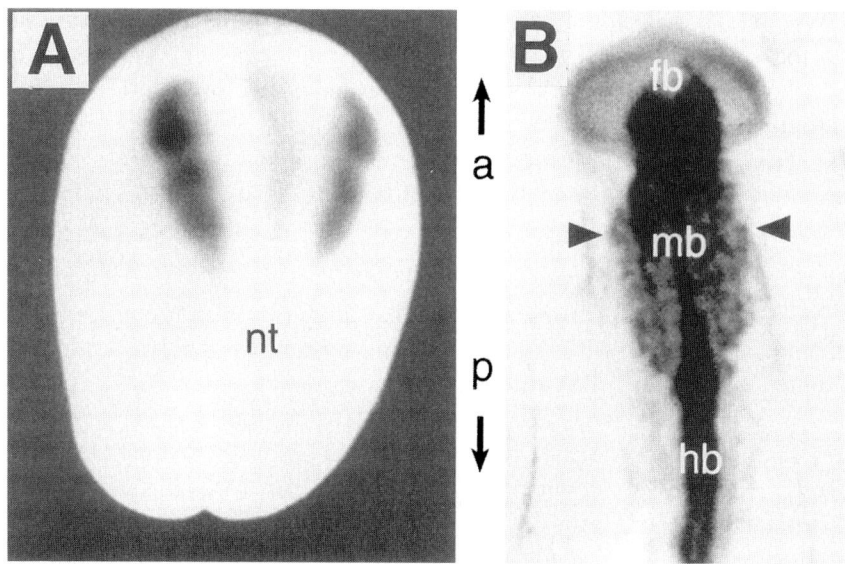

Fig. 3. *Slug* expression pattern in *Xenopus* and chick. The zinc-finger transcription factor Slug is an early marker for the neural crest in *Xenopus* and chick. **(A)** A late-neurula *Xenopus* embryo with *Slug* mRNA expression in the cranial neural crest on both sides of the closing neural tube. The groove down the central portion of the embryo is the forming neural tube. **(B)** An E1.5 (10-somite stage) chick embryo with *Slug* mRNA marking the early-migrating neural crest in the head (arrowheads) and premigratory neural crest at more posterior levels of the neural tube. a, anterior; p, posterior; fb, forebrain; mb, midbrain; hb, hindbrain; nt, neural tube.

of neural crest markers, although *Xenopus* Wnt3a is also expressed before neural tube closure at the edges of the neural plate *(94)*. Furthermore, avian neural crest can be induced in conjugates of epidermis and neural plate without the concomitant expression of either Wnt1 or Wnt3a *(39)*. This suggests that Wnt-1 and -3a are not involved in the initial induction of neural crest. However, Wnt1/3a double knockout mice have a reduction in neurogenic and gliogenic neural crest derivatives, suggesting that fewer neural crest cells emerge in embryos lacking both genes *(97)*. Not all neural crest derivatives are affected, with ventralmost derivatives such as sympathetic ganglia demonstrating normal morphology, whereas dorsal root ganglia are markedly reduced. This is consistent with the possibility that these Wnts play a later maintenance role in neural crest production by the neural tube. Wnts may be involved in the expansion of neural crest progenitors, most likely by regulating the proliferation of the cells after induction has occurred but prior to commencement of emigration *(97)*.

Wnt family members may also be able to control some aspects of neural crest cell fate. In zebrafish experiments, single neural crest cells overexpressing molecules of the Wnt signaling pathway form pigment cells at the expense of neurons or glia. Conversely, overexpressing inhibitors of the pathway biases the neural crest cells to form neurons at the expense of pigment cells *(98)*.

Recent work by Baker et al. *(99)* has put forward a novel model for Wnt function. They demonstrate that expression of *Xenopus* Wnt8, mouse Wnt8, and downstream Wnt targets in frog ectodermal explants can induce expression of the early panneural marker neural cell adhesion molecule (NCAM), without neural induction by BMP antagonists. Additionally, they demonstrate that Wnt signaling components suppress BMP-4 expression in ectoderm explants as assayed by *in situ* hybridization. In fact, Wnt8, and not the BMP antagonist noggin, seems to be capable of blocking BMP-4 expression in the neural plate throughout gastrula stages, suggesting that an early Wnt signal and not a direct BMP antagonist is responsible for the early inhibition of BMP-4 expression in the neural plate. Finally, the authors suggest that there may be parallel pathways for the effects of Wnt signaling in neural induction since inhibition of Wnt8-mediated activation of the neural inducers Xnr3 and *siamois* did not abrogate Wnt8's ability to itself promote neural induction. These results suggest that Wnt signaling may be involved in multiple inductive events in early development. The ramifications of these data for the role of Wnt signaling in neural crest induction are unclear, as these investigators did not explore the effects of the perturbations on neural crest markers. Previous results showing that Wnts could not induce neural crest without a co-expressed neural inducer *(59,87,88)* taken together with the results of Baker et al. *(99)* may indicate that the precise levels of Wnt signaling are critical. Further investigation will be required to determine exactly what role Wnt plays during neural crest development.

Fibroblast Growth Factors

Other molecules expressed in the mesoderm have been shown to have neural crest inducing activities. FGF signaling can induce neural crest markers in frog ectodermal explants when in the presence of BMP antagonists *(59,100,101)*. Overexpression of a dominant-negative FGF receptor can inhibit expression of the early neural crest marker XSlug in whole embryos *(100)*. Other investigators have demonstrated that FGF signaling has a posteriorizing effect on neural tissue *(101–105)*. Indeed, members of the FGF family are spatiotemporally expressed in a way that is consistent with their playing roles in the process of neural and/or neural crest induction *(106–111)*. The results indicate that FGFs may be able to generate both posterior and lateral (i.e., neural crest) fates in the CNS and PNS. The role of FGFs becomes complicated in light of evidence from transgenic frog experiments, however, in which frog embryos expressing a dominant-negative FGF receptor have normally developing posterior neural tissue and border regions including the neural crest, although the investigators did not test a full range of neural crest markers *(112)*. Moreover, FGF-treated neural plate explants do not form neural crest tissue *(100)*. Finally, neural crest induction by FGF may be a secondary result of its ability to induce a member of the Wnt family *(59)*. Thus, FGF signaling is not required for neural crest induction, and the demonstrated effects may be indirect.

Neural Crest Stem Cells

In the past decade, work by several investigators has led to the identification and purification of neural crest stem cells—cells with the potential to self-renew and also to give rise to the diverse population of derivatives that are generated by the neural crest. The first neural crest stem cells were isolated in vitro by clonal analysis of cells that were fractionated from rat neural crest cultures by cell sorting based on expression of a cell surface epitope *(113)*. These cells can be replated to form new stem cells and also can give rise to "blast" cells that are partially restricted to form neurons or glia. These include the sympathoadrenal sublineage, which includes precursors to sympathetic neurons and adrenomedullary cells *(114,115)*, that, in the embryo, appear specified by the time that neural crest-derived cells reach their sites of localization around the dorsal aorta.

Specific molecules can instruct neural crest stem cells to adopt specific fates; for example, glial growth factor (neuregulin) causes the development of glia (Schwann cells) BMP-2 biases clones to develop into neurons (and a small number of smooth muscle cells), and TGF-β1 promotes development of smooth muscle cells *(116–118)*. Thus, it is interesting to note that members of the TGF-β superfamily are not only involved in induction of the neural crest but are also implicated in subsequent cell fate decisions.

Although the neural crest stem cells are very useful in testing the ability of factors to promote certain cell fate decisions, there are possible caveats; for example, the stem cell qualities of the purified cells may have been acquired in vitro and may not reflect an actual state that is present in the embryo. The findings of Frank and colleagues *(34,35)* that neural tubes can give rise to neural crest-like cells that emigrate long after the normal period of neural crest formation suggest that neural crest stem cells may persist within the spinal cord and other sites for long periods. Consistent with this possibility, Morrison et al. *(3)* have recently isolated neural crest stem cells from embryonic rat peripheral nerve. The cells were isolated by fluorescence-activated cell sorting using cell surface epitopes p75 and P0. Under proper culture conditions, these cells self-renew and can differentiate into neurons, glia, and smooth muscle cells within single colonies. The cells are also instructively promoted to form neurons or glia by exposure to either BMP-2 or glial growth factor, respectively, in clonal cultures. An important test of the qualities of these neural crest stem cells is to determine whether newly isolated cells are multipotent when transplanted into an embryo. Indeed, freshly isolated cells that were p75 + /P0 - have stem cell properties and can be back-transplanted into chick embryos, giving rise to both neurons and glia as assayed by differential marker expression *(3)*. By labeling actively dividing cells in embryos with the thymidine analog bromodeoxyuridine, it was shown that endogenous neural crest stem cells persist in the embryo by self-renewing *(3)*.

LINEAGE AND CELL FATE DECISIONS IN THE NEURAL CREST

The existence of neural crest stem cells in the embryo supports the idea that the fate of neural crest cells in vivo is primarily determined by their environment *(119)*. Neural crest cell fate decisions and their relationships to cell lineage have been debated for many years. Although it has been accepted that at least some, if not most, neural crest cells are multipotent, some evidence indicates that other neural crest cells have restricted fates in vivo *(27,28,120)*. However, in these experiments, the potential of the cells has not actually been tested by challenging the cells with all possible factors that might influence cell fate choice. It is obviously difficult to quantify and compare the environment of one cell with another, beginning from their origins in the neural tube and following their migration trajectories through the periphery. In these lineage experiments, single dye-labeled or retrovirally tagged cells often gave rise to clones of progeny with multiple derivatives but sometimes gave rise to clones of only one cell type, suggesting an earlier specification for that progenitor cell. Thus, alternate methods of marking and challenging neural crest cells will be necessary to define the state of multipotency at the single cell level. This is an area in which the neural crest stem cells and their blast cells promise to provide new and important information.

Mechanisms of Neural Crest Diversification

If neural crest cells are truly multipotent and only receive instructions for differentiation when migrating to or reaching their final destinations, then it is interesting to consider how cells are instructed to take on different fates. For example, neural crest cells in the dorsal root ganglia differentiate into both sensory neurons and glia. An asymmetric cell division could produce a blast cell of each type, which could in turn replicate. Alternatively, the progenitor may replicate itself and produce a more restricted daughter cell, which then goes on to form the final derivatives. The latter seems more likely given the ability of neural crest stem cells to self-renew.

Environmental Cues vs Timing of Emigration

Both the environment and the timing of emigration from the neural tube have been proposed to affect the cell fate decisions of the neural crest. A restriction in available cell fate accompanies the time of emigration from the neural tube: the latest migrating cells only populate the dorsal root ganglia as neurons and form melanocytes in the skin and feathers *(34,121)*. However, when transplanted into earlier embryos, neural crest-like cells derived from much older spinal cords were able to migrate more ventrally and make sympathetic and peripheral neurons *(35,122)*. Similarly, in the head, late-migrating cells only formed dorsal derivatives because of the presence, ventrally, of earlier migrating cells; however, they are not restricted in potential *(123)*. Furthermore, the latest migrating cells of the main wave of crest emigration make melanocytes in the skin, but skin culture experiments show that they have the potential to form neurons *(124)*. This suggests that the restriction in available fates in these cases is made by the environment that the cells occupy rather than the time that they emerge from the neural tube (Fig. 1).

Additional evidence for the influence of environment on neural crest cell fate comes from neural crest stem cells, in which single progenitor cells can generate smooth muscle cells when exposed to TGF-β molecules. However, a community effect takes place when denser cultures are exposed to TGF-β molecules, such that either neurons form or cell death occurs, rather than differentiation of smooth muscle cells *(125)*. These data suggest that cell fate in the embryo could also be determined by community effects in which cells respond differently to the same factors depending on the density of neighboring cells *(125)*. Other interesting studies on neural crest stem cells reveal that they can integrate multiple instructive cues and are biased to certain levels of responsiveness based on the growth factors to which they are exposed. If cultures of neural crest stem cells are exposed to saturating levels of both BMP-2 and glial growth factor (neuregulin), BMP-2 appears dominant and neurons differentiate. However, BMP-2 and TGF-β1 seem to be co-dominant *(118)*.

There is evidence, however, that some neural crest cell populations may undergo early fate restrictions. By culturing "early-migrating" and "late-migrating" trunk neural crest cells, Artinger and Bronner-Fraser *(126)* found that the latter are more restricted in their developmental potential than the former; although they can form pigment cells and sensory-like neurons, they fail to form sympathetic neurons. Additionally, late-migrating cells transplanted into an earlier environment can colonize the sympathetic ganglia but failed to form adrenergic cells *(126)*. Thus, the time that a precursor leaves the neural tube may contribute to its potency. Perez et al. *(127)* have provided evidence for early specification of sensory neurons by the basic helix-loop-helix transcription factors neurogenins 1 and 2. These molecules are expressed early in a subset of neural crest cells, and ectopic expression of the molecules biases migrating neural crest cells to localize in the sensory ganglia and express sensory neuron markers.

Another way to account for the process of promoting two different cell fates from one precursor population within a single tissue is the proposal that temporal changes in the target environment bias the cell fate decision *(120)*. This is supported by the fact that first neurons and then glia are born in the dorsal root ganglia (e.g., ref. *128*). The target environment could be influenced to change by early differentiating neural crest cells themselves; for example, some neurons produce glial-promoting factors *(129–133)*. Also, the loss of certain inhibitory glycoconjugates from the extracellular matrix in the dorsolateral migration pathway has been linked to the migration of late-emigrating neural crest cells along this pathway *(134)*, where they are exposed to melanogenic factors and hence adopt a melanocyte fate *(135)*. Thus, there is evidence for the influence of both the timing of emigration and environmental cues in determining neural crest fates.

Progressive Lineage Restriction

It has been proposed that neural crest cells adopt specific fates by progressive lineage restrictions *(11,77,136)*. One way to explain the intermingling of clonally related neurons and glia is that the choice is made stochastically, such that each cell has the capacity to adopt either fate, and environmental factors act by influencing the probability of a fate choice rather than imposing strict commitments *(120)*. Support for the idea of progressive fate restriction comes from the NEP, which can give rise to both CNS- and PNS-type stem cells. PNS stem cells (indistinguishable from neural crest stem cells, as described in ref. *11*) are formed on addition of BMP-2/-4 to the NEP cell cultures *(33)*. BMP-2, a molecule that is known to instruct neural crest stem cells toward an adrenergic neuronal fate, is expressed in the dorsal aorta, near where sympathetic ganglia form *(117,137,138)*. Thus, there is evidence that environmental cues may be able to promote progressive restriction of neural crest cell fates. Many factors act selectively by affecting the proliferation or survival of neural crest derivatives; others act instructively on multipotent progenitors to promote one fate over another. Further work will be required to answer the complex question of how individual cells within the same environment can adopt different fates. The evidence in support of both multipotentiality and lineage restriction may imply that neural crest cells take cues from both the timing of emigration from the neural tube and the environments to which they are exposed in cell lineage decisions. For more discussion on the topic of neural crest diversification, the reader is referred to several recent reviews *(139–141)*.

CONCLUSIONS

The demonstration that multiple molecules from different gene families have the capacity to induce neural crest implies that the mechanism of neural crest induction involves complex and perhaps parallel pathways. It is further interesting to note that the same molecules can have multiple inductive capabilities at different times in development. Although great strides have been made toward understanding the induction and cell fate decisions of the neural crest, many mysteries remain. The field of neural crest research is rich in unanswered questions whose solutions will not only offer deeper understanding of the mechanisms of neural crest development but will also give more general insight into phenomena such as cell migration and differentiation.

ACKNOWLEDGMENTS

We thank Clare Baker and Anne Knecht for invaluable comments on the manuscript and Carole LaBonne for helpful discussions. T.A.M. is a Fellow of the ARCS Foundation.

REFERENCES

1. Le Douarin, N. (1982) *The Neural Crest*, Cambridge University Press, Cambridge.
2. Hall, B. K. and Hörstadius, S. (1988) *The Neural Crest*. Oxford University Press, Oxford.
3. Morrison, S. J., White, P. M., Zock, C., and Anderson, D. J. (1999) Prospective identification, isolation by flow cytometry, and *in vivo* self-renewal of multipotent mammalian neural crest stem cells. *Cell* **96,** 737–749.
4. His, W. (1868) *Untersuchungen über die erste Anlage des Wirbeltierleibes. Die erste Entwicklung des Hühnchens im Ei.* F. C. W. Vogel, Leipzig.
5. Landacre, F. L. (1921) The fate of the neural crest in the head of the Urodeles. *J. Comp. Neurol.* **33,** 1–43.
6. Stone, L. S. (1922) Experiments on the development of the cranial ganglia and the lateral line sense organs in *Amblystoma punctatum. J. Exp. Zool.* **35,** 421–496.
7. Harrison, R. G. (1938) Die Neuralleiste Erganzheft. *Anat. Anz.* **85,** 3–30.
8. Hörstadius, S. (1950) *The Neural Crest*, Oxford University Press, Oxford.
9. Bronner-Fraser, M. (1993) Mechanisms of neural crest cell migration. *Bioessays* **15,** 221–230.
10. Erickson, C. A., and Perris, R. (1993) The role of cell-cell and cell-matrix interactions in the morphogenesis of the neural crest. *Dev. Biol.* **159,** 60–74.

11. Stemple, D. L. and Anderson, D. J. (1993) Lineage diversification of the neural crest: *in vitro* investigations. *Dev. Biol.* **159**, 12–23.
12. Selleck, M. A. J. and Bronner-Fraser, M. (1995) Origins of the avian neural crest: the role of neural plate-epidermal interactions. *Development* **121**, 525–538.
13. Sasai, Y. and De Robertis, E. M. (1997) Ectodermal patterning in vertebrate embryos. *Dev. Biol.* **182**, 5–20.
14. Weinstein, D. C. and Hemmati-Brivanlou, A. (1997) Neural induction in *Xenopus laevis*: evidence for the default model. *Curr. Opin. Neurobiol.* **7**, 7–12.
15. Wilson, P. A. and Hemmati-Brivanlou, A. (1997) Vertebrate neural induction: inducers, inhibitors, and a new synthesis. *Neuron* **18**, 699–710.
16. Chang, C. and Hemmati-Brivanlou, A. (1998) Cell fate determination in embryonic ectoderm. *J. Neurobiol.* **36**, 128–151.
17. Le Douarin, N. M., Fontaine-Perus, J., and Couly, G. (1986) Cephalic ectodermal placodes and neurogenesis. *Trends Neurosci.* **9**, 175–180.
18. Webb, J. F. and Noden, D. M. (1993) Ectodermal placodes: contributions to the development of the vertebrate head. *Am. Zool.* **33**, 434–447.
19. Baker, C. V. H. and Bronner-Fraser, M. (1997) The origins of the neural crest. Part I: Embryonic induction. *Mech. Dev.* **69**, 3–11.
20. LaBonne, C. and Bronner-Fraser, M. (1999) Molecular mechanisms of neural crest formation. *Annu. Rev. Cell Dev. Biol.* **15**, 81–112.
21. Olsson, L. and Hanken, J. (1996) Cranial neural-crest migration and chondrogenic fate in the Oriental fire-bellied toad *Bombina orientalis*: defining the ancestral pattern of head development in anuran amphibians. *J. Morphol.* **229**, 105–120.
22. Bartelmez, G. W. (1922) The origin of the otic and optic primordia. *J. Comp. Neurol.* **34**, 201–232.
23. Holmdahl, D. E. (1928) Die Enstehung und weitere Entwicklung der Neuralleiste (Ganglienleiste) bei Vogeln und Saugetieren. *Z. Mikrosk.-anat. Forsch.* **14**, 99–298.
24. Verwoerd, C. D. A. and van Oostrom, C. G. (1979) Cephalic neural crest and placodes. *Adv. Anat. Embryol. Cell Biol.* **58**, 1–75.
25. Nichols, D. H. (1981) Neural crest formation in the head of the mouse embryo as observed using a new histological technique. *J. Embryol. Exp. Morphol.* **64**, 105–120.
26. Bronner-Fraser, M. (1986) Analysis of the early stages of trunk neural crest migration in avian embryos using the monoclonal antibody HNK-1. *Dev. Biol.* **115**, 44–55.
27. Bronner-Fraser, M. and Fraser, S. (1989) Developmental potential of avian trunk neural crest cells *in situ*. *Neuron* **3**, 755–766.
28. Bronner-Fraser, M. and Fraser, S. E. (1988) Cell lineage analysis reveals multipotency of some avian neural crest cells. *Nature* **335**, 161–164.
29. Collazo, A., Bronner-Fraser, M., and Fraser, S. E. (1993) Vital dye labelling of *Xenopus laevis* trunk neural crest reveals multipotency and novel pathways of migration. *Development* **118**, 363–376.
30. Serbedzija, G. N., Bronner-Fraser, M., and Fraser, S. E. (1992) Vital dye analysis of cranial neural crest cell migration in the mouse embryo. *Development* **116**, 297–307.

31. Serbedzija, G. N., Bronner-Fraser, M., and Fraser, S. E. (1994) Developmental potential of trunk neural crest cells in the mouse. *Development* **120,** 1709–1718.

32. Artinger, K. B., Chitnis, A. B., Mercola, M., and Driever, W. (1999) Zebrafish *narrowminded* suggests a genetic link between formation of neural crest and primary sensory neurons. *Development* **126,** 3969–3979.

33. Mujtaba, T., Mayer-Proschel, M., and Rao, M. S. (1998) A common neural progenitor for the CNS and PNS. *Dev. Biol.* **200,** 1–15.

34. Sharma, K., Korade, Z., and Frank, E. (1995) Late-migrating neuroepithelial cells from the spinal cord differentiate into sensory ganglion cells and melanocytes. *Neuron* **14,** 143–152.

35. Korade, Z. and Frank, E. (1996) Restriction in cell fates of developing spinal cord cells transplanted to neural crest pathways. *J. Neurosci.* **16,** 7638–7648.

36. Ruffins, S., Artinger, K., and Bronner-Fraser, M. (1998) Early migrating neural crest cells can form ventral neural tube derivatives when challenged by transplantation. *Dev. Biol.* **203,** 295–304.

37. Rollhäuser-ter Horst, J. (1980) Neural crest replaced by gastrula ectoderm in Amphibia. *Anat. Embryol.* **160,** 203–211.

38. Moury, J. D. and Jacobson, A. G. (1990) The origins of neural crest cells in the axolotl. *Dev. Biol.* **141,** 243–253.

39. Dickinson, M. E., Selleck, M. A. J., McMahon, A. P., and Bronner-Fraser, M. (1995) Dorsalization of the neural tube by the non-neural ectoderm. *Development* **121,** 2099–2106.

40. Mancilla, A., and Mayor, R. (1996) Neural crest formation in *Xenopus laevis*: mechanisms of *Xslug* induction. *Dev. Biol.* **177,** 580–589.

41. Scherson, T., Serbedzija, G., Fraser, S., and Bronner-Fraser, M. (1993) Regulative capacity of the cranial neural tube to form neural crest. *Development* **118,** 1049–1061.

42. Sechrist, J., Nieto, M. A., Zamanian, R. T., and Bronner-Fraser, M. (1995) Regulative response of the cranial neural tube after neural fold ablation: spatiotemporal nature of neural crest regeneration and up-regulation of *Slug*. *Development* **121,** 4103–4115.

43. Hunt, P., Ferretti, P., Krumlauf, R., and Thorogood, P. (1995) Restoration of normal Hox code and branchial arch morphogenesis after extensive deletion of hindbrain neural crest. *Dev. Biol.* **168,** 584–597.

44. Suzuki, H. R. and Kirby, M. L. (1997) Absence of neural crest cell regeneration from the postotic neural tube. *Dev. Biol.* **184,** 222–233.

45. Couly, G., Grapin-Botton, A., Coltey, P., and Le Douarin, N. M. (1996) The regeneration of the cephalic neural crest, a problem revisited — the regenerating cells originate from the contralateral or from the anterior and posterior neural fold. *Development* **122,** 3393–3407.

46. Dale, L., Howes, G., Price, M. J., and Smith, J. C. (1992) Bone morphogenetic protein 4: a ventralizing factor in early *Xenopus* development. *Development* **115,** 573–585.

47. Fainsod, A., Steinbeisser, H., and De Robertis, E. M. (1994) On the function of *BMP-4* in patterning the marginal zone of the *Xenopus* embryo. *EMBO J.* **13,** 5015–5025.

48. Hemmati-Brivanlou, A. and Thomsen, G. H. (1995) Ventral mesodermal patterning in *Xenopus* embryos: expression patterns and activitie of BMP-2 and BMP-4. *Dev. Gen.* **17,** 78–89.

49. Lamb, T. M., Knecht, A. K., Smith, W. A., Stachel, S. E., Economides, A. N., Stahl, N., Yancopoulos, G. D., and Harland, R. M. (1993) Neural induction by the secreted polypeptide noggin. *Science* **262,** 713–718.

50. Zimmerman, L. B., De Jesús-Escobar, J. M., and Harland, R. M. (1996) The Spemann organizer signal noggin binds and inactivates bone morphogenetic protein 4. *Cell* **86,** 599–606.

51. Sasai, Y., Lu, B., Steinbesser, H., Geissert, D., Gont, L. K., and De Robertis, E. M. (1994) *Xenopus chordin*: a novel dorsalizing factor activated by organizer-specific homeobox genes. *Cell* **79,** 779–790.

52. Piccolo, S., Sasai, Y., Lu, B., and De Robertis, E. M. (1996) Dorsoventral patterning in *Xenopus*: inhibition of ventral signals by direct binding of chordin to BMP-4. *Cell* **86,** 589–598.

53. Hemmati-Brivanlou, A., Kelly, O. G., and Melton, D. A. (1994) Follistatin, an antagonist of activin is expressed in the Spemann organizer and displays direct neuralizing activity. *Cell* **77,** 283–295.

54. Fainsod, A., Deißler, Yelin, R., Marom, K., Epstein, M., Pillemer, G., Steinbeisser, H., and Blum, M. (1997) The dorsalizing and neural inducing gene *follistatin* is an antagonist of *BMP-4*. *Mech. Dev.* **63,** 39–50.

55. Raven, C. P., and Kloos, J. (1945) Induction by medial and lateral pieces of the archenteron roof with special reference to the determination of the neural crest. *Acta. Neerl. Morphol.* **5,** 348–362.

56. Knecht, A. K., Good, P. G., Dawid, I. B., and Harland, R. M. (1995) Dorsal-ventral patterning and differentiation of noggin-induced neural tissue in the asbence of mesoderm. *Development* **121,** 1927–1936.

57. Wilson, P. A., Lagna, G., Suzuki, A., and Hemmati-Brivanlou, A. (1997) Concentration-dependent patterning of the *Xenopus* ectoderm by BMP4 and its signal transducer Smad1. *Development* **124,** 3177–3184.

58. Marchant, L., Linker, C., Ruiz, P., Guerrero, N., and Mayor, R. (1998) The inductive properties of mesoderm suggest that the neural crest cells are specified by a BMP gradient. *Dev. Biol.* **198,** 319–329.

59. LaBonne, C. and Bronner-Fraser, M. (1998) Neural crest induction in *Xenopus*: evidence for a two signal model. *Development* **125,** 2403–2414.

60. Nguyen, V. H., Schmid, B., Trout, J., Connors, S. A., Ekker, M., and Mullins, M. C. (1998) Ventral and lateral regions of the zebrafish gastrula, including the neural crest progenitors, are established by a bmp2b/swirl pathway of genes. *Dev. Biol.* **199,** 93–110.

61. Hammerschmidt, M., Serbedzija, G. N., and McMahon, A. P. (1996) Genetic analysis of dorsoventral pattern formation in the zebrafish: requirement of a BMP-like ventralizing activity and its dorsal repressor. *Genes Dev.* **10,** 2452–2461.

62. Kishimoto, Y., Lee, K. H., Zon, L., Hammerschmidt, M., and Schulte-Merker, S. (1997) The molecular nature of zebrafish *swirl*: BMP2 function is essential during early dorsoventral patterning. *Development* **124,** 4457–4466.

63. Delot, E., Kataoka, H., Goutel, C., Yan, Y.-L., Postlethwait, J., Wittbrodt, J., and Rosa, F. M. (1999) The BMP-related protein Radar: a maintenance factor for dorsal neuroectoderm cells? *Mech. Dev.* **85,** 15–25.

64. Hogan, B. L. M. (1996) Bone morphogenetic proteins: multifunctional regulators of vertebrate development. *Genes Dev.* **10,** 1580–1594.

65. Winnier, G., Blessing, M., Labosky, P. A., and Hogan, B. L. (1995) Bone morphogenetic protein-4 is required for mesoderm formation and patterning in the mouse. *Genes Dev.* **9,** 2105–2116.

66. Matzuk, M. M., Lu, N., Vogel, H., Sellheyer, K., Roop, D. R., and Bradley, A. (1995) Multiple defects and death in mice deficient in follistatin. *Nature* **374,** 360–363.

67. McMahon, J. A., Takada, S., Zimmerman, L. B., Fan, C. M., Harland, R. M., and McMahon, A. P. (1998) Noggin-mediated antagonism of BMP signaling is required for growth and patterning of the neural tube and somite. *Genes Dev.* **12,** 1438–1452.

68. Lee, K. J. and Jessell, T. M. (1999) The specification of dorsal cell fates in the vertebrate central nervous system. *Annu. Rev. Neurosci.* **22,** 261–294.

69. Liem, K. F., Tremmi, G., Roelink, H., and Jessell, T. M. (1995) Dorsal differentiation of neural plate cells induced by BMP-mediated signals from epidermal ectoderm. *Cell* **82,** 969–979.

70. Schultheiss, T. M., Burch, J. B., and Lassar, A. B. (1997) A role for bone morphogenetic proteins in the induction of cardiac myogenesis. *Genes Dev.* **11,** 451–462.

71. Watanabe, Y. and Le Douarin, N. M. (1996) A role for BMP-4 in the development of subcutaneous cartilage. *Mech. Dev.* **57,** 69–78.

73. Streit, A., Lee, K. J., Woo, I., Roberts, C., Jessell, T. M., and Stern, C. D. (1998) Chordin regulates primitive streak development and the stability of induced neural cells, but is not sufficient forneural induction in the chick embryo. *Development* **125,** 507–519.

73. Basler, K., Edlund, T., Jessell, T., and Yamada, T. (1993) Control of cell pattern in the neural tube: regulation of cell differentiation by *dorsalin-1,* a novel TGFβ family member. *Cell* **73,** 687–702.

74. Selleck, M. A., Garcia-Castro, M. I., Artinger, K. B., and Bronner-Fraser, M. (1998) Effects of shh and noggin on neural crest formation demonstrate that BMP is required in the neural tube but not ectoderm. *Development* **125,** 4919–4930.

75. Pera, E., Stein, S., and Kessel, M. (1999) Ectodermal patterning in the avian embryo: epidermis versus neural plate. *Development* **126,** 63–73.

76. Liem, K. F., Tremml, G., and Jessell, T. M. (1997) A role for the roof plate and its resident TGFbeta-related proteins in neuronal patterning in the dorsal spinal cord. *Cell* **91,** 127–138.

77. Anderson, D. J. (1993) Molecular control of cell fate in the neural crest: the sympathoadrenal lineage. *Annu. Rev. Neurosci.* **16,** 129–158.

78. Varley, J. E., McPherson, C. E., Zou, H., Niswander, L., and Maxwell, G. D. (1998) Expression of a constitutively active type I BMP receptor using a retroviral vector promotes the development of adrenergic cells in neural crest cultures. *Dev. Biol.* **196,** 107–118.

79. Schneider, C., Wicht, H., Enderich, J., Wegner, M., and Rohrer, H. (1999) Bone morphogenetic proteins are required *in vivo* for generation of sympathetic neurons. *Neuron* **24,** 861–870.

80. Wilson, P. A. and Hemmati-Brivanlou, A. (1995) Induction of epidermis and inhibition of neural fate by Bmp-4. *Nature* **376,** 331–333.

81. Mitani, S. and Okamoto, H. (1991) Inductive differentiation of two neural lineages reconstituted in a microculture system from *Xenopus* early gastrula cells. *Development* **112,** 21–31.

82. Bang, A. G., Papalopulu, N., Kintner, C., and Goulding, M. D. (1997) Expression of Pax-3 is initiated in the early neural plate by posteriorizing signals produced by the organizer and by posterior non-axial mesoderm. *Development* **124,** 2075–2085.

83. Bonstein, L., Elias, S., and Frank, D. (1998) Paraxial-fated mesoderm is required for neural crest induction in *Xenopus* embryos. *Dev. Biol.* **193,** 156–168.

84. Wodarz, A. and Nusse, R. (1998) Mechanisms of Wnt signaling in development. *Annu. Rev. Cell Dev. Biol.* **14,** 59–88.

85. Seiber-Blum, M. (1998) Growth factor synergism and antagonism in early neural crest development. *Biochem. Cell Biol.* **76,** 1039–1050.

86. Vaccarino, F. M., Schwartz, M. L., Raballo, r., Rhee, J., and Lyn-Cook, R. (1999) Fibroblast growth factor signaling regulates growth and morphogenesis at multiple steps during brain development. *Curr. Top. Dev. Biol.* **46,** 179–200.

87. Saint-Jeannet, J.-P., He, X., Varmus, H. E., and Dawid, I. B. (1997) Regulation of dorsal fate in the neuraxis by Wnt-1 and Wnt-3a. *Proc. Natl. Acad. Sci. USA* **94,** 13,713–13,718.

88. Chang, C. and Hemmati-Brivanlou, A. (1998) Neural crest induction by Xwnt7B in *Xenopus*. *Dev. Biol.* **194,** 129–134.

89. Mayor, R., Morgan, R., and Sargent, M. G. (1995) Induction of the prospective neural crest of *Xenopus*. *Development* **121,** 767–777.

90. Christian, J. L., McMahon, J. A., McMahon, A. P., and Moon, R. T. (1991) Xwnt-8, a *Xenopus* Wnt-1/int-1-related gene responsive to mesoderm-inducing growth factors, may play a role in ventral mesodermal patterning during embryogenesis. *Development* **111,** 1045–1055.

91. Hume, C. R., and Dodd, J. (1993) Cwnt-8C: a novel Wnt gene with a potential role in primitive streak formation and hindbrain organization. *Development* **119,** 1147–1160.

92. Hollyday, M., McMahon, J. A., and McMahon, A. P. (1995) *Wnt* expression patterns in the chick embryo nervous system. *Mech. Dev.* **52,** 9–25.

93. Wolda, S. L., Moody, C. J., and Moon, R. T. (1993) Overlapping expression of *Xwnt-3a* and *Xwnt-1* in neural tissue of *Xenopus laevis* embryos. *Dev. Biol.* **155,** 46–57.

94. McGrew, L. L., Hoppler, S., and Moon, R. T. (1997) Wnt and FGF pathways cooperatively pattern anteroposterior neural ectoderm in *Xenopus*. *Mech. Dev.* **69,** 105–114.

95. Roelink, H. and Nusse, R. (1991) Expression of two members of the Wnt family during mouse development-restricted temporal and spational pattern in the neural tube. *Genes Dev.* **5,** 381–388.

96. Parr, B. A., Shea, M. J., Vassileva, G., and McMahon, A. P. (1993) Mouse *Wnt* genes exhibit discrete domains of expression in the early embryonic CNS and limb buds. *Development* **119,** 247–261.

97. Ikeya, M., Lee, S. M., Johnson, J. E., McMahon, A. P., and Takada, S. (1997) Wnt signaling required for expansion of neural crest and CNS progenitors. *Nature* **389,** 966–970.

98. Dorsky, R. I., Moon, R. T., and Raible, D. W. (1998) Control of neural crest cell fate by the Wnt signalling pathway. *Nature* **396,** 370–373.

99. Baker, J. C., Beddington, R. S. P., and Harland, R. M. (1999) Wnt signaling in *Xenopus* embryos inhibits BMP4 expression and activates neural development. *Genes Dev.* **13,** 3149–3159.

100. Mayor, R., Guerrero, N., and Martinez, C. (1997) Role of FGF and noggin in neural crest induction. *Dev. Biol.* **189,** 1–12.

101. Kengaku, M. and Okamoto, H. (1993) Basic fibroblast growth factor induces differentiation of neural tube and neural crest lineages of cultured ectoderm cells from *Xenopus* gastrula. *Development* **119,** 1067–1078.

102. Cox, W. G. and Hemmati-Brivanlou, A. (1995) Caudalization of neural fate by tissue recombination and bFGF. *Development* **121,** 4349–4358.

103. Lamb, T. M. and Harland, R. M. (1995) Fibroblast growth factor is a direct neural inducer, which combined with noggin generates anterior-posterior neural pattern. *Development* **121,** 3627–3636.

104. Launay, C., V., F., DL., S., and JC, B. (1996) A truncated FGF receptor blocks neural induction by endogenous *Xenopus* inducers. *Development* **122,** 869–880.

105. Xu, R. H., Kim, J., Taira, M., Sredni, D., and Kung, H. (1997) Studies on the role of fibroblast growth factor signaling in neurogenesis using conjugated/aged animal caps and dorsal ectoderm-grafted embryos. *J. Neurosci.* **17,** 6892–6898.

106. Tannahill, D., Isaacs, H. V., Close, M. J., Peters, G., and Slack, J. M. W. (1992) Developmental expression of the *Xenopus int-2* (FGF-3) gene: activation by mesodermal and neural induction. *Development* **115,** 695–702.

107. Isaacs, H. V., Tannahill, D., and Slack, J. M. W. (1992) Expression of a novel FGF in the *Xenopus* embryo. A New candidate inducing factor for mesoderm formation and anteroposterior specification. *Development* **114,** 711–720.

108. Mahmood, R., Kiefer, P., Guthrie, S., Dickson, C., and Mason, I. (1995) Multiple roles for FGF-3 during cranial neural development in the chicken. *Development* **121,** 1399–1410.

109. Riese, J., Zeller, R., and Dono, R. (1995) Nucleo-cytoplasmic translocation and secretion of fibroblast growth factor-2 during avian gastrulation. *Mech. Dev.* **49,** 13–22.

110. Bueno, D., Skinner, J., Abud, H., and Heath, J. K. (1996) Spatial and temporal relationships between Shh, Fgf4, and Fgf8 gene expression at diverse signalling centers during mouse development. *Dev. Dyn.* **207,** 291–299.

111. Storey, K. G., Goriely, A., Sargent, C. M., Brown, J. M., Burns, H. D., Abud, H. M., and Heath, J. K. (1998) Early posterior neural tissue is induced by FGF in the chick embryo. *Development* **125,** 473–484.

112. Kroll, K. L. and Amaya, E. (1996) Transgenic *Xenopus* embryos from sperm nuclear transplantations reveal FGF signaling requirements during gastrulation. *Development* **122,** 3173–3183.

113. Stemple, D. L. and Anderson, D. J. (1992) Isolation of a stem cell for neurons and glia derived from the mamalian neural crest. *Cell* **71,** 973–985.

114. Doupe, A. J., Landis, S. C., and Patterson, P. H. (1985)Environmental influences in the development of neural crest derivatives: glucocorticoids, growth facors and chromaffin cell plasticity. *J. Neurosci.* **5,** 2119–2142.

115. Doupe, A. J., Patterson, P. H., and Landis, S. C. (1985) Small intensely fluorescent (SIF) cells in culture: role of glucocorticoids and growth factors in their development and phenotypic interconversions with other neural crest derivatives. *J. Neurosci.* **5,** 2143–2160.

116. Shah, N. M., Marchionni, M. A., Isaacs, I., Stroobant, P. W., and Anderson, D. J. (1994) Glial growth factor restricts mammalian neural crest stem cells to a glial fate. *Cell* **77,** 349–360.

117. Shah, N. M., Groves, A. K., and Anderson, D. J. (1996) Alternative nerual crest cell fates are instructively promoted by TGFβ superfamily members. *Cell* **85,** 331–343.

118. Shah, N. M. and Anderson, D. A. (1997) Integration of multiple instructive cues by neural crest stem cells reveals cell-intrinsic biases in relative growth factor responsiveness. *Proc. Natl. Acad. Sci. USA* **94,** 11369–11374.

119. Le Douarin, N. M. (1986) Cell line segregation during peripheral nervous system ontogeny. *Science* **231,** 1515–1522.

120. Frank, E. and Sanes, J. R. (1991) Lineage of neurons and glia in chick dorsal root ganglia: Analysis *in vivo* with a recombinant retrovirus. *Development* **111,** 895–908.

121. Serbedzija, G. N., Fraser, S. E., and Bronner-Fraser, M. (1989) A vital dye analysis of the timing and pathways of avian trunk neural crest cell migration. *Development* **106,** 809–816.

122. Weston, J. A. and Butler, S. L. (1966) Temporal factors affecting localization of neural crest cells in the chicken embryo. *Dev. Biol.* **14,** 246–266.

123. Baker, C. V. H., Bronner-Fraser, M., Le Douarin, N. M., and Teillet, M.-A. (1997) Early- and late-migrating cranial neural crest cell populations have equivalent developmental potential *in vivo*. *Development* **124,** 3077–3087.

124. Richardson, M. K. and Sieber-Blum, M. (1993) Pluripotent neural crest cells in the developing skin of the quail embryo. *Dev. Biol.* **157,** 348–358.

125. Hagedorn, L., Suter, U., and Sommer, L. (1999) P0 and PMP22 mark a mutipotent neural crest-derived cell type that displays community effects in response to TGFβ family factors. *Development* **126,** 3781–3794.

126. Artinger, K. B. and Bronner-Fraser, M. (1992) Partial restriction in the developmental potential of late emigrating avian neural crest cells. *Dev. Biol.* **149,** 149–157.

127. Perez, S. E., Rebelo, S., and Anderson, D. J. (1999) Early specification of sensory neuron fate revealed by expression and function of neurogenins in the chick embryo. *Development* **126,** 1715–1728.

128. Carr, V. M. and Simpson, S. B. Jr. (1978) Proliferative and degenerative events in the early development of chick dorsal root ganglia. *J. Comp. Neuro.* **182,** 727–740.

129. Marchionni, M. A., Goodearl, A. D., Chen, M. S., Bermingham-McDonogh, O., Kirk, C., Hendricks, M., Danehy, F., Misumi, D., Sudhalter, J., Kobayashi,

K., Wroblewski, D., Lynch, C., Baldassare, M., Hiles, I., Davis, J. B., Hsuan, J. J., Totty, N. F., Otsu, M., McBurney, R. N., Waterboy, M. D., Stroobarb, P., and Gwynne, D. (1993) Glial growth factors are alternatively spliced erbB2 ligands expressed in the nervous system. *Nature* **362**, 312–318.

130. Orr-Urtreger, A., Trakhtenbrot, L., Ben-Levy, R., Wen, D., Rechavi, G., Lonai, P., and Yarden, Y. (1993) Neural expression and chromosomal mapping of Neu differentiation factor to 8p12-p21. *Proc. Nat. Acad. Sci. USA* **90**, 1867–1871.

131. Meyer, D. and Birchmeier, C. (1995) Multiple essential functions of neuregulin in development. *Nature* **378**, 386–390.

132. Lemke, G. (1996) Neuregulins in development. *Mol. Cell. Neurosci.* **7**, 247–262.

133. Meyer, D., Yamaai, T., Garratt, A., Riethmacher-Sonnenberg, E., Kane, D., Theill, L. E., and Birchmeier, C. (1997) Isoform-specific expression and function of neuregulin. *Development* **124**, 3575–3586.

134. Oakley, R. A., Lasky, C. J., Erickson, C. A., and Tosney, K. W. (1994). Glycoconjugates mark a transient barrier to neural crest migration in the chicken embryo. *Development* **120**, 103–114.

135. Perris, R., von Boxberg, Y., and Lofberg, J. (1988) Local embryonic matrices determine region-specific phenotypes in neural crest cells. *Science* **241**, 86–89.

137. Anderson, D. J. (1999) Lineages and transcription factors in the specification of vertebrate primary sensory neurons. *Curr. Op. Neurobiol.* **9**, 517–524.

137. Bitgood, M. J. and McMahon, A. P. (1995) *Hedgehog* and *Bmp* genes are coexpressed at many diverse sites of cell-cell interaction in the mouse embryo. *Dev. Biol.* **172**, 126–138.

138. Lyons, K. M., Hogan, B. L. M., and Robertson, E. J. (1995) Colocalization of BMP2 and BMP7 RNAs suggests that these factors cooperatively mediate tissue interaction during murine development. *Mech. Dev.* **50**, 71–83.

140. Ito, K. and Sieber-Blum, M. (1993) Pluripotent and developmentally restricted neural-crest-derived cells in posterior visceral arches. *Dev. Biol.* **156**, 191–200.

140. LaBonne, C. and Bronner-Fraser, M. (1998) Induction and patterning of the neural crest, a stem cell-like precursor population. *J. Neurobiol.* **36**, 175–189.

141. Groves, A. and Bronner-Fraser, M. (1999) Neural crest diversification. *Curr. Top. Dev. Biol.* **43**, 221–258.

8

Neural Progenitor Cells of the Adult Human Brain

Steven A. Goldman

Over the past two decades, studies of cell genesis in the adult vertebrate brain have revealed the persistence of neural progenitor cells in the neuroepithelial lining of the cerebral ventricles, and in contiguous granule cell populations such as the olfactory stream and dentate gyrus. Competent progenitor cells have now been identified in fish, reptiles, birds, rodents, monkeys, and humans *(1,2)*. These progenitors, in particular those able to give rise to neurons, reside in the ventricular lining *(3–7)*, within which they appear to be largely subependymal in origin *(8,9)*, although multipotential neurogenic ependymal cells have also been reported *(10)*. Whether subependymal or ependymal, these progenitor cells extend throughout the adult ventricular system *(11–13)*, persist throughout adult life *(14,15)*, and may include or derive from multipotential founders *(16–18)*.

Together, these observations have pointed to the existence in adults of a ventricular zone (VZ) precursor cell population that remains neurogenic in selected regions, such as the avian neostriatum and rodent olfactory bulb, but which more typically becomes quiescent unless activated *(16,16a)*, then yielding only glia or short-lived neuronal progeny *(10,16b)*. Importantly, these neural progenitors persist and remain abundant in the normal human brain, as well as in experimental animals *(5,19–21)*. We review here the identification, initial isolation, relative distributions, and lineage competence of three major progenitor cell phenotypes that have been isolated from the adult human forebrain: the ventricular zone neural progenitor cell, the hippocampal neuronal progenitor, and the white matter glial progenitor *(19–21)*. Each of these cell types has recently been isolated from adult human brain tissue, and each may now be enriched in bulk to near purity. The acquisition of these cells should allow us to advance our understanding of their lineage potential and growth factor dependence, and to thereby better utilize resident progenitor cell populations of the adult human brain.

From: *Stem Cells and CNS Development*
Edited by: M. S. Rao © Humana Press Inc., Totowa, NJ

NEURONAL PRECURSOR CELLS RESIDE IN THE ADULT HUMAN FOREBRAIN VZ

In adult primates, the forebrain VZ, composed of the apposing ependymal and subependymal cell layers, continues to harbor dividing cells, predominantly if not exclusively within the subependymal cell population *(22,23)*. The distribution of these mitotic subependymal cells is roughly analogous to that of the adult rodent forebrain *(12,24,25)*, in which subependymal cell division is followed by the migration of neuronal daughter cells rostrally to the olfactory bulb *(7,26)*, and—at least developmentally—to the subgranular zone of the hippocampus. On the basis of these observations, we postulated that the adult human brain might retain a reservoir of such subependymal progenitor cells, which cease generating neurons in vivo yet retain the capacity for neurogenesis in vitro.

To test this possibility, we sought evidence of neurogenesis in cultures of adult human temporal lobe *(5)*. Both explants and dissociates were prepared from fresh brain, obtained during temporal resection; this was dissected into cortical, subcortical, and periventricular zone samples and cultured under conditions permissive for neuronal differentiation from the adult rat VZ. These human VZ cultures gave rise to neurons identified both antigenically (*see* Fig. 1) and physiologically. Only explants containing VZ generated such neuronal outgrowth, whereas neocortical explants failed to do so. Furthermore, when dissociates of adult temporal VZ were exposed to [^{3}H]thymidine, antigenically verified neurons that incorporated thymidine were found, indicating that these cells arose from precursor division in vitro (Fig. 1). Importantly, these new VZ-derived neurons were functionally as well as antigenically neuronal: When VZ outgrowths were loaded with fluo-3 and depolarized during confocal imaging, neurons showed rapid, 4–10-fold elevations in [Ca^{2+}]i in response to 60 mM K$^+$, responses typical of neuronal voltage-gated calcium channels. In contrast, glial responses to K$^+$ were minimal. Thus, progenitor cells derived from the adult human VZ exhibited mitotic neurogenesis, and their daughters developed mature neuronal function *(5,27,28)*.

PROGENITOR CELLS MAY BE IDENTIFIED HISTOLOGICALLY IN THE ADULT HUMAN SUBEPENDYMA

Study of the neural and more-committed neuronal progenitor cell populations of the adult central nervous system (CNS) was hampered for decades, by the lack of any available antigenic markers by which these cells might be specifically identified. The identification of nestin protein as an intermediate filament expressed at high levels by neuroepithelial cells aided and accelerated the study of these cells *(29)*, despite nestin's lack of absolute

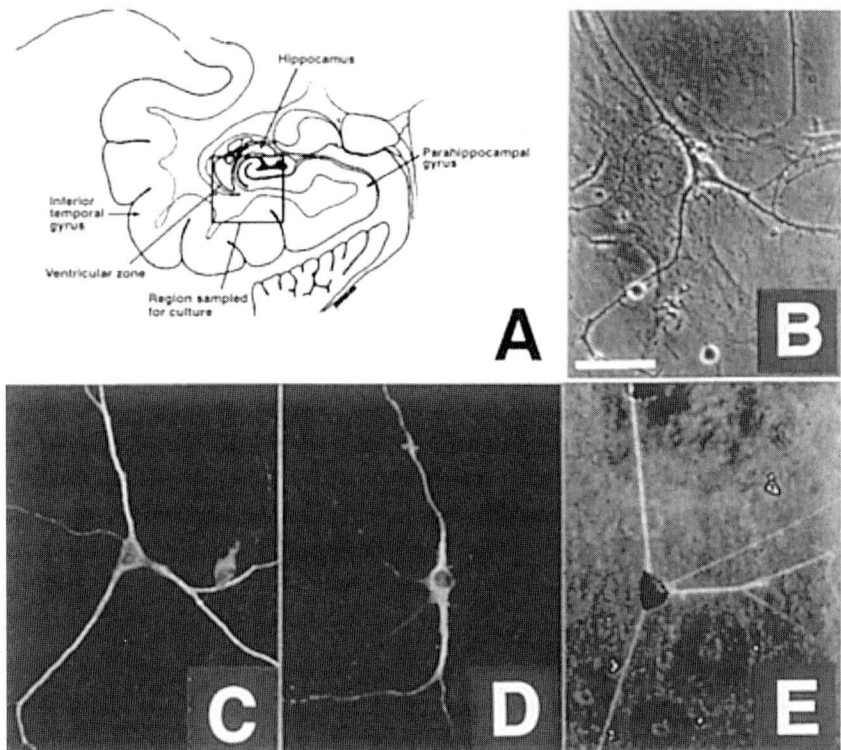

Fig. 1. The adult human temporal lobe provides an accessible source of progenitor cells. (**A**) Samples were taken during temporal lobectomy, typically either for decompressive resection or refractory epilepsy. The borders of a typical temporal lobe resection are outlined; each was dissected into cortical, subcortical, and periventricular portions, the latter including the ependyma and adjacent subventricular tissue. When the hippocampus was included in the resection, it was dissected from the temporal lobe, and the dentate gyrus then dissected clean of its overlying ventricular wall. (**B**) Outgrowth from an adult VZ explant, in which a presumptive neuron is seen upon a layer of flat substrate cells at 19 DIV. (**C**) MAP-2$^+$ neuron, found in a subcortical culture at 18 DIV. (**D**) N-CAM$^+$ neuron in an SZ dissociate at 12 DIV. (**E**) MAP-5$^+$ cell that incorporated [^{3}H]thymidine in vitro, suggesting its origin from precursor cell mitosis. Scale bar = 50 µm. (Adapted with permission from ref. *52*.)

phenotypic specificity *(30)*. In addition, recent discoveries of RNA binding proteins specific for neural phenotype have led to the identification and development of several new probes for neurons and their progenitors. Musashi protein is one such RNA binding protein, expressed only by mitotic, uncommitted progenitors in development, and by VZ cells and parenchymal astrocytes in adulthood *(31)*. Musashi was first identified in *Drosophila* and *Xenopus (32,33)*, in which it is expressed by CNS stem cells and their mitotic daughters. In mammalian development, musashi expression is limited to cycling cells in the ventricular and subventricular zones, and diminishes rapidly with cell migration. It is not expressed by neurons or oligodendrocytes, but is by astrocytes *(34)*. In adults, its expression is limited largely to the ventricular and olfactory subependyma, a distribution pattern similar to that of nestin *(29)*; its sequence is highly conserved, allowing antibodies against mouse musashi to identify human precursor cells *(27)*.

The Hu proteins constitute another such family of neuronal RNA binding proteins *(35)*; three of its four known members appear to be expressed only by neurons, perhaps at different phases of neuronal ontogeny. As a result, the anti-Hu monoclonal antibody 16A11, which recognizes a conserved epitope on the Hu proteins HuC, HuD, and Hel-N1, recognizes only neurons and their committed progenitors *(36–38)*. On the basis of these studies, we immunostained tissue sections of the adult human temporal VZ to quantify and map the distribution of musashi and Hu-defined neural and neuronal progenitor cells, respectively *(27)* (Fig. 2). Each of the five patient samples that we assessed included >5000 scored ependymal and subependymal cells, counted in at least three sections from each patient; the patients ranged in age from 9 to 46 yr. Overall, $7.8 \pm 2.2\%$ of subependymal cells expressed Hu, and $6.2 \pm 2.6\%$ musashi. Together then, almost 15% of temporal VZ cells expressed one or the other of these markers. Although seemingly high, this estimate of the frequency of progenitor cells and their derivatives in the adult subependyma must be viewed in the context of the patchy distribution and evanescent thinness of the adult VZ, which is but a noncontiguous cellular monolayer along the adult human temporal horn. As a result, the actual incidence of VZ progenitor cells in the resected brain tissue is low, probably less than 1 in 10^6 cells overall *(28)*.

ADULT HUMAN SUBEPENDYMAL PRECURSOR CELLS RESPOND TO FGF-2 AND BDNF WITH EXPANSION, NEURONAL DIFFERENTIATION AND LONG-TERM SURVIVAL

In rodents, the proliferation of these adult VZ precursor cells is promoted by fibroblast growth factor-2 (FGF-2) *(18,39,40)*, as the differentiation, maturation, and survival of their neuronal daughters is supported by brain-derived neurotrophic factor (BDNF) *(12,41)*. On this basis, we sequentially treated explants of the adult human temporal VZ with FGF-2 followed by BDNF, and found that neurogenesis could indeed be induced and supported with this combination of agents *(28)*. Neuronal number and survival in explants

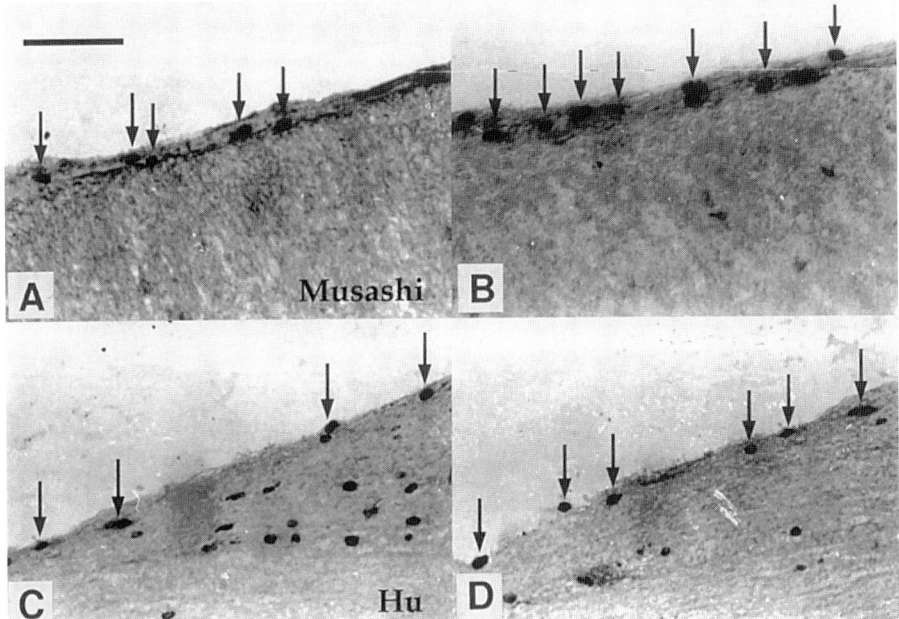

Fig. 2. Musashi and Hu proteins recognize uncommitted progenitor cells and their neuronal daughters in the adult temporal ventricular zone. These sections were taken from the ependyma/subependyma lining of the lateral ventricle, deep to the inferior temporal gyrus, in a 27-yr-old man with mesial temporal sclerosis. **(A,B)** Scattered islands of ventricular cells, generally subependymal, immunoperoxidase stained for musashi protein, an RNA binding protein of neural progenitors. musashi expression was limited to the VZ of these adult human temporal resections. **(C,D)** Loose aggregates of adult SZ cells also expressed Hu, a triad of early, neuron-specific RNA binding proteins recognized by monoclonal antibody 16A11. Scale bar = 50 μm. (Reproduced with permission from ref. *27*.)

raised for 1 wk in 20 ng/mL FGF-2, followed by 8 wk in 40 ng/mL BDNF, were both substantially greater than in unsupplemented plates or those given *either* FGF-2 *or* BDNF. After 9 wk in vitro, many explants raised in FGF-2/BDNF exhibited elaborate networks of scores of healthy neurons (Fig. 3). These cells expressed MAP-2 and displayed sharp calcium increments to K$^+$- depolarization, suggesting their functional maturation. Many had incorporated [^{3}H]thymidine during their first week in vitro, indicating their genesis during their inital week in FGF-2, 8 wk earlier. No surviving neurons were noted beyond 2 wk in plates not treated with FGF-2 and BDNF. These data indicated that serial application of FGF-2 and BDNF allowed the generation of complex networks of new neurons by subependymal explants of the adult human brain *(27,28)*.

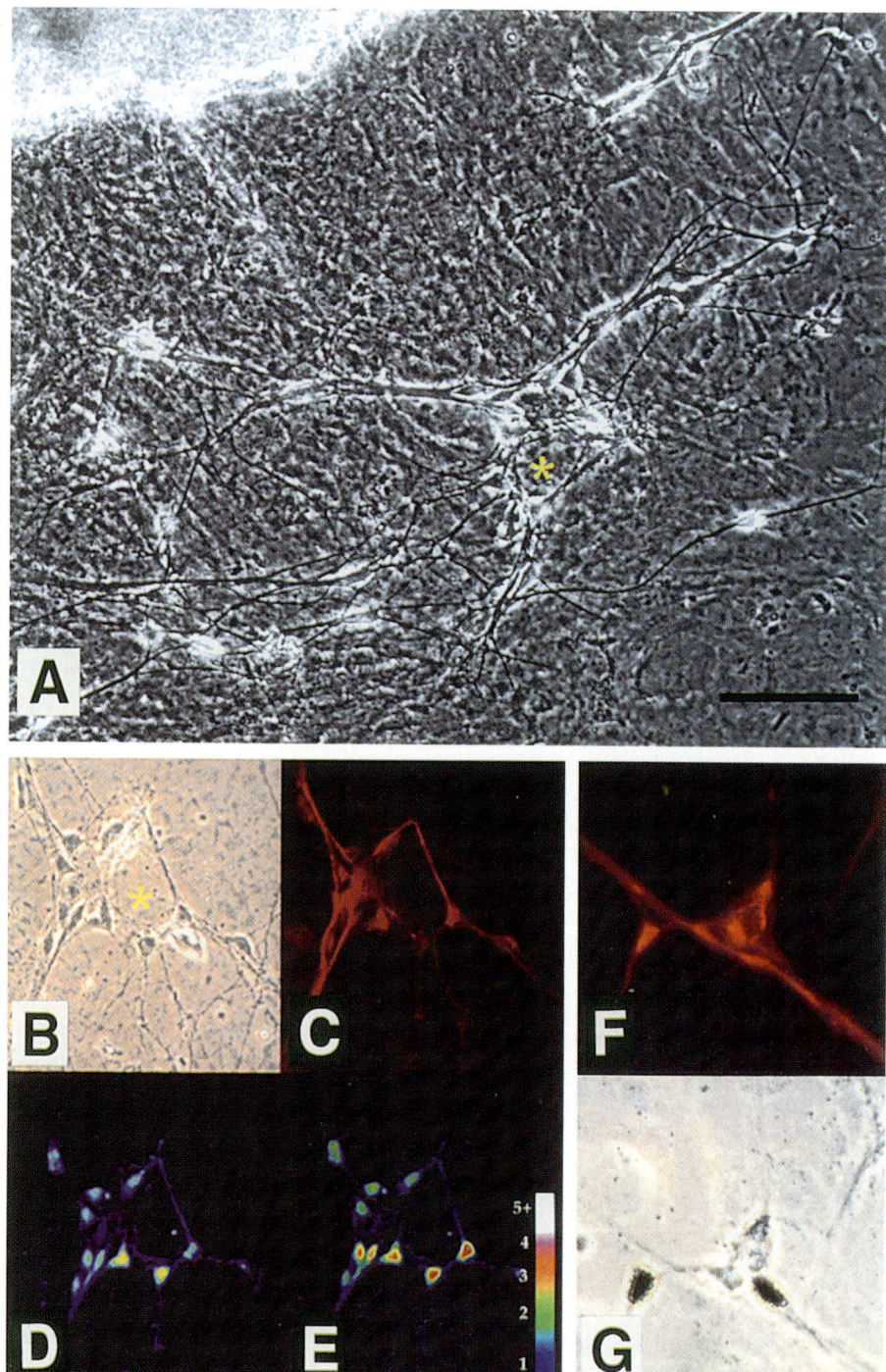

THE INACCESSIBILITY OF HUMAN NEURAL PRECURSORS HAS ENCOURAGED STRATEGIES FOR THEIR EXPANSION

The harvest of primary neuronal precursor cells from the adult human forebrain has been limited by the low yields attending its enzymatic dissociation and the difficulty in recognizing and purifying surviving precursor cells as such. To improve the yield of adult-derived neural progenitors, several groups have taken the approach of raising neural cell lines derived from single precursors, exposed continuously to mitogens in serum-deficient culture. Although first established for use in the adult rodent brain *(16,42,43)*, the propagation of clonally derived neurospheres was extended to that of fetal human neural progenitor cells as well *(44–46)*. On that basis, a recent study tackled the vexing issue of propagating and expanding adult human subependymal progenitor cells, by preparing neurosphere cultures of dissociated adult temporal VZ and hippocampus *(47)*. The authors found that mRNA transcripts for neuronal, oligodendrocytic, and astrocytic genes could be detected in single neurospheres, suggesting the existence of a multipotential progenitor cell in these clonally derived cell masses. The latter conclusion must be qualified, however, since neither the antigenicity nor the functional competence of the cells generated by these adult-derived human neurospheres were assessed in this initial report. Nonetheless, the generation of multiple phenotypes in these cultures suggested that the residual neuronal progenitor cells found in adult human VZ explant cultures *(5,27)*, might be derivatives of a resident multipotential neural stem cell.

It is important to note that despite the abundance of cells obtained through the sustained propagation of VZ cells in vitro, this strategy for preparing engraftable neural progenitors has a number of limitations. First, the directed differentiation of these cells into desired terminally differentiated phenotypes remains an unrealized goal, despite some progress in modifying

---

Fig. 3. Adult human ventricular zone progenitor cells can be stimulated to expand in vitro, and generate functional neurons. Cultures from this patient were grown in added FGF-2 for 1 wk and in BDNF thereafter. **(A)** A field of neurons that have arisen from a neocortical explant after 9 wk in vitro; highly neuritic neurons lay on ependymal cells and glia. **(B)** Higher magnification (asterisk). **(C)** MAP-2 staining confirms the neuronal identity of these cells. **(D)** Laser-scanning confocal microscopy at 488 nm images the basal calcium signal of the neuronal outgrowth, after loading with fluo-3. **(E)** A sharp increase in fluo-3 fluorescence, typical of neuronal calcium responses, to K^+- depolarization. **(F,G)** Of 11 MAP-2[+] cells in **B**, 4 incorporated [³H]thymidine[+] during their first week in vitro. Scale bar = 50 µm. (Reproduced with permission from ref. *27*.)

or biasing the phenotypic choice and fate of these cells by defined neurotrophic and gliotrophic agents *(12,41,48,49,50a–52)*. Second, the lineage potential, transformation state, and karyotype of these lines all remain uncertain; after prolonged passage at high split ratios, the antigenic expression patterns of repetitively passaged precursors often manifest mixed lineages *(40)*. Indeed, under the stress of prolonged mitogenic stimulation in serum-free culture, neither the clonality nor karyotypic integrity of these cells can be assumed. Although the reversion to a mixed-antigenic phenotype may represent the emergence to a stem cell phenotype *(53)*, it might also manifest the degradation of committed lineages into lines that are at best unrepresentative of their founders and, at worst, transformed neuroectodermal blasts. Third, it remains unclear whether neurons generated from repetitively passaged precursors retain the normal characteristics of neuronal electrophysiologic function, although recent observations of action potential generation from extensively expanded human fetal progenitor cell lines have been reassuring in this regard *(46)*.

HUMAN NEURAL PROGENITOR CELLS MAY ACT AS SUCH UPON XENOGRAFT

A number of recent studies have reported that fetal human neural progenitors, derived from abortuses and expanded in vitro as neurospheres, may terminally differentiate and histologically integrate when xenografted to both the prenatal and adult rodent brain *(46,54,55)*. Similarly, v-myc-immortalized human neural stem cells have been shown to differentiate into all three major neural lineages upon perinatal engraftment into the rodent brain *(56)*. Together, these studies have indicated that neural stem cells may act as such in vivo as well as in vitro, generating multiple lineages in context-dependent fashions. These studies have tremendously advanced our conception of the potential therapeutic roles of neural stem cells in both structural repair and enzymatic repletion. Nonetheless, because of the possibility of phenotypic degradation attending either immortalization or expansion, it has been unclear whether prolonged mitogen-assisted expansion from single isolated precursors will prove to be an acceptable strategy for propagating therapeutically sound neural progenitor cells.

With this caveat in mind, we postulated that native progenitor cells might be more likely than their expanded counterparts to generate functionally sound neurons upon eventual transplantation. Indeed, we had already found that neurons generated directly from freshly cultured VZ explants exhibited characteristic neuronal responses to both depolarization and excitatory transmitters *(27,57)*. However, the theoretical advantages of using directly harvested progenitor cells had been limited in practice by our lack of means for specifically identifying and harvesting these cells from donor brain tissues. As a result, relatively large amounts of scarce human fetal tissues have been required to generate enough fetal progenitor cells for engraftment, whereas no adult tissues have yet yielded progenitor cells in sufficient numbers or purity to allow their direct implantation. To redress this problem of enriching scarce progenitor cells from much larger populations of brain cells, we therefore established a means of identifying and selecting neural progenitor cells on the basis of their expression of fluorescent transgenes driven by cell-specific promoter sequences.

GFP Allows Neuronal Precursors to be Identified in Cultures of Both the Embryonic and Adult VZ/VZ

To recognize live neuronal precursors as such, we chose to use neural promoters to drive the expression of the gene encoding green fluorescent protein (GFP) *(58)*. GFP is a coelenterate protein that fluoresces upon blue excitation, with little toxicity; it has evolved into an effective transcriptional reporter in live cells *(59)*. To identify neuronal progenitor cells while they were still alive, as opposed to in fixed histologic sections, we developed constructs of a mutant GFP optimized for human codon usage (hGFP) *(60)*, placed under the control of the early neuronal P/Tα1-tubulin promoter *(61,62)*. In accord with the neuronal specificity of Tα1-tubulin promoter expression *(61)*, P/Tα1-tubulin-driven hGFP was strongly expressed by precursors and young neurons, but not by glia *(63,64)*. In both fetal and adult-derived cultures, P/Tα1-driven GFP fluorescence was specific to neurons and their committed precursors (dividing as well as postmitotic) and remained bright up to 14 d after transfection *(64)*. This observation established that GFP, when expressed under the control of cell-specific regulatory elements, might be used as a reporter of phenotype in live cells.

Promoter-Defined Phenotypic Restriction of GFP Permits Fluorescence-Activated Sorting of Neural Progenitor Cells

To separate and harvest native neural progenitor cells from brain tissue, we then capitalized on the ability of promoters, such as that for Tα1-tubulin, to direct fluorescent gene expression to preselected progenitor cell phenotypes. As an initial proof-of-principle, we chose to separate P/Tα1:hGFP-defined VZ neuronal progenitor cells from the larger brain cell population, by using fluorescence-activated cell sorting (FACS). To this end, we transfected cultured monolayer dissociates of VZ cells with P/Tα1:GFP and then sorted the transfectants based on GFP fluorescence *(64)*. This technique allowed both a high degree of enrichment of neuronal progenitor cells and as a virtual abolition of glial contaminants. By this means, we were able to isolate embryonic chick and rat neural precursor cells and to observe their subsequent neuronal maturation (Fig. 4).

With the advent of improved methods for dissociating adult forebrain tissue, we were able to extend this approach to include the isolation and purification of neuronal progenitors from the adult brain as well as from fetuses. Using tissue dissociation conditions optimized for use with adult brain parenchyma, we found that P/Tα1:hGFP$^+$ cells could be readily identified in, and sorted from, transfected dissociates of the adult rat VZ (65). These sorted progenitors were mitotically competent upon initial harvest and indeed continued to express P/Tα1:hGFP fluorescence while still dividing. Nonetheless, most of them rapidly matured as neurons when raised in serum-containing media: Most of these cells began to express the early neuronal proteins Hu and TuJ1/ βIII-tubulin within the first week in vitro. Fewer than 5% expressed astrocytic markers, compared with over half of the cells in matched unsorted culture, or in cultures that were sorted after transfection with DNA bearing GFP under the control of the constitutive CMV promoter. Together, these results indicated that the use of an early neuron-selective promoter to drive phenotype-specific GFP expression permitted the targeted extraction of neuronal progenitor cells from the adult as well as from the fetal rat brain.

Mitotic Neuronal Progenitor Cells Can Be Selected From Adult Human VZ

Based on the successful selection of neuronal progenitor cells from the rat VZ using P/Tα1:hGFP-based FACS, we next sought to separate progenitors from the adult human VZ (Fig. 5) (19). To this end, the temporal ventricular wall was dissected from an initial sample of temporal lobes resected from four adult patients undergoing therapeutic lobectomy. These samples were dissociated and the cultured cells transduced with either P/Tα1:hGFP or E/nestin:EGFP plasmid DNA. A week later, the cells were redissociated, selected via FACS on the basis of neural promoter-driven GFP expression, and replated (Figs. 5 and 6). Most of these cells expressed the early neuronal protein βIII-tubulin upon FACS; within the week thereafter, most matured as morphologically evident neurons that co-expressed βIII-tubulin and MAP-2. Many of these neurons had incorporated bromodeoxyuridine in the days before FACS, indicating their mitogenesis in vitro (Fig. 7).

The Nestin Enhancer Directs GFP to Neural Progenitor Cells in the Adult Human VZ

The Tα1-tubulin promoter gave us a means by which to identify and isolate neuronal progenitor cells on the basis of their transcriptional activation. We next sought to identify reagents by which uncommitted neural

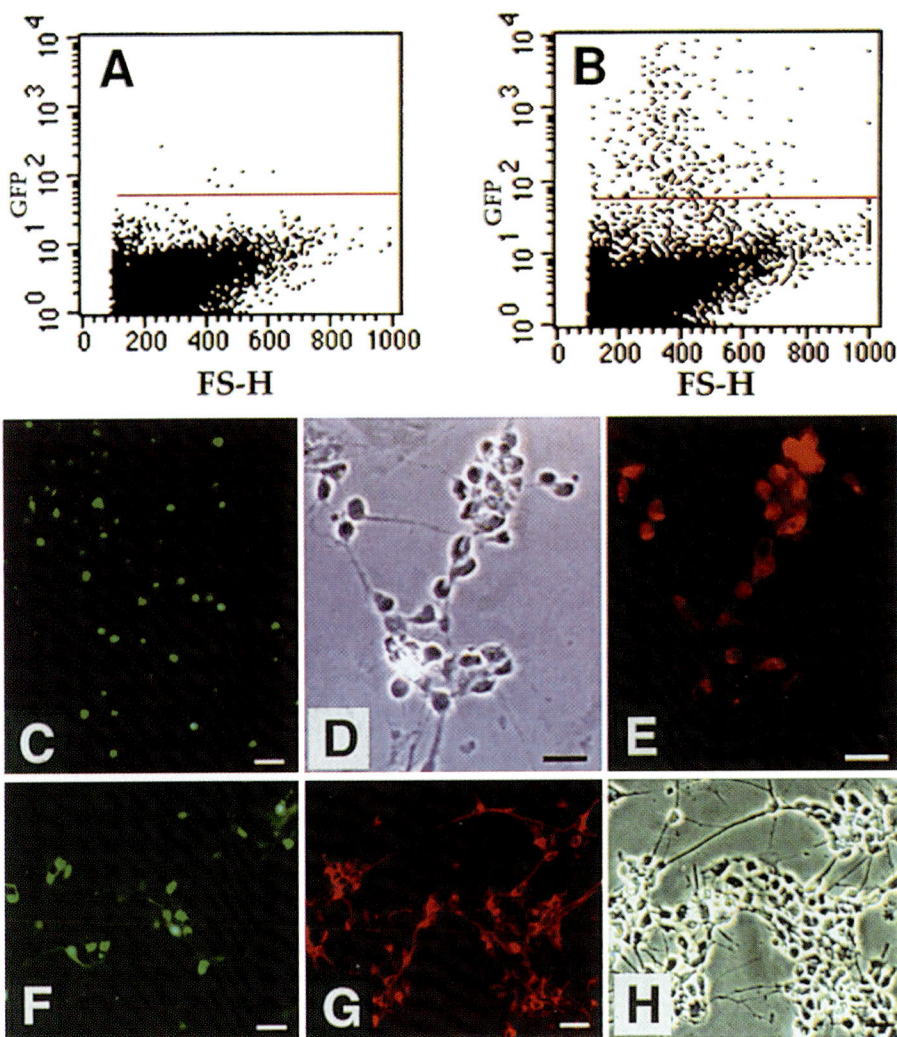

Fig. 4. Neuronal precursors may be separated from developing forebrain by fluorescence-activated cell sorting on the basis of Tα1-tubulin promoter-driven GFP. (**A,B**) GEP fluorescence intensity vs forward scatter (cell size). (**A**) E14 forebrain cells transfected with Tα1:*lacZ*, sorted 36 hlater. (**B**) Matched cells separated after transfection with Tα1:*GFP*. Each graph shows 10^4 cells; in B, 4.2% of the cells were separated on the basis of Tα1:GFP. (**C**) Tα1:GFP-sorted cells, 16 h after FACS. (**D**) 2 d later, the Tα1:GFP-sorted cells appeared neuronal. (**E**) Stained for nestin. (**F**) Similiar enrichment was seen for three markers: nestin, Hu, and βIII-tubulin. Two days post FACS, <1% cells expressed GFAP; <0.1% were O4⁺. Scale bars = C: = 50 μm; D, E: = 25 μm; F–H: = 75 μm. (Reproduced with permission from ref. *64*.)

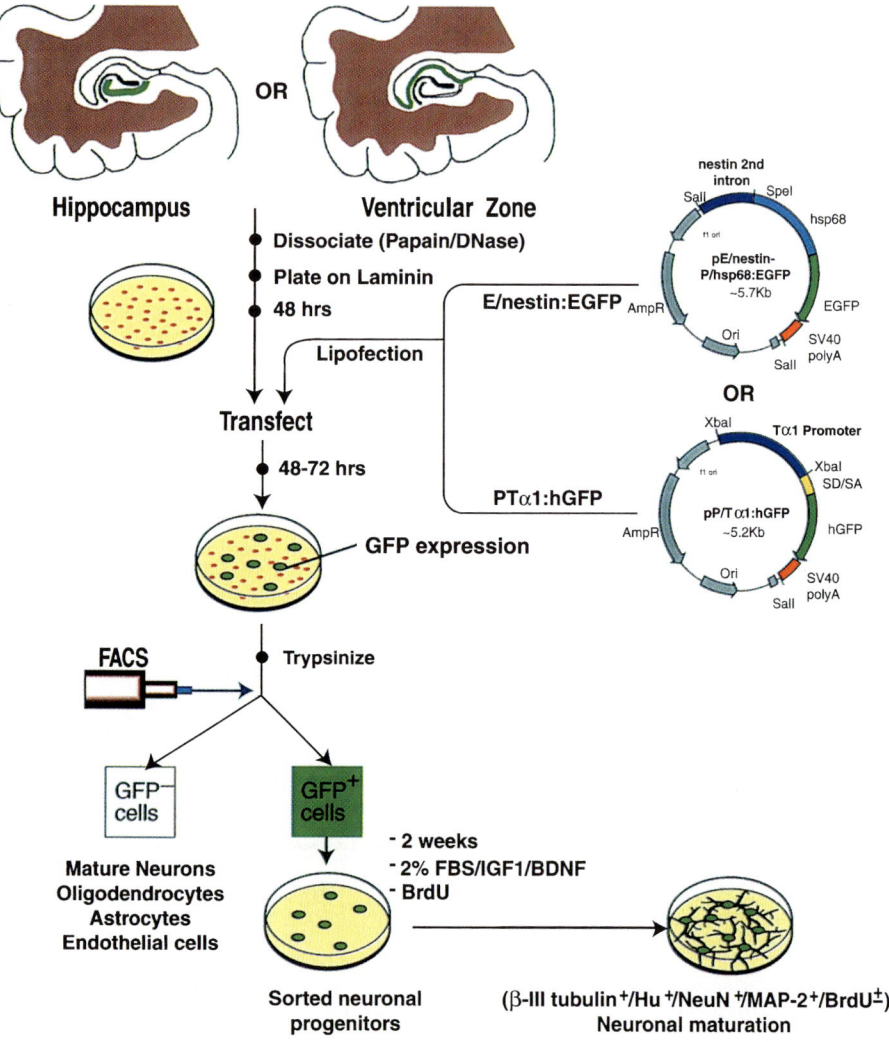

Fig. 5. Separation of neural progenitor cells from the adult human brain via promoter-defined GFP-based FACS. This schematic outlines alternative strategies by which FACS has been used to separate neural progenitor cells from the adult human temporal lobe. Both P/Tα1:hGFP and E/nestin:EGFP selection plasmids have been used to separate neuronal and less committed neural progenitor cells, respectively, from the adult VZ and hippocampus. Both regions of the brain are outlined, and each selection plasmid is schematized. In addition, white matter derived from these samples has been transfected with DNA encoding hGFP placed under the control of the early oligodendrocytic CNP2 promoter (see text); use of essentially the same logic and experimental protocols as employed for isolating the E/nestin and P/Tα1 tubulin-defined progenitor pools, has allowed the isolation and sorting of P/CNP2-defined oligodendrocytic progenitor cells from the adult white matter as well. (Adapted with permission from refs. 19 and 20).

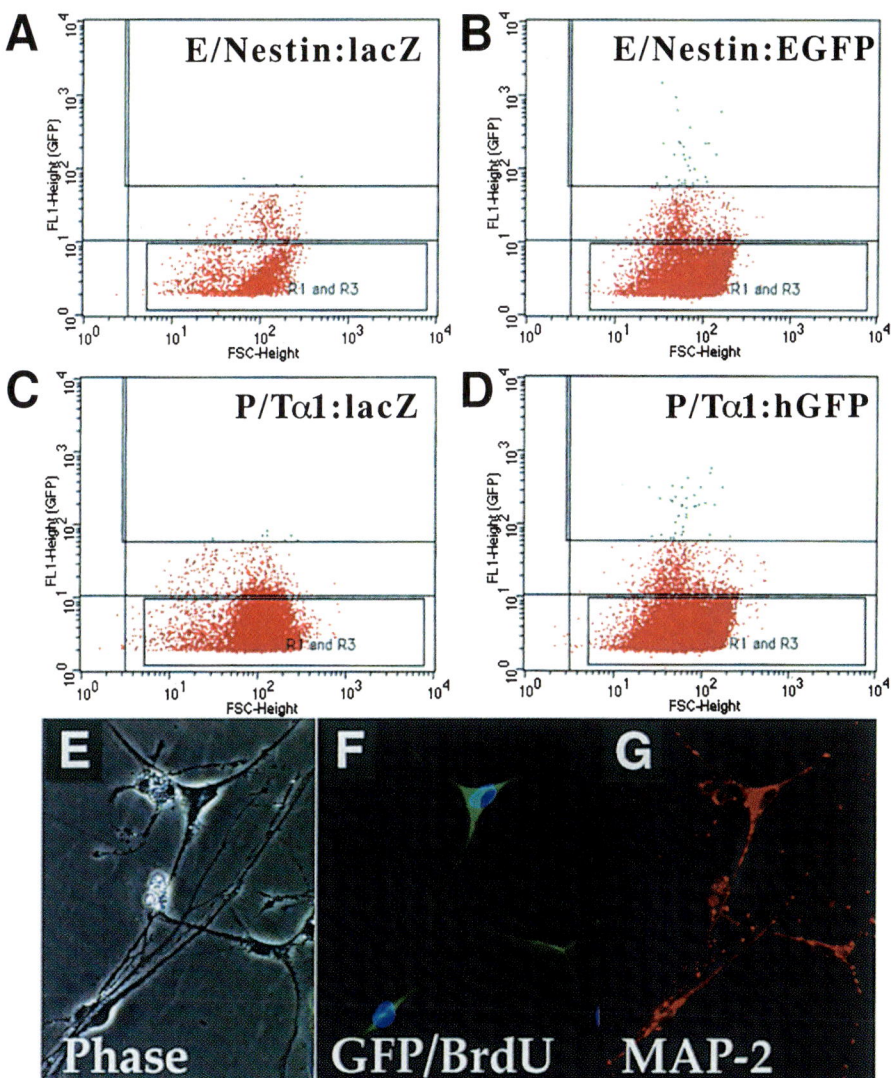

Fig. 6. The Tα1-tubulin promoter identifies neurons arising from mitotic progenitor cells in the adult human ventricular zone. Cultured cells derived from the adult VZ were transfected with pP/Tα1:hGFP after 2 DIV. This plasmid identifies neuronal precursor cells and their young daughters; in vitro, its expression persists during early neuronal maturation. (**A–C**) A young bipolar neuron (**A**, *phase contrast*), that is expressing P/Tα1:hGFP (**B**, *green*) after 5 DIV. The cell co-expressed neuronal βIII-tubulin (**C**, *red*) and incorporated BrdU (**B**, *blue*), indicating its genesis in the days before. (**D–F**) A cluster of neurons (**D**, *phase*) in a matched culture of adult VZ cells, stained for βIII-tubulin (**F**, *red*) at 14 DIV. (**E**) These cells expressed P/Tα1:hGFP (*green*), and each incorporated BrdU (*blue*) to which they were exposed during the first week in culture, indicating their in vitro mitogenesis. (**G**) These more mature neurons were photographed after 14 DIV, 10 d after transfection of an adult VZ culture with P/Tα1:hGFP. Scale bar = 25 μm. (Reproduced with permission from ref. *19*.)

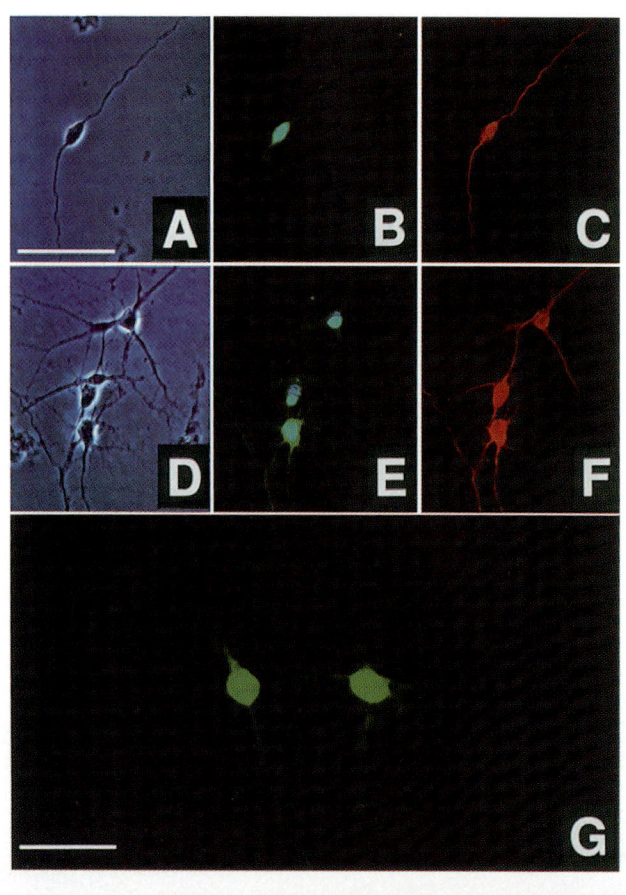

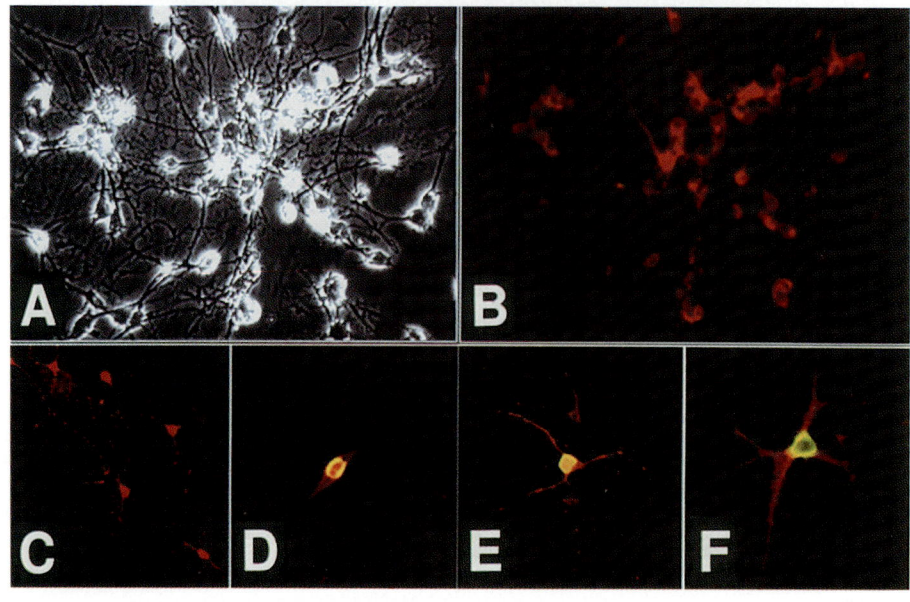

progenitor and stem cells might be similarly identified and isolated. To this end, we used the neuroepithelial-selective enhancer for the immature neural filament protein nestin *(66)*. Briefly, the evolutionarily conserved region of the second intron of rat nestin gene, which is sufficient to direct gene expression to neuroepithelial progenitors *(66,67)*, was placed at the upstream site of the mouse basal heat shock protein promoter (hsp68). P/hsp68 was used since it is expressed only when placed downstream of a strong enhancer *(68)*. This was fused to the cassette of the EGFP coding region, flanked by the SV40 polyA, yielding E/nestin: P/hsp68:EGFP. We transfected this selection cassette (designated E/nestin:EGFP) into both fetal rat and human VZ and hippocampal cell cultures, and found that in each setting, the E/nestin:P/hsp68 regulatory sequence directed GFP expression in a select fraction of the cell population. These E/nestin:EGFP+ cells either expressed nestin protein alone, without concurrent neuronal or

---

Fig. 7. (**Opposite page, top**) Both E/nestin:EGFP and P/Tα1:hGFP-defined neural progenitors may be enriched from the adult human ventricular zone using GFP-based FACS. (**A,B**) VZ cells derived from a 20-yr-old man, transfected with either E/nestin:*lacZ* (*A*, a nonfluorescent control), or E/nestin:EGFP (**B**). For both sorts, GFP fluorescence intensity (FL1) was plotted against cell size (forward scatter, FCS). In this representative run, among the E/nestin:EGFP-transfected cells, 0.10% achieved an arbitrary threshold of fluorescence intensity, which was calibrated to be met by <0.01% of control cells. (**C,D**) VZ cells from the same patient transfected with either the control plasmid P/Tα1:*lacZ* (**C**), or the fluorescence expression construct P/Tα1:hGFP (**D**). Among the PTα1:hGFP-transfected cells, 0.12% achieved threshold and were gated for separation. (**E,G**) FACS based on P/Tα1:hGFP expression yielded mitotic cells able to give rise to neurons in vitro. (**E**) Representative cluster of cells sorted from an adult VZ culture. These cells were sorted 3 d after transfection with pP/Tα1:hGFP, and fixed a week later at 10 DIV. (**F**) Residual P/Tα1:hGFP fluorescence (*green*) contrasts with the (*blue*) BrdU-immunolabeled nuclei of cells that arose through in vitro mitogenesis. (**G**) The P/Tα1:hGFP+/BrdU+ cells expressed the mature neuronal protein MAP-2 (*red*). Scale bar = 25 µm. (Reproduced with permission from ref. *19*.)

Fig. 8. (**Opposite page, bottom**) Adult human hippocampus harbors mitotic neuronal progenitor cells. (**A,B**), A monolayer dissociate of adult human dentate gyrus, removed from a 33-yr-old man after temporal lobectomy. (**A**) *phase* and (**B**) *fluorescence* images at 7 DIV of a cluster of adult dentate neurons, labeled with the antineuronal antibody MAP-2 (**B**, *red*). (**C**) Hippocampal culture derived from a 35-yr-old, immunostained for βIII-tubulin. This culture was exposed to BrdU in vitro, and then fixed and stained for BrdU as well as bIII-tubulin. (**D–F**) TuJ1+ (*red*)/ BrdU+ (*green*) Neurons, generated by mitotic neurogenesis from hippocampal progenitors. Scale bar = 30 µm. (Adapted with permission from ref. *19*.)

glial antigenic expression, or concurrently with immature neuronal anti-gens such as βIII-tubulin. Furthermore, E/nestin:EGFP did not yield GFP fluorescence in cultures of human astrocytes, fibroblasts, or endothelial cells. Thus, the E/nestin:EGFP cassette appeared to restrict GFP expres-sion to neural progenitors — which included bipolar A2B5$^+$ cells that may have been of early oligodendroglial or neuronal lineage — and immature but distinct Hu$^+$/TuJ1$^+$ neurons. The neuroepithelial cell-specific expres-sion of this construct has also been confirmed in transgenic mice (Kawaguchi, et al., in press).

Using dissociates of the adult human ventricular zone transfected with plasmid DNA bearing either the E/nestin:P/hsp68:EGFP or P/Tα1:hGFP selection cassettes, we were able to identify and isolate distinct populations of mitotic neural progenitor cells from the adult human ventricular epithelium. In particular, we used P/Tα1:hGFP plasmid DNA to report neuronally committed progenitors and their daughters in the adult VZ, and E/nestin:EGFP to similarly identify and sort a nominally less committed population of neural precursor cells from the VZ (19). Together, these experiments established the feasibility of prospectively identifying and isolating viable populations of persistent neural progenitor cells from the adult human forebrain. This in turn led us to seek and isolate competent progenitor cells in other less characterized regions of the human brain. These other populations of progenitor cells, which include the neuronal progenitors in the hippocampal dentate gyrus, and glial progenitors in the subcortical white matter, may be more common, more accessible, and clinically valu-able than their parental—and prototypic—VZ progenitor cells.

NEURONAL PROGENITOR CELLS CAN BE SELECTED FROM THE ADULT HUMAN HIPPOCAMPUS

Past studies have suggested the persistence of neuronal progenitor cells in the dentate gyrus of the adult mammalian hippocampus. Histological evidence of dividing hippocampal progenitor cells has been found in adult animals ranging from chickadees to humans (69–75). In rodents, hippocam-pal neurogenesis can be modulated by stress (76), enrichment (77), exercise (78), and learning (79). Furthermore, among primates, both adult macaques (74,80) and humans (72) exhibit histological evidence of neurogenesis in the dentate gyrus. Indeed, the dentate gyrus remains the only site of persis-tent neurogenesis thus far noted in the adult human brain, although postnatal neuronal addition to the frontal neocortex has been noted in both young children (81,82) and rhesus monkeys (79). In culture, neurogenic hippoc-ampal progenitor cells have been found in suspension cultures derived from both adult rats *(83)* and humans *(47)*; these can expand in response to FGF-2, include multipotential founders *(83)*, and are capable of heterotopic integration into other regions of granular neurogenesis, such as the olfactory subependyma *(84)*.

Yet despite the widespread incidence of hippocampal neurogenesis in adult animals, human hippocampal progenitor cells had not previously been isolated. As a result, no assessment of the abundance, factor-responsiveness, or regenerative capacity of these cells has been possible. To identify and extract neuronal progenitors from the adult human hippocampus, we therefore transfected VZ-free dissociates of surgically resected adult human hippocampus with plasmid DNA bearing the gene for hGFP, placed under the control of regulatory sequences for the genes encoding either $T\alpha 1$ tubulin *(61,62,64)* or nestin *(29,66)*. These constructs each recognized a population of cells in adult hippocampal dissociates, which divided in vitro and gave rise to antigenically and functionally appropriate neurons (Fig. 8) *(21)*. In the presence of FGF-2, both the $T\alpha 1$:hGFP and E/nestin:EGFP-defined cells incorporated BrdU from the culture media, and both matured to express typical neuronal antigens, including βIII-tubulin, MAP-2, Hu, and NeuN. Thus, adult hippocampal cultures harbored a pool of dividing cells, whose progeny became neurons, and which could be identified while alive by their transcriptional activation of the nestin and $T\alpha 1$-tubulin regulatory sequences.

Using FACS, we then isolated both E/nestin:EGFP$^+$ and P/Ta1:hGFP$^+$ hippocampal cells, enriching each to near purity. The progenitor cell pools thereby obtained were still able to generate neurons, which matured as such not only morphologically and antigenically, but also physiologically (Fig. 9): Neurons arising in FACS-purified cultures of P/$T\alpha 1$:hGFP hippocampal cells developed depolarization-induced calcium elevations of >300%, typical of neuronal voltage-gated calcium channels. In addition, patch clamp analysis revealed that they had fast sodium channels, and responded as neurons with rapid sodium currents to voltage-stepping. Together, these observations indicated that the adult human hippocampus harbors mitotically competent progenitor cells, which can be expanded in vitro to give rise to antigenically appropriate, functionally competent neurons. By using FACS based on GFP expression driven by the $T\alpha 1$-tubulin promoter and nestin enhancer, these cells may be specifically targeted and extracted from hippocampal tissue, yielding viable, highly enriched populations of hippocampal progenitor cells.

A DISTINCT POOL OF OLIGODENDROCYTE PROGENITORS RESIDES IN ADULT HUMAN WHITE MATTER

Given the encouraging selectivity of neuronal and uncommitted neural progenitor enrichment, based respectively on the $T\alpha 1$-tubulin promoter and

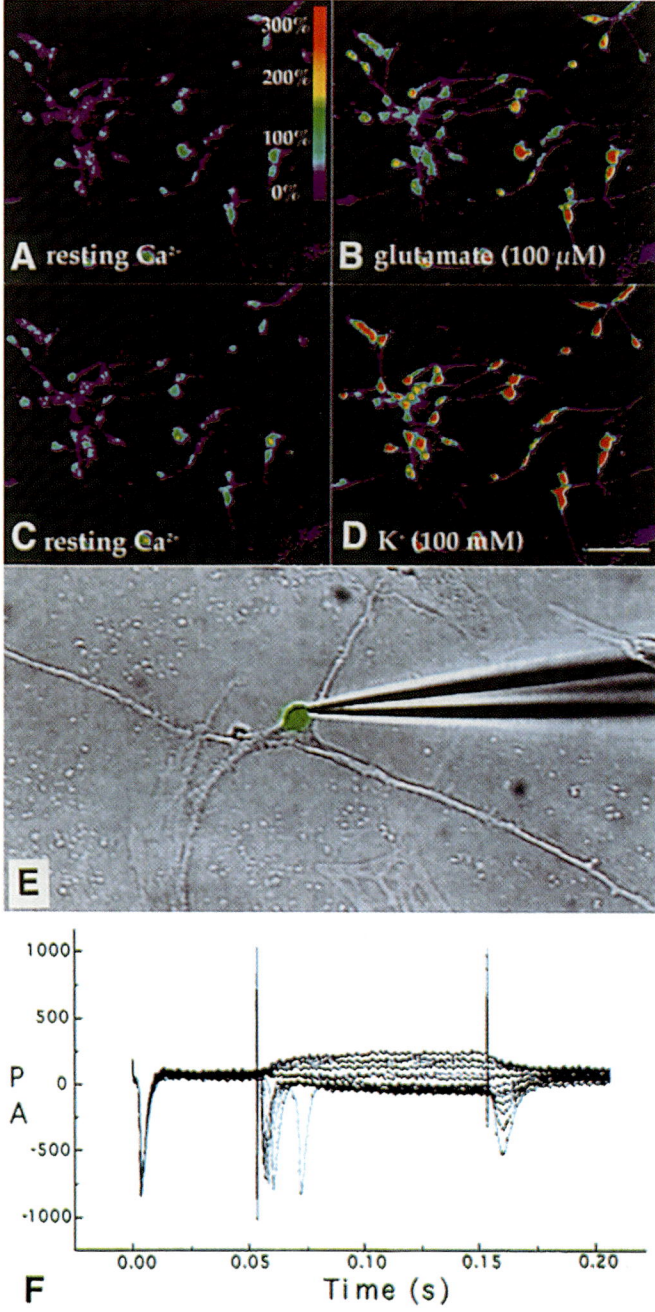

Fig. 9. P/Tα1:GFP-sorted human hippocampal cells develop into physiologically mature neurons. (**A–D**) P/Tα1:hGFP-sorted progenitors developed neuronal

the nestin enhancer, we next attempted cell type-specific enrichments based on other phenotype-selective regulatory sequences. We began by attempting to identify progenitor cells of the adult human white matter, whose existence in rats had been previously established in culture *(85)*, and validated by retroviral lineage analysis *(86,87)*. Yet, despite these prior studies of mitotic oligodendrocyte progenitors in the adult rat, no analogous phenotype had ever been isolated from adult human brain tissue *(94,95)*. To this end, we constructed plasmid vectors containing the early promoter for the oligodendrocyte protein cyclic nucleotide phosphodiesterase (P/CNP2; Dr. P. Braun, Montreal), which we placed 5' to the coding region for hGFP. When we transfected this construct into dissociates of the adult human capsular white matter, we observed that P/CNP2:hGFP was expressed initially by only a single, morphologically and antigenically discrete class of bipolar cells *(20)*. These cells were mitotically competent, in that they incorporated BrdU from their culture medium. In a sample of 30 cultures derived from three patients, $55 \pm 14.8\%$ of the P/CNP2:hGFP$^+$ cells incorporated BrdU by 7 DIV. Cell division persisted in low-serum base medium containing FGF-2, platelet-derived growth factor (PDGF) and NT-3. Of the P/CNP2:hGFP$^+$ cells, most initially expressed the early oligodendrocytic marker A2B5 but failed to express the more differentiated markers O4, O1, or galactocerebroside. Some expressed astrocytic glial fibrillary acidic protein (GFAP), but none expressed neuronal markers when initially identified by their GFP fluorescence in mixed, unsorted cultures. When FACS was used to purify these P/CNP2:hGFP$^+$ cells), most (>90%) were found to mature as oligodendrocytes (Fig. 10), progressing through a stereotypic sequence of A2B5, O4, O1, and galactocerebroside expression, as in development *(88)*. By 3 wk after FACS, $74.1 \pm 7.7\%$ of the P/CNP2:hGFP$^+$ sorted cells expressed

---

Fig. 9. *(continued)* Ca^{2+} responses to depolarization. (**A**) P/Ta1:hGFP-sorted cells loaded with the calcium indicator dye fluo-3, 10 d after FACS; these have matured uniformly into fiber-bearing cells of neuronal morphology. (**B**) The same field after exposure to glutamate. (**C**) Upon return to baseline after media wash. (**D**) After exposure to a depolarizing stimulus of 60 mM KCl. The neurons displayed rapid, reversible, >300% elevations in cytosolic calcium in response to K$^+$, consistent with the activity of neuronal voltage-gated calcium channels. Whole cell patch-clamp revealed voltage-gated sodium currents in P/Ta1:hGFP$^+$ dentate neurons. (**E**) Representative cell 14 d after P/Tα1:hGFP-based FACS. Identified visually as a progenitor-derived neuron on the basis of its residual GFP expression, the cell was patch-clamped in a voltage-clamped configuration and its responses to current injection recorded. (**F**) Fast negative deflections noted after depolarized voltage steps are typical of the voltage-gated sodium currents of mature neurons. Scale bar = 50 μm. (Reproduced with permission from ref. *21*.)

oligodendrocytic CNP protein; $66.3 \pm 6.8\%$ were O4$^+$ (Fig. 11) Of these, most matured as galactocerebroside$^+$ oligodendrocytes under our culture conditions *(20)*.

By cytometry based on P/CNP2-driven GFP, these oligodendrocyte progenitor cells were not rare, comprising over 0.4% of the sorted white matter cell pool *(20)*. In fact, correcting for our average plasmid transfection efficiency of roughly 10%, we would estimate that roughly 4% of successfully dissociated white matter cells might be competent progenitors, as defined by P/CNP2:hGFP.

Importantly, some development of non-oligodendrocytic phenotypes was also noted in the sorted P/CNP2:hGFP$^+$ cultures: By 4 d after FACS, $6.5 \pm 5.4\%$ of the sorted cells expressed GFAP, and $11 \pm 5\%$ were GFAP$^+$ by 3 wk in vitro. Even more remarkably, $7.7 \pm 4.4\%$ of P/CNP2:hGFP-sorted cells matured into Hu or TuJ1$^+$ neurons in the week after FACS separation, even though no neurons whatsoever—as defined by Hu and/or TuJ1/βIII-tubulin-immunoreactivity—had been observed either prior to FACS, or at the same timepoint in unsorted control cultures. Both the P/CNP2:hGFP$^+$/ TuJ1$^+$/$\pm$Hu$^+$ neurons and the P/CNP2:hGFP$^+$/GFAP$^+$ astrocytes were confirmed as true positives, in that they continued to express P/CNP2-driven GFP. These findings suggested that CNP2-defined white matter progenitors, despite their apparent commitment to become oligodendrocytes in native cultures of adult white matter, were able to generate neurons and astrocytes as well, once sorted to low-density culture apart from other cell types. Thus, these cells appeared to retain multilineage potential (Fig. 12) and as such may comprise a pool of human parenchymal stem cells, perhaps akin to those previously harvested from FGF-2-treated rat cortical parenchyma *(40,50a)* and optic nerve *(96)*. These results argue strongly that the local environment of the adult human brain may serve to restrict the phenotypic potential of its residual progenitors, and that the removal of these cells from the paracrine influences of the local environment may permit them to revert to less restricted programs of differentiation.

CONCLUSIONS

Neural precursor cells persist within subependymal and dentate granule cell regions of the adult human forebrain and include distinct populations of neuronal and uncommitted precursors. These may be identified, separated, and enriched from the adult brain, based on their expression of fluorescent transgenes driven by cell-specific promoters. At first glance, these appear to represent distinct cell populations. However, the broadened lineage potential of the adult white matter progenitor cells in the setting of low-density, pure culture suggests that each of these progenitor phenotypes may in fact be relatively plastic in its autonomous lineage potential and that phenotype and fate

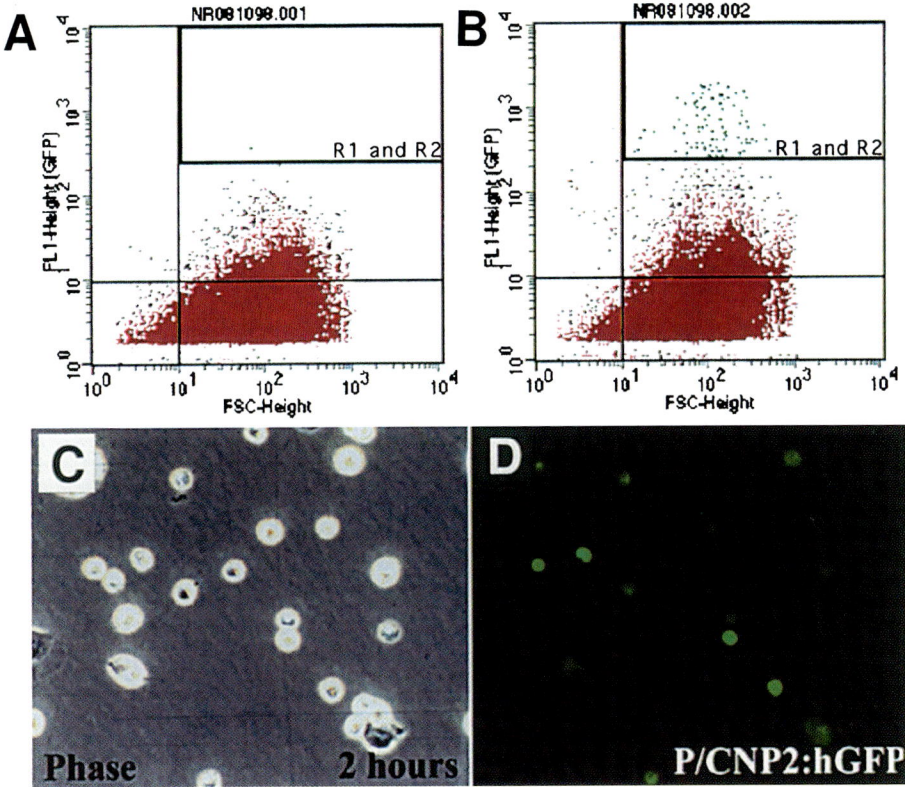

Fig. 10. A distinct pool of mitotically competent oligodendrocyte progenitor cells may be isolated from adult human white matter. (**A,B**) Representative sort of a human white matter sample, derived from the frontal lobe of a 42-yr-old woman during repair of an intracranial aneurysm. This plot shows 50,000 cells (sorting events) with their GFP fluorescence intensity (FL1), plotted against their forward scatter (FSC, a measure of cell size). (**A**) Plot obtained from a nonfluorescent P/hCNP2:*lac*Z-transfected control. (**B**) Corresponding result from a matched culture transfected with P/hCNP2:hGFP. The boxed area (R1+R2) includes those P/hCNP2:hGFP⁺ cells recognized and separated on the basis of their fluorescence emission. The many cells thereby recognized in the P/hCNP2:hGFP-transfected sample (**B**) contrasts to the rare cells so identified in the nonfluorescent P/hCNP2:*lac*Z-transfected control (**A**). (**C,D**) Phase and fluorescence images of P/hCNP2:hGFP⁺ cells 2 h after sorting. Scale bar = 20 μm. (Reproduced with permission from ref. *20*.)

may rather be dictated by local environmental constraints to which each cell must necessarily succumb. Our ability to harvest these cells opens new possibilities for their functional analysis, for designing strategies intended to support

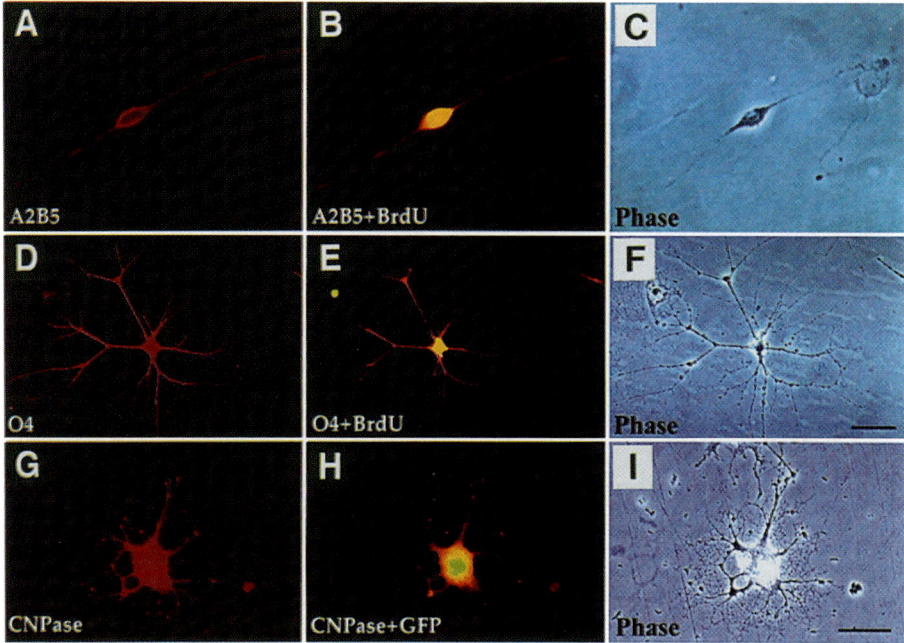

Fig. 11. Isolated P/CNP2:hGFP⁺ subcortical progenitor cells divide to give rise to oligodendrocytes. Dissociates of the adult capsular white matter harbored a population of cells that could be recognized by P/CNP2:hGFP and that were able to divide to give rise to oligodendrocytes. These cells could be selected and isolated on the basis of P/CNP2:hGFP-based FACS; the sorted cells continued to divide after FACS, and their daughters expressed oligodendrocytic markers. (A–C) A bipolar A2B5⁺/ BrdU⁺ cell, 48 h after FACS. (D–F) Within 3 wk the initially bipolar cells matured as fibrous, O4⁺ cells. These cells incorporated BrdU in vitro, indicating their origin from replicating A2B5⁺ cells. (G–I) A multipolar oligodendrocyte expressing CNPase protein, still actively expressing GFP fluorescence 3 wk after P/CNP2:hGFP-based FACS. Scale bars = 20 μm. (Reproduced with permission from ref. 20.)

these cells in the setting of injury and disease, and for formulating approaches intended to enhance their endogenous proliferation and recruitment.

Having established the persistence and abundance of mitotically competent neural progenitor cells in the adult human CNS, our goal is to utilize these cells for neuronal and oligodendrocytic replacement in the damaged CNS. Axiomatically, this may be achieved by activating endogenous neural progenitors (16b,89,90,97), or by implanting exogenous precursors (55,56,91–93); both are feasible therapeutic options for functional repair. To utilize adult neural progenitors in clinical therapeutics, we need to establish whether different adult progenitor populations are lineage restricted or

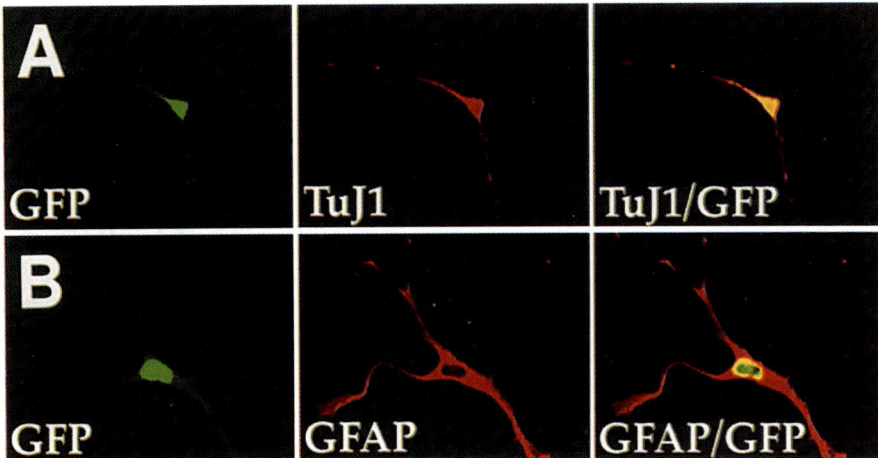

Fig. 12. White matter precursor cells may constitute a pool of multipotential progenitors. By a week after FACS, some P/hCNP2:hGFP-sorted cells were noted to mature into either TuJ1[+] neurons (**A**) or GFAP[+] astrocytes (**B**). Both the TuJ1[+] (*red* in **A**) and GFAP[+] (*red* in **B**) cells were confirmed visually as expressing P/hCNP2:hGFP (*green*). No such neuronal differentiation of CNP2:hGFP-identified cells was ever noted in unsorted plates, within which these cells generally matured as oligodendrocytes, and much less so as astrocytes. This suggests that P/hCNP2-defined progenitors may retain some degree of multilineage potential, which may be selectively exercised in the low-density, homogeneous environment of a sorted cell pool, in which paracrine influences upon differentiation are minimized. (Reproduced with permission from ref. *20*.)

not, whether some or all are innately multipotent, the extent to which they may self-renew, the likelihood of their phenotypic degradation or transformation, and how different regionally defined progenitor phenotypes might differ from one another in their migration competence and humoral regulation. Only by so understanding the biology of adult human neuronal progenitors will we be able to establish a rational and informed foundation for their therapeutic use. Once at that point, we may then begin to consider how best to train newly generated networks of neurons and glia to assume lost functions, in diseases as diverse as stroke, trauma, demyelinating disease, and the degenerative dementias.

ACKNOWLEDGMENTS

I would like to thank my collaborators in the studies reviewed here, in particular Drs. Neeta Singh Roy, Abdellatif Benraiss, Su Wang, Hong Wu, David Pincus, H. Michael Keyoung, Richard A. R. Fraser, Robert Goodman, William T. Couldwell, Michael Gravel, Peter Braun, William K. Rashbaum,

Hideyuki Okano, and Maiken Nedergaard. This work was supported by NINDS grants R01NS33106, R01NS29813, and R01NS39559, and by grants from the Mathers Charitable Foundation, Project ALS, the Heredity Disease Foundation, the Human Frontiers Scientific Program, and the National Multiple Sclerosis Society.

REFERENCES

1. Goldman, S. (1998) Adult neurogenesis: From canaries to the clinic. *J. Neurobiol.* **36,** 267–286.
2. Goldman, S. A. and Luskin, M. B. (1998) Strategies utilized by migrating neurons of the postnatal vertebrate forebrain. *Trends Neurosci.* **21,** 107–114.
3. Goldman, S. A. and Nottebohm, F. (1983) Neuronal production, migration, and differentiation in a vocal control nucleus of the adult female canary brain. Proc. Natl. Acad. Sci. **80,** 2390–2394.
4. Goldman, S. A., Zukhar, A., Barami, K., Mikawa, T., and Niedzwiecki, D. (1996) Ependymal/subependymal cells of the postnatal and adult songbird brain generate both neurons and non-neuronal siblings, *in vitro* and *in vivo*. *J. Neurobiol.* **30,** 505–520.
5. Kirschenbaum, B., Nedergaard, M., Preuss, A., Barami, K., Fraser, R. A., and Goldman, S. A. (1994) In vitro neuronal production and differentiation by precursor cells derived from the adult human forebrain. *Cereb. Cortex* **4,** 576–589.
6. Lois, C. and Alvarez-Buylla, A. (1993) Proliferating subventricular zone cells in the adult mammalian forebrain can differentiate into neurons and glia. *Proc. Natl. Acad. Sci. USA* **90,** 2074–2077.
7. Luskin, M. B. (1993) Restricted proliferation and migration of postnatally generated neurons derived from the forebrain subventricular zone. *Neuron* **11,** 173–189.
8. Chiasson, B., Tropepe, V., Morshead, C., and van der Kooy, D. (1999) Adult mammalian forebrain ependymal and subependymal cells demonstrate proliferative potential, but only subependymal cells have neural stem cell characteristics. *J. Neurosci.* **19,** 4462–4471.
9. Doetsch, F., Caille, I., Lim, D., Garcia-Verdugo, J., and Alvarez-Buylla, A. (1999) Subventricular zone astrocytes are neural stem cells in the adult mammalian brain. *Cell* **97,** 703–716.
10. Johansson, C., Momma, S., Clarke, D., Risling, M., Lendahl, U., and Frisen, J. (1999) Identification of a neural stem cell in the adult mammalian central nervous system. *Cell* **96,** 25–34.
11. Doetsch, F. and Alvarez-Buylla, A. (1996) Network of tangential pathways for neuronal migration in the adult mammalian brain. *Proc. Natl. Acad. Sci. USA* **93,** 14,895–14,900.
12. Kirschenbaum, B. and Goldman, S. A. (1995) Brain-derived neurotrophic factor promotes the survival of neurons arising from the adult rat forebrain subependymal zone. *Proc. Natl. Acad. Sci. USA* **92,** 210–214.
13. Weiss, S., Dunne, C., Hewson, J., Wohl, C., Wheatley, M., Peterson, A. C., and Reynolds, B. A. (1996) Multipotent CNS stem cells are present in the adult mammalian spinal cord and ventricular neuroaxis. *J. Neurosci.* **16,** 7599–7609.

14. Goldman, S., Kirschenbaum, B., Harrison-Restelli, C., and Thaler, H. (1997) Neuronal precursor cells of the adult rat ventricular zone persist into senescence, with no change in spatial extent or BDNF response. *J. Neurobiol.* **32,** 554–566.

15. Kuhn, G., Dickinson-Anson, H., and Gage, F. (1996) Neurogenesis in the dentate gyrus of the adult rat: Age related decrease of neuronal progenitor proliferation. *J. Neurosci.* **16,** 2027–2033.

16. Morshead, C. M., Reynolds, B. A., Craig, C. G., McBurney, M. W., Staines, W. A., Morassutti, D., et al. (1994) Neural stem cells in the adult mammalian forebrain: a relatively quiescent subpopulation of subependymal cells. *Neuron* **13,** 1071–1082.

16a. Morshead, C. and van der Kooy, D. (1992) Postmitotic death is the fate of constitutively proliferating cells in the subependymal layer of the adult mouse brain. *J Neurosci* **12,** 249–256.

16b. Craig, C. G., Tropepe, V., Morshead, C. M., Reynolds, B. A., Weiss, S., and van der Kooy, D. (1996) In vivo growth factor expansion of endogenous subependymal neural precursor cell populations in the adult mouse brain. *J. Neurosci.* **16,** 2649–2658.

17. Reynolds, B. A. and Weiss, S. (1992) Generation of neurons and astrocytes from isolated cells of the adult mammalian central nervous system. *Science* **255,** 1707–1710.

18. Richards, L. J., Kilpatrick, T. J., and Bartlett, P. F. (1992) De novo generation of neuronal cells from the adult mouse brain. *Proc. Natl. Acad. Sci. USA* **89,** 8591–8595.

19. Roy, N., Benraiss, A., Wang, S., Fraser, R. A. R., Goodman, R., Couldwell, W. T., et al. (2000) Promoter-targeted selection and isolation of neural progenitor cells from the adult human ventricular zone. *J. Neurosci. Res.* **59,** 321–331.

20. Roy, N., Wang, S., Benraiss, A., Fraser, R., Gravel, P., Braun, P., and Goldman, S. (1999) Identification, isolation and enrichment of oligodendrocyte progenitor cells from the adult human subcortical white matter. *J. Neurosci.* **19,** 9986–9995.

21. Roy, N., Wang, S., Jiang, L., Kang, J., Restelli, C., Fraser, R., et al. (2000) In vitro neurogenesis by neural progenitor cells isolated from the adult human hippocampus. *Nature Med.* **6,** 271–277.

22. Gould, E., Beylin, A., Tanapat, P., Reeves, A., and Shors, T. (1999) Learning enhances adult neurogenesis in the adult hippocampal formation. *Nature Neurosci.* **2,** 260–265.

23. Kaplan, M. (1983) Proliferation of subependymal cells in the adult primate CNS: differential uptake of thymidine by DNA-labeled precursors. *J. Hirnforsch.* **23,** 23–33.

24. Altman, J. and Das, G. D. (1966) Autoradiographic and histological studies of postnatal neurogenesis. I. A longitudinal investigation of the kinetics, migration and transformation of cells incorporating tritiated thymidine in neonate rats, with special reference to postnatal neurogenesis in some brain regions. *J. Comp. Neurol.* **127,** 337–390.

25. Smart, I. (1961) The subependymal layer of the mouse brain and its cell production as shown by radioautography after thymidine injection. *J. Comp. Neurol.* **116,** 325–347.

26. Lois, C. and Alvarez-Buylla, A. (1994) Long-distance neuronal migration in the adult mammalian brain. *Science* **264,** 1145–1148.

27. Pincus, D., Goodman, R., Fraser, R., Nedergaard, M., and Goldman, S. (1998) Neural stem and progenitor cells: a strategy for gene therapy and brain repair. *Neurosurgery* **42,** 858–868.

28. Pincus, D. W., Harrison, C., Goodman, R. R., Edgar, M., Keyoung, H., Fraser, R. A., et al. (1998) FGF2/BDNF-associated maturation of new neurons generated from adult human subependymal cells. *Ann. Neurol.* **43,** 576–585.

29. Frederiksen, K. and McKay, R. D. (1988) Proliferation and differentiation of rat neuroepithelial precursor cells in vivo. *J. Neurosci.* **8,** 1144–1151.

30. Clark, S., Shetty, A., Bradley, J., and Turner, D. (1994) Reactive astrocytes express the embryonic intermediate neurofilament nestin. *Neuroreport* **5,** 1885–1888.

31. Sakakibara, S., Imai, T., Hamaguchi, K., Okabe, M., Aruga, J., Nakajima, K., et al. (1996) Mouse-Musashi-1, a neural RNA-binding protein highly enriched in the mammalian CNS stem cell. *Dev. Biol.* **176,** 230–242.

32. Nakamura, M., Okano, H., Blendy, J., and Montell, C. (1994) Musashi, a neural RNA-binding protein required for Drosophila adult external sensory organ development. *Neuron* **13,** 67–81.

33. Richter, K., Good, P., and Dawid, I. (1990) A developmentally regulated, nervous system specific gene in *Xenopus* encodes a putative RNA-binding protein. *New Biol.* **2,** 556–565.

34. Sakakibara, S. and Okano, H. (1997) Expression of neural RNA-binding proteins in the postnatal CNS: implications of their roles in neuronal and glial cell development. *J. Neurosci.* **17,** 8300–8312.

35. Szabo, A., Dalmau, J., Manley, G., Rosenfeld, M., Posner, J., and Furneaux, H. (1991) HuD, a paraneoplastic encephalomyelitis antigen, contains RNA binding domains and is homologous to elav and sex-lethal. *Cell* **67,** 325–333.

36. Barami, K., Iversen, K., Furneaux, H., and Goldman, S. A. (1995) Hu protein as an early marker of neuronal phenotypic differentiation by subependymal zone cells of the adult songbird forebrain. *J. Neurobiol.* **28,** 82–101.

37. Marusich, M., Furneaux, H., Henion, P., and Weston, J. (1994) Hu neuronal proteins are expressed in proliferating neurogenic cells. *J. Neurobiol.* **25,** 143–155.

38. Marusich, M. and Weston, J. (1992) Identification of early neurogenic cells in the neural crest lineage. *Dev. Biol.* **149,** 295–306.

39. Gritti, A., Parati, E. A., Cova, L., Frolichsthal, P., Galli, R., Wanke, E., et al. (1996) Multipotential stem cells from the adult mouse brain proliferate and self-renew in response to basic fibroblast growth factor. *J. Neurosci.* **16,** 1091–1100.

40. Palmer, T. D., Ray, J., and Gage, F. H. (1995) FGF-2-responsive neuronal progenitors reside in proliferative and quiescent regions of the adult rodent brain. *Mol. Cell Neurosci.* **6,** 474–486.

41. Ahmed, S., Reynolds, B. A., and Weiss, S. (1995) BDNF enhances the differentiation but not the survival of CNS stem cell-derived neuronal precursors. *J. Neurosci.* **15,** 5765–5778.

42. Reynolds, B. A., Tetzlaff, W., and Weiss, S. (1992) A multipotent EGF-responsive striatal embryonic progenitor cell produces neurons and astrocytes. *J. Neurosci.* **12,** 4565–4574.

43. Vescovi, A. L., Reynolds, B. A., Fraser, D. D., and Weiss, S. (1993) bFGF regulates the proliferative fate of unipotent (neuronal) and bipotent (neuronal/ astroglial) EGF-generated CNS progenitor cells. *Neuron* **11,** 951–966.

44. Carpenter, M., Cui, X., Hu, Z., Jackson, J., Sherman, S., Seiger, A., and Wahlberg, L. (1999) In vitro expansion of a multipotent population of human neural progenitor cells. *Exp. Neurol.* **158,** 265–278.

45. Svendsen, C., Caldwell, M., Shen, J., ter Borg, M., Rosser, A., Tyers, P., et al. (1997) Long-term survival of human central nervous system progenitor cells transplanted into a rat model of Parkinson's disease. *Exp. Neurol.* **148,** 135–146.

46. Vescovi, A., Parati, E., Gritti, A., Polin, P., Ferrario, M., Wanke, E., et al. (1999) Isolation and cloning of multipotential stem cells from the embryonic human CNS and establishment of transplantable human stem cells lines by epigenetic stimulation. *Exp. Neurol.* **156,** 71–83.

47. Kukekov, V., Laywell, E., Susdlov, O., Davies, K., Scheffler, B., Thomas, B., et al. (1999) Multipotent stem/progenitor cells with similar properties arise from two neurogenic regions of adult human brain. *Exp. Neurol.* **156,** 333–344.

48. Jiang, W., McMurtry, J., Niedzwiecki, D., and Goldman, S. A. (1998) Insulin-like growth factor-1 is a radial cell-associated neurotrophin that promotes neuronal recruitment from the adult songbird ventricular zone. *J. Neurobiol.* **36,** 1–15.

49. Johe, K. K., Hazel, T. G., Muller, T., Dugich-Djordjevic, M. M., and McKay, R. D. (1996) Single factors direct the differentiation of stem cells from the fetal and adult central nervous system. *Genes Dev.* **10,** 3129–3140.

50. Marmur, R., Kessler, J., Zhu, G., Gokhan, S., and Mehler, M. (1998) Differentiation of oligodendroglial progenitors derived from cortical multipotential cells requires extrinsic signals including activation of gp130/LIF receptors. *J. Neurosci.* **18,**

50a. Marmur, R., Mabie, P., Gokhan, S., Song, Q., Kessler, J., and Mehler, M. (1998) Isolation and developmental characterization of cerebral cortical multipotent progenitors. *Dev. Biol.* **204,**

51. Rao, M. and Mayer-Proschel, M. (1997) Glial-restricted precursors are derived from multipotential neuroepithelial stem cells. *Dev. Biol.* **188,** 48–63.

52. Wagner, J., Akerud, P., Castro, D., Holm, P., Canals, J., Snyder, E., Perlmann, T., and Arenas, E. (1999) Induction of a midbrain dopaminergic phenotype in Nurr1-overexpressing neural stem cells by type 1 astrocytes. *Nature Biotechnol.* **17,** 653–659.

53. Weiss, S., Reynolds, B. A., Vescovi, A. L., Morshead, C., Craig, C. G., and van der Kooy, D. (1996) Is there a neural stem cell in the mammalian forebrain? *Trends Neurosci.* **19,** 387–393.

54. Brustle, O., Choudhary, K., Karram, K., Huttner, A., Murray, K., Dubois-Dalcq, M., and McKay, R. (1998) Chimeric brains generated by intraventrciualr transplantation of fetal human brain cells into embryonic rats. *Nature Biotech.* **16,** 1040–1044.

55. Fricker, R., Carpenter, M., Winkler, C., Greco, C., Gates, M., and Bjorklund, A. (1999) Site-specific migration and neuronal differentiation of human neural progenitor cells after transplantation in the adult rat brain. *J. Neurosci.* **19,** 5990–6005.

56. Flax, J., Aurora, S., Yang, C., Simonin, C., Willis, A., Sidman, R., et al. (1998) Engraftable human neural stem cells respond to developmental cues, replace neurons, and express foreign genes. *Nature Biotech.* **16,** 1033–1039.

57. Goldman, S. A. and Nedergaard, M. (1992) Newly generated neurons of the adult songbird brain become functionally active in long-term culture. *Dev. Brain Res.* **68,** 217–223.

58. Chalfie, M., Tu, Y., Euskirchen, G., Ward, W., and Prasher, D. (1994) Green fluorescent protein as a marker for gene expression. *Science* **263,** 802–805.

59. Cheng, L., Fu, J., Tsukamoto, A., and Hawley, R. (1996) Use of green fluorescent protein variants to monitor gene transfer and expression in mammalian cells. *Nature Biotech* **14,** 606–609.

60. Levy, J., Muldoon, R., Zolotukhin, S., and Link, C. (1996) Retroviral transfer and expression of a humanized, red-shifted green flourescent protein gene into human tumor cells. *Nature Biotech.* **14,** 610–614.

61. Gloster, A., Wu, W., Speelman, A., Weiss, S., Causing, C., Pozniak, C., et al. (1994) The Tα1 α-tubulin promoter specifies gene expression as a function of neuronal growth and regeneration in transgenic mice. *J. Neurosci.* **14,** 7319–7330.

62. Miller, F., Naus, C., Durand, M., Bloom, F., and Milner, R. (1987) Isotypes of α-tubulin are differentially regulated during neuronal maturation. *J. Cell Biol.* **105,** 3065–3073.

63. Goldman, S., Nedergaard, M., Harrison-Restelli, C., Jiang, W., Keyoung, H. M., Leventhal, C., et al. (1997) Neural precursors and neuronal production in the adult mammalian forebrain. *Ann. N.Y. Acad. Sci.* **835,** 30–55.

64. Wang, S., Wu, H., Jiang, W., Isdell, F., Delohery, T., and Goldman, S. A. (1998) Identification and enrichment of forebrain neuronal precursor cells by fluorescence-activated sorting of ventricular zone cells transfected with GFP regulated by the Tα1 tubulin promoter. *Nature Biotechnol.* **16,** 196–201.

65. Wang, S., Roy, N., Benraiss, A., Harrison-Restelli, C., and Goldman, S. (2000) Promoter-based isolation and purification of mitotic neuronal progenitor cells from the adult mammalian ventricular zone. *Dev. Neurosci.* **22,** 167–176.

66. Zimmerman, L., Lendahl, U., Cunningham, M., McKay, R., Parr, B., Gavin, B., Vassileva, G., and McMahon, A. (1994) Independent regulatory elements in the nestin gene direct transgene expression to neural stem cells and muscle precursors. *Neuron* **12,** 11–24.

67. Lothian, C. and Lendahl, U. (1997) An evolutionarily conserved region in the second intron of the human nestin gene directs gene expression to CNS progenitor cells and to early neural crest cells. *Eur. J. Neurosci.* **9,** 452–462.

68. Rossant, J., Zirngibl, R., Cado, D., Shago, M., and Giguere, V. (1991) Expression of a retinoic acid response element-hsp/lacZ transgene defines specific domains of transcriptional activity during mouse embryogenesis. *Genes Dev.* **5,** 1333–1344.

69. Altman, J. and Das, G. D. (1965) Autoradiographic and histological evidence of postnatal hippocampal neurogenesis in rats. *J. Comp. Neurol.* **124,** 319–335.

70. Barnea, A. and Nottebohm, F. (1994) Seasonal recruitment of hippocampal neurons in adult free-ranging black-capped chickadees. *Proc. Natl. Acad. Sci. USA* **91,** 11,217–11,221.

71. Bayer, S., Yackel, J., and Puri, P. (1982) Neurons in the rat dentate gyrus granular layer substantially increase during juvenile and adult life. *Science* **216,** 890–892.

72. Eriksson, P., Perfilieva, E., Bjork-Eriksson, Alborn, A., Nordberg, C., Peterson, D., and Gage, F. (1998) Neurogenesis in the adult human hippocampus. *Nature Med.* **4,** 1313–1317.

73. Gould, E., McEwen, B., Tanapat, P., Galea, L., and Fuchs, E. (1997) Neurogenesis in the dentate gyrus of the adult tree shrew is regulated by psychosocial stress and NMDA receptor activation. *J. Neurosci.* **17,** 2492–2498.

74. Gould, E., Tanapat, P., McEwen, B., Flugge, G., and Fuchs, E. (1998) Proliferation of granule cell precursors in the dentate gyrus of adult monkeys is diminished by stress. *Proc. Natl. Acad. Sci. USA* **95,** 3168–3171.

75. Kaplan, M. S. and Hinds, J. W. (1977) Neurogenesis in the adult rat: electron microscopic analysis of light radioautographs. *Science* **197,** 1092–1094.

76. Gould, E., Cameron, H., Daniels, D., Wooley, C., and McEwen, B. (1992) Adrenal hormones suppress cell division in the adult rat dentate gyrus. *J. Neurosci.* **12,** 3642–3650.

77. Kempermann, G., Kuhn, H., and Gage, F. (1997) More hippocampal neurons in adult mice living in an enriched environment. *Nature* **386,** 493–495.

78. van Praag, H., Kempermann, G., and Gage, F. (1999) Running increases cell proliferation and neurogenesis in the adult mouse dentate gyrus. *Nature Neurosci.* **2,** 266–270.

79. Gould, E., Reeves, A., Graziano, M., and Gross, C. (1999) Neurogenesis in the neocortex of adult primates. *Science* **286,** 548–552.

80. Kornack, D. and Rakic, P. (1999) Continuation of neurogenesis in the hippocampus of the adult macaque monkey. *Proc. Natl. Acad. Sci. USA* **96,** 5768–5773.

81. Shankle, W., Landing, B., Rafii, M., Schiano, A., Chen, J., and Hara, J. (1998) Evidence for a postnatal doubling of neuron number in the developing human cerebral cortex between 15 months and 6 years. *J. Theor. Biol.* **191,** 115–140.

82. Shankle, W., Rafii, M., Landing, B., and Fallon, J. (1999) Approximate doubling of numbers of neurons in postnatal human cerebral cortex and in 35 specific cytoarchitectural areas from birth to 72 months. *Pediatr. Dev. Pathol.* **2,** 244–259.

83. Palmer, T., Takahashi, J., and Gage, F. (1997) The adult rat hippocampus contains primordial neural stem cells. *Molec. Cell Neurosci.* **8,** 389–404.

84. Suhonen, J., Peterson, D., Ray, J., and Gage, F. (1996) Differentiation of adult hippocampus-derived progenitors into olfactory neurons in vivo. *Nature* **383,** 624–627.

85. Wolswijk, G. and Noble, M. (1989) Identification of an adult-specific glial progenitor cell. *Development* **105,** 387–400.

86. Gensert, J. M. and Goldman, J. E. (1997) Endogenous progenitors remyelinate demyelinated axons in the adult CNS. *Neuron* **19,** 197–203.

87. Gensert, J. M. and Goldman, J. E. (1996) In vivo characterization of endogenous proliferating cells in adult rat subcortical white matter. *Glia* **17,** 39–51.

88. Noble, M. (1997) The oligodendrocyte-type 2 astrocyte lineage: In vitro and in vivo studies on development, tissue repair and neoplasia, in *Isolation, Characterization and Utilization of CNS Stem Cells* (Gage, Y. C. F., ed.), Springer-Verlag, Berlin, pp. 101–128.

89. Benraiss, A., Lerner, K., Cherminiecki, E., Roh, D., Hackett, N., Crystal, R., and Goldman, S. (1999) In vivo transduction of the adult rat ventricular zone with an adenoviral BDNF vector substantially increases neurogenesis and neuronal recruitment to the rat olfactory bulb. *Soc. Neurosci. Abstr.* **25,** 1028.

90. Zigova, T., Peneca, V., Wiegand, S., and Luskin, M. (1998) Intraventricular administration of BDNF increases the number of newly generated neurons in the adult olfactory bulb. *Molec. Cell. Neurosci.* **11,** 234–245.

91. Brustle, O., Jones, K., Learish, R., Karram, K., Choudhary, K., Wiestler, O., et al. (1999) Embryonic stem cell derived glial precursors: a source of myelinating transplants. *Science* **285,** 754–756.

92. Zhang, S.-C., Ge, B., and Duncan, I. (1999) Adult brain retains the potential to generate oligodendroglial progenitors with extensive myelination capacity. *Proc. Natl. Acad. Sci. USA* **96,** 4089–4094.

93. Zigova, T., Betarbet, R., Minnen, M., Bakay, R., Wiegand, S., Lindsay, R., and Luskin, M. B. (1997) Transplantation of neuronal progenitor cells from the neonatal subventricular zone into the neonatal and adult striatum of the rat. *Abstracts of the 6th International Neural Transplantation Meeting.*

94. Gogate, N., Verma, L. Zhou, J., et al. (1994) Plasticity in the adult human oligodendrocyte lineage. *J. Neurosci.* **14,** 4571–4587.

95. Armstrong, R., Dorn, H., Kufta, C., Fredman, E., and Dubois-Dalcq, M. (1992) Pre-oliodendrocytes from adult human CNS. *J. Neurosci.* **12,** 1538–1547.

96. Kondo, T. and Raff, M. (2000) Oligodendrocyte precursor cells reprogrammed to become multipotent CNS stem cells. *Science* **289,** 1754–1756.

97. Kuhn, H., Winkler, J., Kempermann, G., Thal, L., and Gage, F. (1997) EGF and FGF-2 have different effects on neural progenitors in the adult rat brain. *J. Neurosci.* **17,** 5820–5829.

9

ES Cells and Neurogenesis

John W. McDonald

INTRODUCTION

The scientific potential of embryonic stem (ES) cells was harnessed in 1981, when ES cells were first derived from mice, marking the birth of transgenic animal technology (*1,2*; Table 1). However, the concept of a pluripotent embryonic cell has its origins much earlier. ES cell development evolved from work on mouse teratocarcinomas, tumors that arise in the gonads of a few inbred strains. Classical work on teratocarcinomas established their origins from germ cells and provided the concept of a stem cell, then termed embryonal carcinoma cells or EC cells *(130–132)*. The developmental capacity of EC cells became apparent following successful generation of chimeric mice by blastocyst injection of EC cells. The ability to prodice teratocarcinomas from ectopically transplanted blastocysts lead to the reasoning and demonstration the pluripotent cells could be derived directly from blastocysts *(1,2)*. The transgenic revolution dramatically impacted on neurosciences in the 1990s with the development of ES cell technology for studies at the cellular level. Genetic flexibility is at the heart of this new technology: double-allele gene knockins and knockouts, with and without inducible promoter systems, are now readily carried out. Such genetic approaches provide powerful tools for exploring cultured neural cell systems, especially with advances in methods for deriving differentiated neural cells from ES cells in culture. Thus, ES cell-derived cultures can now reliably serve as replacements for primary culture systems. The ES cell system offers many additional advantages that are illustrated and discussed throughout this review.

ES cell-like pluripotent stem cells have been derived from multiple vertebrate species, including mice *(1,2)*, nonhuman primates *(3,4)*, and, most recently, humans *(5–7)*. The ES cell is, developmentally speaking, the earli-

From: *Stem Cells and CNS Development*
Edited by: M. S. Rao © Humana Press Inc., Totowa, NJ

Table 1
Development of the ES Cell Field in Neuroscience

Date	Observation	Ref.
1981	**Derivation of ES cells from murine blastocyts (4–6 day embryos).**	1
	Derivation of ES cells from murine preimplantation blastocyst. Subcutaneous injection of ES cells into athymic mice produced teratomas containing three germ layers when examined 6 wk later. Culturing whole EBs produced some neuron-like cells over a period of 6 wk.	2
1988	**Definitive demonstration of neuronal differentiation from murine ES cells in culture.** NGF accelerated the generation of neurons (identified by silver staining) from ES cells plated as whole EBs in culture.	41
1992	**Methods developed for isolating and culturing murine ES cells from PGCs.**	12,112
1994	**Early example of the strength of the in vitro ES cell system for analyzing targeted mutations.** By using in vitro differentiation, this study identified the defect in GATA-1 ES cells as a block late in the erythroid lineage. GATA-1 (–/–) mice fail to generate mature erythroid cells but do generate white blood cells. Therefore, the in vitro system was key in determining the stage-specific role of the GATA-1 defect.	113
	Demonstration of the strength of the in vitro ES cell neural differentiation system by applying double-allele gene inactivation and subsequent "rescue" of the mutant cells. Homologous recombination was used to both inactivate T3R-α alleles in murine ES cells (to evaluate the role of unliganded T3R-α in early development and on RA-stimulated neural development). In T3R-α (–/–) ES cells, there was increased basal and RA-induced expression of the endogenous RA-responsive genes, RAR-β and alkaline phosphatase. This effect was rescued by co-transfection of T3R-α1 but not by co-transfection of the T3R-α variant c-erbAα2.	114
1995	**Derivation of ES cell-like cells from a nonhuman primate (*Rhesus*, Old World species).** When transplanted into leg muscle or testis of SCID mice, the cells formed teratomas were formed with cellular representation of all three germ layers (examined 8–15 wk post transplantation).	3
1996	**Transplantation of murine ES cells into the CNS.** Neurons developed after transplantation of RA-induced ES cells into the quinolinic acid-lesioned striatum.	57
1997	**Demonstration of the utility of double-allele gene knockouts in ES cells for the study of cell biology.** ES cells with double-allele deletions of the FGF-4 gene exhibited altered growth and survival of ES-derived neural cells. Note that embryos were arrested shortly after implantation. FGF-4 (–/–) ES cells formed teratomas with neural components 30, 35, FGF-4 (–/–) 47, and 61 d after subcutaneous transplantation in 129/SvJ mice. Addition of FGF-4 in culture dramatically enhanced survival of ES FGF-4 (–/–)-derived cells, where death rate was greater than that of FGF (+/+) ES cells.	110

Table 1 (cont.)
Development of the ES Cell Field in Neuroscience

Date	Observation	Ref.
1998	**Postmitotic neurons and glia were generated after rhesus ES cells** were transplanted into immunocompromised host muscle (SCID mice) to form forming teratomas. At 5–12 wk posttransplantation, all the teratomas exhibited neural differentiation. Some aspects of normal differentiation are recapitulated in teratomas from rhesus ES cells.	101
	Demonstration of genetic selection of lineage-restricted neural progenitors from ES cells. ES cells were engineered using homologous recombination to contain a bifunctional selection marker/reporter gene β*geo* integrated into the *Sox-2* gene. When induced with the RA (4-/4+ protocol), ~50% of the dissociated cultured cells expressed β-galactosidase activity and *Sox-2* immunoreactivity. Addition of G418 to cultures selected for ~90% expression of β-galactosidase activity and *Sox-2* immunoreactivity. Further differentiation produced mature neurons. Therefore, genetic selection can be applied to select pure populations of NRP cells.	58
1998	**Demonstration of utility of double-allele targeted ES-cell culture system for analyzing effects of specific genes on neuronal development.** GD3S (–/–) ES cells, double-targeted disruptions of GD3 synthase gene, were deficient in GD3S activity, and neurons could still differentiate normally. Therefore, the previous hypothesis that b-series gangliosides are necessary for neuronal differentiation was disproved.	115
	Application of "stem-cell selection" using ES cells engineered to express resistance to antibiotic in their undifferentiated state. The CGR8 ES cell line was transfected with the Oct4neofos construct. The Oct3/4 gene was linked to lacZ/neo (OKO160). Undifferentiated Oct4neofoss11 cells expressed low levels of neomycin phosphotransferase under control of the Oct4 promoter element [stem cell-specific transcription factor Oct4 (71)]. Differentiated progeny did not express the transgene and therefore were killed by the selection agent G418.	70
	Demonstration of the utility of double-allele gene knockouts in ES cells for analyzing cellular mechanisms. Caspase 9 (–/–) ES cells were derived from caspase 9 (–/+) ES cells cultured at an increased concentration of G418. Caspase 9 (–/–) ES cells and ES-derived thymocytes and splenocytes are resistant to apoptosis induced by UV and γ-irradiation. Note that caspase 9 (–/–) was early postnatal lethal (most mice died by PND 3).	111
	Derivation of ES cell-like cells from human blastocyst. Transplanting each of the five cell lines into SCID mice produced teratomas containing all three germ layers and formed rosettes of neuroepithelium and neural ganglia (after 4 mo).	5

(continued)

Table 1 (cont.)
Development of the ES Cell Field in Neuroscience

Date	Observation	Ref.
	Derivation of pluripotent EG cells from human PGCs. Cultures of PGCs produced cells from all three embryonic germ layers, including neuroepithelia.	6
1999	**Demonstration that ES cell transplantation can enhance recovery of lost functions.** Transplanting dissociated 4–/4+ RA-stimulated EBs, but not vehicle medium or neocortical cells from adult mice, into the spinal cord 9 d after contusion injury improved functional hindlimb locomotion. ES cell-derived neural cells integrated, migrated 1 cm, and differentiated into neurons, oligodendrocytes, and astrocytes.	76
		76
2000	**Demonstration of the usefulness of double-allele Smad4 gene inactivation on cellular mechanisms of TGF-β signaling.** Smad4 (–/–) ES cells were produced by targeted disruption and were used to demonstrate a variable requirement for Smad4 in TGF-β signaling. Typically, Smad4 (–/–) mutant mice die before d 7.5 of embryogenesis.	116
	Demonstration that neural progenitors can be isolated from human ES cells and can be induced to generate mature neurons.	7

aOutlined are the key events in the evolution of the ES cell system in the neurosciences. The list is not meant to be all-inclusive, it provides highlights.

Abbreviations: EB, embryoid body; ES, embryonic stem; FGF, fibroblast growth factor; GD3S, GD3 synthase; NGF, nerve growth factor; NRP, neuronal restricted precursor; PGC, primordial germ cells; PND, postnatal day; RA, retinoic acid; RAR, retinoic acid receptor; SCID, severe combined immunodeficiency; TGF-β, transforming growth factor-β; T3R-α, T3 receptor-α; +, positive reactivity; (–/–), double-allele gene inactivation; 4–/4+ RA, induction protocol that exposes EBs to RA only in the last 4 d of an 8-d protocol; UV, ultraviolet.

est form of stem cell and is theoretically capable of producing all subsequent stages of stem cells that are described in this book. It is the only type of stem cell presently available that can be proved to be genetically normal. This can be shown by reinserting an ES cell into a blastocyst, thereby generating a normal chimeric organism *(8,9)*. In vivo, ES cells incorporated into a host blastula will integrate into the germline, contributing to virtually every somatic cell population. ES cells are able produce all cell types that make up an organism, a property termed totipotency. Furthermore, ES cells can replicate, without apparent limit, in culture. Even after 250 replications in culture, ES cells remain totipotent *(10)*. They represent only one known immortal stem cell. Although there are differences across cell lines, the typical doubling time for murine ES cells is approx 14 h. Thus, one T25 flask of ES cells, passaged at a 1:5 ratio every other day, can yield 625 flasks at the end of 1 wk (8 d). The potential for a practically unlimited supply of cells is quickly apparent.

Totipotent Cell Types

A distinction needs to be drawn between ES cells and embryonic germ (EG) cells, two of the earliest pluripotent stem-cell types. ES cells are derived from the inner cell mass of the preimplantation embryo *(1,2)*, whereas EG cells are derived from primordial germ cells (PGCs), the progenitors of germ cells (sperm and ova; *11,12*). Both ES and EG cells are pluripotent, demonstrating germline transmission in chimeric animals *(13,14)*, and they share several cytologic characteristics, such as high levels of intracellular alkaline phosphatase and similar expression of cell-surface glycolipids and glycoproteins *(15–18)*. These characteristics do not make stem cells pluripotent, however. Additional necessary characteristics include normal and stable karyotypes, the ability to be passaged continuously, growth as multicellular colonies, and the capacity to differentiate into the cell types found in all three embryonic germ layers. Ultimately, contribution to all cell lineages in a chimeric animal (including the germline) is the definitive test of pluripotent developmental potential. However, this test is not practically or ethically possible for every species. To date, this level of stringency has only been achieved in the mouse, but stong evidence also exists in the following vertebrates: mice, swine, birds, rats, rabbits, mink, and fish. Table 1 provides a historical timeline for the derivation of ES cells and their development as research tools.

Other types of cell lines exhibit some level of pluripotency, but these are not discussed in this chapter because excellent reviews of such transformed or tumor-derived cells already exist *(19–23)*. Later stage precursor cells derived from the embryonic, postnatal, and adult CNS, although relevant to this chapter, are discussed elsewhere in this volume.

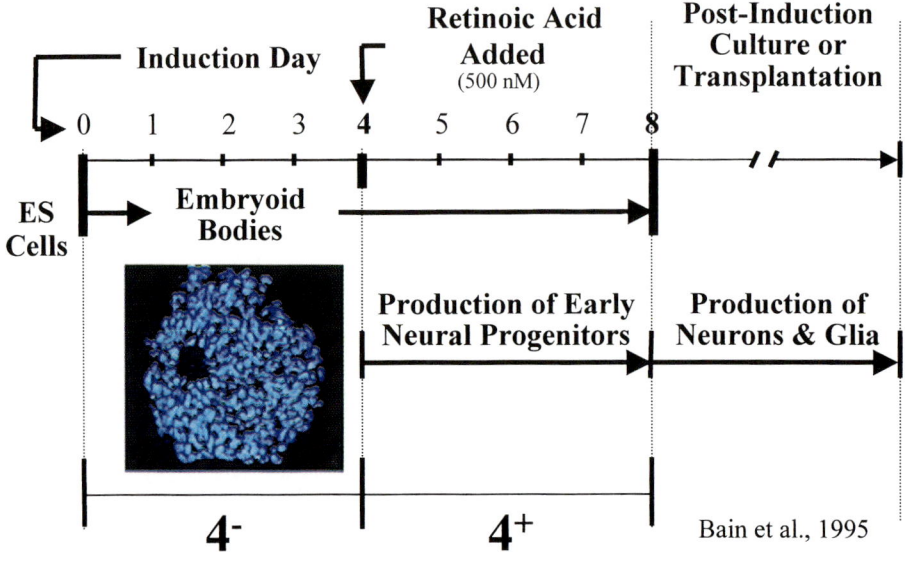

Fig. 1.

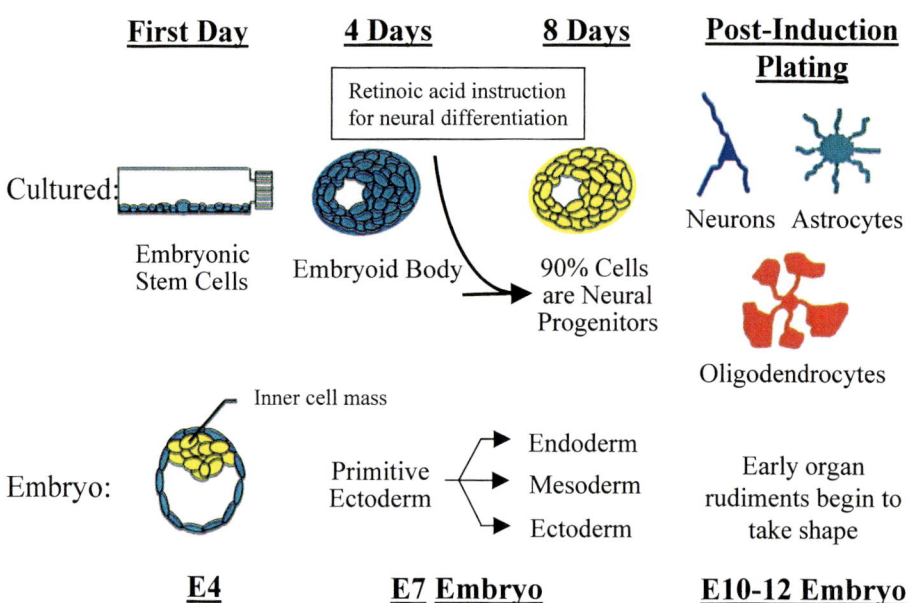

Fig. 2.

THE IN VITRO ES CELL SYSTEM

Mouse ES cells typically are passaged in the presence of leukemia inhibitory factor (LIF) and β-mercaptoethanol — agents that suppress spontaneous differentiation *(24)*. Although murine ES cells respond to the agents, nonhuman primate and human ES cells do not respond to LIF; therefore, it is difficult to restrain their spontaneous differentiation and passage them in an undifferentiated state. Upon withdrawal of LIF and β-mercaptoethanol, murine ES cells, cultured either in suspension in nonadhesive dishes or in "hanging drops," form aggregates containing hundreds of cells (Figs. 1 and 2). These aggregates have been termed embryoid bodies (EBs; Fig. 3). Removal of LIF and production of cytokines by neighboring ES cells in these clusters permit differentiation to occur, which can be confirmed by immunostaining for lineage-specific markers. Early studies have demonstrated that cell types corresponding to multiple body tissues can be derived when murine ES cells are either induced to differentiate or allowed to spontaneously differentiate in culture. These cell types include skeletal myocytes *(25)*, vascular endothelial cells *(26,27)*, hematopoietic lineage cells *(26,28–30)*, and cardiomyocytes *(31,32)*. To date, the greatest progress has occurred in deriving hematopoietic and cardiomyocyte systems (for review see refs. *33* and *34*).

During ES cell differentiation, tissue-specific genes, proteins, ion channels, and receptors are differentially expressed in a pattern closely resembling that observed during normal mouse embryogenesis *(25,34–40)*. Of those listed, the hematopoietic system is best defined, and studies with a number of gene and protein expression markers have shown that the temporal profile of

---

Fig. 1. Schematic comparison of ES-cell differentiation and early embryo development. We begin with the ES cells that are similar to the cells of the 4-d mouse embryo. Day 4 of ES-cell differentiation in culture is similar to the time when the cells of an embryo chose among three different fates: to become endodermal, mesodermal, or ectodermal cells. When cultured ES cells are induced to differentiate into the three principal types of nerve cells, the early rudiments of organs are forming in the human embryo. The recapitulation of development is commonly held to be an important feature of the successful regeneration that we are attempting to achieve after spinal cord injury. (Reproduced with permission from ref. *43*.)

Fig. 2. Schematic description of ES-cell differentiation. The method of differentiation used in our studies was developed by Bain et al. *(43)*. ES cells are grown in clusters called embryoid bodies. The blue inset shows the individual cells of the embryoid body. After 4 d (4–), the cells are exposed to retinoic acid for 4 d. This process, termed the 4 – /4 + protocol, instructs the ES cells to become neural lineage cells. At the 4 – /4 + stage, cells are plated or transplanted.

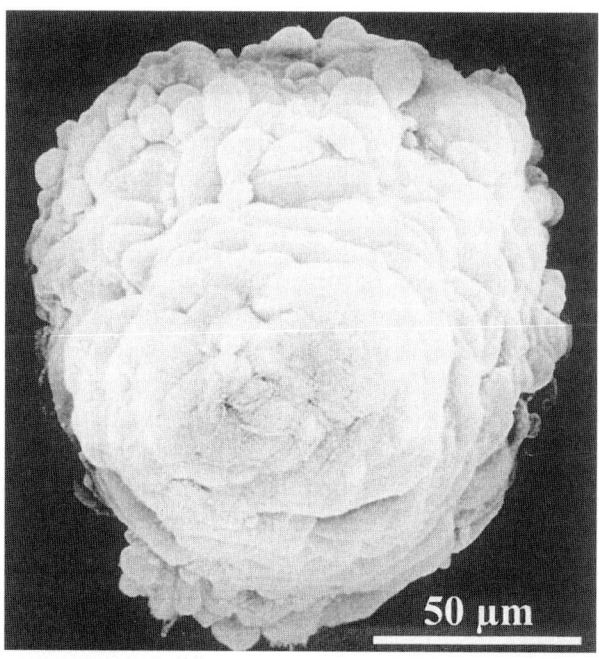

Fig. 3. Scanning EM shows a 4 – /4 + stage EB, characterized as floating clusters of undifferentiated cells. Most EB cells are NEPs, nestin +, and do not express markers of differentiated neural cells. Scale bar = 50 μm.

differentiation from ES cells into blood cells largely correlates with embryogenesis *(34)*. The times of onset and duration of the various stages of hematopoietic differentiation in EBs or cultures derived from EBs are roughly as follows (days of differentiation in a 24-d period): primitive erythroid (d 4–10), definitive erythroid and myeloid (d 5–24), mast cell (d 10–24), and lymphoid (d 14–24). Likewise, the first sign of hematopoiesis in the mouse embryo is the appearance of erythroid cells within the blood islands of the yolk sac (d 7.5 of gestation). These primitive erythroid cells disappear between d 10 and 11 of gestation. Definitive erythroid and myeloid cells first appear in the yolk sac shortly after primitive erythroid cells emerge. Subsequent development of mast-cell and lymphoid-cell lineages occurs in the embryo at times akin to those seen in EB development.

Generally, the in vitro ES-cell system obeys the rules of development that prevail in the normal embryo. In the absence of specific inducers, neural lineage cells represent only a small fraction of the total cells *(39,41)*. However, exposure to cytokines, homeobox gene products, retinoic acid (RA), dimethyl sulfoxide, and a variety of other factors alters the timing of differentiation or restricts or supports it further down particular paths.

ES CELLS CAN GIVE RISE TO ALL THREE PRINCIPAL NEURAL LINEAGES

The neural ES cell field is still young, but recent interest has sparked innovative studies. ES cells can be induced to differentiate into the three principal neural cell types: astrocytes, oligodendrocytes, and neurons (*35,37,41–50*; Table 2; Figs. 4–6). However, methods for isolating enriched populations of these cells were not developed until the late 1990s (see below). Unlike mesodermal cell types, neuronal cell types cannot be produced efficiently from EBs without additional stimuli or selection procedures. For example, RA strongly promotes neural gene expression and represses mesodermal gene expression in cultured mouse EBs *(51)*.

Under these selected conditions, it is possible to derive neuroepithelial precursor cells (NEPs; immunopositive [+] for markers to nestin or epithelial neuronal cell adhesion molecule [E-NCAM]), glial-restricted precursors (GRPs; A2B5 + or NG2 +), neuronal cells (microtubule-associated protein [MAP]-2 +; Fig. 4), astrocytes (glial fibrillary acidic protein [GFAP] +; Fig. 5), and oligodendrocytes (NG2 +, O4+, O1+, galactocerebroside [Gal-C]+, myelin basic protein [MBP] +) (Table 2; Fig. 6). ES cell-derived neurons express proteins or gene products characteristic of primary neurons: those responsible for general neuronal functions (β-tubulin III, neurofilament subunits), neuronal surface markers (NCAM), transmitter synthesizing enzymes (glutamic acid decarboxylase [GAD], tyrosine hydroxylase [TH], choline acetyltransferase [ChAT]), transmitter receptor subunits (glutamate receptor [GluR], γ-aminobutyric acid [GABA]-R), and neurotransmitters (glutamate, GABA). Furthermore, ES cell-derived neurons exhibit electrophysiologic responses (e.g., to glutamate and GABA) and intracellular calcium fluxes similar to those of neurons derived from primary cells (Table 2).

During normal development, motor neurons are among the first neurons to appear. In vertebrates, their appearance requires induction from the notocord and floor plate, which is mediated by sonic hedgehog (Shh), a secreted glycoprotein *(52)*. Shh inhibits Pax7 expression, giving rise to central nervous system (CNS) progenitors of neurons characteristic of the ventral spinal cord. A gradient in Shh activity also regulates Pax6 expression in progenitor cells and influences the identity of developing neurons (e.g., interneurons vs motoneurons; *53,54*). RA also is thought to ventralize neuronal development by promoting generation of neurons characteristic of the ventral spinal cord. This hypothesis comes from the observation that RA may select for hindbrain neural phenotypes because BF-1 (expressed selectively in anterior regions of the CNS) is absent in RA-treated cultures *(55)*.

There is also mounting evidence for the idea that of multiple neuronal phenotypes can develop from ES cells. These include GABAergic, gluta-

Table 2
Published Studies of ES Cell-Derived Neural Cells

Date	Observation	Cell line	RA induction	Neurons	Glia	Ref.
1981	**Growth of whole murine EBs in vitro produced some neuron-like cells over a period of 6 wk.** First derivation of ES cells from murine preimplantation blastocyst. SQ injection of ES cells into athymic mice produced teratomas containing three germ layers when examined 6 wk later.	Murine ICRxSWR/J	No	Neuron-like cells in vitro	NE	2
1988	**First demonstration of definitive neuronal differentiation from murine ES cells in vitro.** NGF accelerated the generation of neurons from ES cells plated as whole EBs in vitro. One day after EB plating, 44% of EBs treated with NGF contained neuron-like cells, compared to only 8% of control EBs. NGF treatment also eliminated undifferentiated ES cells by 8 DIV, whereas ES cells persisted in control cultures even at 9 DIV.	Murine BLC 6 (129/Sv Gat)	No; whole EBs plated	Silver stain +	NE	41
1992	**Neuron-like cells derived from putative bovine ES cell-like cells.** First derivation of bovine ES cell-like cells from blastocysts.	Bovine Two lines	No	Rare neuron-like cells present after spontaneous differentiation	NE	98
	First demonstration of neurons differentiated from mink ES cell-like cells. First derivation of mink ES cells from blastocysts. Subsequent work by the same group has characterized an additional 10 ES cell lines in vitro and in vivo (e.g., teratoma formation 90).	Mink, 10 cell lines: MES1–10	No	Neural cells present in 2/10 lines (MES8 and -9) after DIV NF+	NE	89

Year	Description	Source	Differentiation	Cell type		Ref
1993	**Derivation of putative ES cell-like cells from rabbit preimplantation embryos.** Able to be passed continuously (more than 1 yr), maintain normal karyotype, form EBs, and differentiate into cells characteristic of the three principle germ layers.	Rabbit GM3	No	Neural crest	NE	87
	Differentiation of neuron-like cells from porcine and ovine ES cell-like cells.	Porcine, ovine PICM-8-12, 20	No/yes RA	Neuron-like cells differentiated from ES cells cultured on feeder layers.	NE	117,118
1994	**Development of neurons after in vitro differentiation of porcine ES cells.** First establishment that porcine (*Sus scrofa*) "true" ES cells can be derived from preimplantation embryos and can develop into normal chimeras.	Porcine	Yes RA	Neurons	NE	96
1994	**Demonstration of the strength of the in vitro ES-cell neural differentiation system by applying double-allele gene inactivation and subsequent "rescue" of the mutant cells.** Demonstration that T3R-α gene induction is essential for RA stimulation of neuronal development from murine ES cells. Homologous recombination techniques were used both to inactivate T3R-α gene alleles in murine ES cells (to evaluate the role of unliganded T3R-α in early development) and on RA-stimulated neural development. In T3R-α (−/−) ES cells, increased basal and RA-induced expression of endogenous RA-responsive genes, RAR-β and alkaline phosphatase.	Murine CC1.2	Yes RA D1–5	Neurons NF-M+	NE	114

(continued)

Table 2
Published Studies of ES Cell-Derived Neural Cells

Date	Observation	Cell line	RA induction	Neurons	Glia	Ref.
	This effect was rescued by co-transfection of T3R-α1 but not the T3R-α variant c-$erbA\alpha2$. Therefore, it was concluded that T3R-α inhibits the RA response.					
1995	**Demonstration that RA strongly induces neural cell differentiation in ES cell aggregates.** Neuron-like cells comprised 38% of cells in 4-/4+ RA-induced cultures but only about 0% in 4-/4- cultures (2 DIV after dissociation). **First demonstration of functional neurons derived from murine ES cells in vitro.** First demonstration of ES cell-derived astrocytes in culture. 4-/4+ RA-induced ES cells produced β- tubulin III and NF-M+ neurons, as well as cells that expressed gene products for NF-L, glutamate receptor subunits (GluR$_{1-4,6}$), Brn-3, GAD$_{65}$, GAD$_{67}$, and GFAP. A noted absence of BF-1 expression suggested that RA may select for hindbrain neural phenotypes because BF-1 is expressed selectively in anterior regions of the CNS (55). Neurons generated action potentials, glycine-expressed TTX-sensitive Na+ channels, voltage-gated K$^+$ channels, and Ca^{2+} channels and were sensitive to kainate, NMDA, GABA, or glycine.	Murine D3 CCE	Yes 4-/4+ s RA a EBs	β-tubulin III+, NFM + RT-PCR; glutamate receptor subunits (GluR$_{1-4,6}$), Brn-3, GAD$_{67}$, GAD$_{65}$ No BF-1, TH expression Physiologic properties: responses to kainite, NMDA, GABA, glycine	GFAP + RT-PCR; GFAP	43

Year	Description	Species (cell line)		Markers / characterization		Ref
	Neural tube-like structures developed in rhesus primate ES cell-derived teratomas. First derivation of ES cell-like cells from nonhuman primate (*Rhesus*, Old World species). When transplanted into leg muscle or testis of SCID mice, ES cells formed teratomas containing cells of all three germ layers (examined 8–15 wk post-transplantation).	Nonhuman primate R278.5	No	Neural tube-like structures	NE	3
	First demonstration of ES cell-derived oligo-dendrocytes in cultures of RA-induced murine ES cells. Demonstrated nestin+ neural precursors in cultures of murine ES cells induced with RA (2+/2– as EBs, then dissociated) that were capable of generating mature neurons (GABAergic and cholinergic) and glia (oligodendrocytes and astrocytes). O4+ cells were less than 1% of cultured cells. GFAP+ were 75% of cultured cells. MAP- 2+ neurons were 25% of cultured cells. Voltage-dependent channels observed in voltage-clamp studies.	Murine CGR8 (129 Sv)	Yes 2+/2– RA as EBs	Neuron-like cells: N-CAM+, nestin+ GAD+, AChE activity+ MAP-2+, MAP-5+NFH+, synaptophysin+ Physiology: see column 1	GFAP +, O4+	45
1995	**Demonstration of the value of ES cell-derived in vitro systems for analyzing neuronal function and development at the cellular level.** RA induction for 2 d enhanced neuronal numbers but did not alter phenotypic fate. Complex electrophysiologic and immunocytologic properties of postmitotic neurons were evident, and the sequence expression of voltage-gated and receptor-operated ion channels paralleled that of previous studies in primary cultures of rat neurons (120).	Murine BLC6	Yes 4–/2+	Synaptophysin+, synaptobrevins+, NF-L,-M,-H+, Synaptic vesicle protein2+, N-CAM+, GAD+ Physiology: voltage-dependent (K^+,	GFAP+	35

(continued)

Table 2
Published Studies of ES Cell-Derived Neural Cells

Date	Observation	Cell line	RA induction	Neurons	Glia	Ref.
				Na^+, Ca^{2+}) and receptor-operated (GABA$_A$, glycine, AMPA, NMDA) ionic channels, Ca^{2+}-dependent GABA release		
1996	**Demonstration of acquisition of neuronal polarity, synapse formation, and functional synaptic transmission in ES cell-derived neurons in vitro.** Within 14–21 DIV, RA-induced (4–/4+) ES cell-derived neurons formed excitatory synapses, mediated by glutamate receptors, or inhibitory synapses, mediated by receptors for GABAor glycine. Both NMDA and non-NMDA receptors contributed to the excitatory postsynaptic responses. Most synaptic connections were excitatory (~80%); the minority were inhibitory. Only glycinergic inhibitory synapses were observed, and no GABA-ergic synapses were found. Most ES cell-derived neurons displayed spontaneous activity.	Murine D3	Yes 4+4– RA EBs	GAP-43+axons MAP-2 dendrites synaptophysin+ SV2+ synapsin+ Physiology: see column 1	NE	47

Description	Cell source	EB formation / Day	Markers		Reference
First study to transplant murine ES cells into the CNS. Development of neurons after transplantation of RA-induced ES cells. RA induction enhanced neuronal production and differentiation in culture.	Murine D3 and E14TG2a	Yes 4 + RA; Yes 4+	AChE+, GABA+, NSE+, Thy1.2+, III-β-tubulin+, GABA+,GAD+, NF+III-β-tubulin+, A2B5+; GFAP+	NE	57
Demonstration that RA promotes neural gene expression neural and represses mesodermal gene expression in mouse ES cells in vitro. 4-/4+ RA treatment of EBs enhanced expression of NF-L, NF-M, GAD_{65}, GAD_{67}, Wnt-1, and MASH-1. In EBs not treated with RA (4-/4-), these genes were not expressed, with the exception of low levels of NF-L. RA downregulated expression of the mesodermal genes Brachyury, cardiac actin, and ζ-globin. During RA treatment, sequential neural gene activation was observed in the following order: Wnt-1, then MASH-1, then NFs, then GAD only with the appearance of mature neurons.	Murine D3	Yes 4-/4+	RT-PCR: enhanced expression with RA treatment — NF-L, NF-M, GAD_{65}, GAD_{67}, Wnt-1, MASH-1	NE	51
Murine EG cells: following induction and differentiation protocols similar to previous ES-cell experiments, EG-1 cells had the capacity to differentiate into cardiac muscle, skeletal muscle, and neuronal cells in a process very analogous to ES-cell differentiation.	Murine EG-1 (129/Sv) D3 ES cell	Hanging drop	NF-M+, neurocan+ RT-PCR: NFM, neurocan β-tubulin		

(continued)

Table 2
Published Studies of ES Cell-Derived Neural Cells

Date	Observation	Cell line	RA induction	Neurons	Glia	Ref.
1996	**Differentiation of neuronal precursors and glia from avian ES cells.** First derivation of "true" ES cells from avian blastocysts. Previous work had shown that similarly isolated cells could produce chimeric offspring in chickens, but the cells were never cultured to verify the other features of ES cells (*121*). The chicken ES cells could be maintained in culture long-term, and they exhibited morphology, development, cytokine dependence, and high telomerase activity consistent with ES cells. The ES cells were shown to form cell types from all three major germ layers. Injection of ES cells into host blastocysts produced chimeric chickens. Anti-RA antibody prevented spontaneous differentiation, indicating that early differentiating ES cells release RA.	Avian CEC QEC	Anti-RA Antibody	N-CAM+	GFAP+	83
	First demonstration that bFGF could select for a highly enriched population of ES cell-derived NEPs. Nestin-immunoreactive cells could develop into glia and neurons (multiple neuronal phenotypes) in culture. Further demonstrated using TEM that ES cell-derived neurons could form synapses in vitro, and could respond physiologically to glutamate and GABA. With the bFGF, ITSFn induction system, more	Murine J1 CJ7 R1 D3	No bFGF ITSFn Yes-RA in	GABA+, glutamine+ MAP-2+, NF-M+, synapsin-1+ *No ChAT+* RT-PCR:GAD$_{65}$, AMPAR, NMDA-R1, NMDA-R2A, B, D Physiology:	GFAP+, O4+	46

Year	Description	Cell line/organism	Differentiation	Markers/Methods		Ref
	than 95% cells were nestin+ and more than 60% were MAP-2+.		B27 suppl.	responses to glutamate and GABA	NE	
	Differentiation of neuron-like cells derived from fish ES cells. First derivation of ES cell-like cells from fish (medakafish; *Oryzias latipes*) blastocysts. Chimera development not attempted/demonstrated.	Fish MES 1–3	Yes RA in EBs	Neuron-like cells	NE	81
1997	**Characterization of Ca²⁺ channel development in murine ES cell-derived neurons.** Development of G protein-mediated Ca²⁺ channel regulation in murine ES cell-derived neurons. Somatostatin efficiently suppressed L- and N-type Ca²⁺ channels in immature and mature neurons. In contrast, inhibition of L- and N-type channels by baclofen was rarely observed at the early stage, and was confined to N-type channels. The findings suggest that specific neurotransmitters such as somatostatin regulate voltage-gated Ca²⁺ channels via G proteins during the early stages of neurogenesis, thus providing a mechanism for the epigenetic control of neuronal differentiation.	Murine BCL-6	Yes 2+/2– hanging drop/suspension culture	NF-L+, -M+, -H+, synaptophysin+ RT-PCR: synaptophysin, β-tubulin, $G\alpha_{01}$, $G\alpha_{02}$ Physiology: whole-cell Ca²⁺ {I(Ca)} currents: somatostatin and baclofen reversibly inhibited I(Ca) e.g., GABAergic and glutamatergic	NE	36

(continued)

Table 2
Published Studies of ES Cell-Derived Neural Cells

Date	Observation	Cell line	RA induction	Neurons	Glia	Ref.
	Isolation of porcine EG cells (D24-25 embryos) that were capable of differentiating into neuron-like cells in vitro. EG cells injected into host blastocysts were able to contribute to chimeric piglets.	Porcine PEGC142 PEGC273 PEGC367 PEGC62	Yes	Neuron-like cells and neural rosettes formed in vitro	NE	97
1997	**Demonstration of utility of double-allele gene knockouts in ES cells for the study of cell biology.** ES cells with double-allele deletion of FGF-4 gene altered growth and survival of ES cell-derived neural cells. Note that FGF-4 (−/−) embryos were arrested shortly after implantation. FGF-4 (−/−) ES cells formed teratomas with neural components 30, 35, 47, and 61 d after subcutaneous transplantation in 129/SvJ mice. Addition of FGF-4 in culture dramatically enhanced survival of ES FGF-4 (−/−)-derived cells, which otherwise died at a faster rate than FGF (+/+) ES cells.	Murine R1(129) and D3, GB3,FD6 HD3,1C3	Yes	Mature neural tissue observed in teratomas created SQ	NE	110
	Embryonic intraventricular transplantation of RA-induced ES cells generated neurons, astrocytes, and oligodendrocytes that integrated with host tissues. The appearance of the differentiated transplanted cells temporally appeared in correlation with the normal postnatal development of each cell	Murine J1	No	NeuN+,MAP2, TH+ Neuroepithelium formed in ventricles	CNPase+ Oligos GFAP+	75

type. Four-day EBs were plated in culture for 5–12 d in ITSFn medium (46) prior to transplantation.

	Demonstration that RA-mediated *Pax6* expression and markedly enriched the yield of neurons and glia from murine ES cells.	Murine R1 MPI-II *Pax6*/lacZ	Yes 4-d-RA as EBs	Neuron-like cells NFM+	GFAP+ 119
1998	**Postmitotic neurons and glia were generated after rhesus ES cells were transplanted into immuno-compromised host muscle (SCID mice) where they formed teratomas.** At 5–12 wk posttransplantation, neural differentiation was seen in all teratomas: neural tube-like structures, embryonic ganglia, dispersed neurons, and brain-like gray matter. Neurons were NF+. Axon tracts were common features in teratomas more than 6 wk old. Increased astrocyte differentiation (GFAP+) was observed as teratomas aged. Therefore, some aspects of normal differentiation were recapitulated in teratomas from rhesus ES cells. Group has isolated 7 rhesus ES cell lines in total (*102*).	Nonhuman primate R278.5	No	Intramuscle teratomas: neural tube-like structures, embryonic ganglia, neurons, brain-like gray matter NF+	Intra-muscle teratomas: GFAP+ 101
	First demonstration of normal developmental characteristics of neuromuscular junctions from murine ES cells in vitro. Colocalization of agrin, synaptophysin, and AChE in mixed muscle/neurons derived from ES cells. The temporal pattern of expression of striated muscle and neuronal markers closely match similar expression in vivo during development.	Murine D3 and BLC6	No 5D EBs	Cholinergic cells NF-L+ and synaptophysin (5D+2); NF-M, RT-PCR: NF-L, NF-M, NF-H (5D+6); synaptophysin (5D	NE 37

(continued)

Table 2
Published Studies of ES Cell-Derived Neural Cells

Date	Observation	Cell line	RA induction	Neurons	Glia	Ref.
1998	**First demonstration of genetic selection of lineage-restricted neural progenitors from ES cells.** These ES cells contained a bifunctional selection marker/reporter gene βgeo integrated into the *Sox-2* gene by homologous recombination. When induced with RA using the 4−/4+ protocol, approx 50% of the dissociated cultured cells expressed β-galactosidase activity and Sox-2 immunoreactivity. Further addition of G418 resulted in cultures in which more than 90% of the cells expressed β-galactosidase activity and Sox-2 immunoreactivity. Of the selected cells, 46% were Pax6+, 35% were Pax3+, 24% were Mash1+, 14% were Mash4A+, 30% were Delta1+, and 3% were Islet1+. Further differentiation produced mature neurons expressing GABA, glutamate, NF, and MAP immunoreactivity. Therefore, genetic selection can be used to obtain pure populations of neural restricted precursor cells, in this case NRPs.	Murine E14TG2a CGR8 CCE-Sox2	Yes 4−/4+ RA	GABA+, glutamine+, GFAP+ β-tubulin III+, MAPs+, synapsin-1+, NFL+, NFH+, nestin+	EBs); neurocan (5D + 2); tau (5D + 6); NF-H (5D + 18)	58
	First demonstration of the utility of double-allele gene targeted ES cell culture system for analyzing the	Murine J1	Yes 4−/4+	NFM+, GAP-43+,	NE	115

Description	Cells		Method	Markers		Ref
effects of specific genes on neuronal development. GD3S (−/−) ES cells, double-targeted gene disruptions of GD3 synthase gene, were deficient in GD3S activity, and neurons could still differentiate normally. Therefore, the previous hypothesis that b-series gangliosides are necessary for neuronal differentiation was disproved.				MAP-2+	GFAP+	49
Demonstration of the utility of double-allele gene knockouts for studing cellular mechanisms in vitro. Loss of β1-integrin function accelerated neuronal differentiation. ES cells with double-allele deletion of β1-integrin differentated into neurons at an accelerated rate (expression of neuron-specific genes and increased numbers of neurons) but neurite outgrowth was retarded.	Murine D3 G119(β1+/−) G201 G100(β1−/−)	No and Yes	Hanging drop and mass cultures of EBs	Multiple neuron-specific (NFM+) and glial-specific GABA release RT-PCR: expression of NF-L and NF-M, NF-H, synaptophysin, neurocan, tau		
Neuroepithelium and neuroganglia present in teratomas derived from human ES cell-like cells. First-time ES cell-like cells were derived from human blastocyst. Transplanting each of the five cell lines into SCID-beige mice produced teratomas containing all three germ layers and rosettes of neuroepithelium and neural ganglia (after 4 mo).	Human H1, H7,H9, H13,H14	No	Neuroepithelium and neuroganglia in teratomas	NE		5

(*continued*)

Table 2
Published Studies of ES Cell-Derived Neural Cells

Date	Observation	Cell line	RA induction	Neurons	Glia	Ref.
	Demonstration of neurons in EBs derived from human EG cells. First isolation of pluripotent EG cells from human PGCs; cultures of PGCs produced cells from all three derivatives of all three embryonic germ layers, including neuroepithelia.	Human BF1, KF1, R13, R14, R15, BF2	No	NF+ in cultured EBs	S-100+ in cultured EBs	6
1998	**RA induction favors differentiation of ventral CNS neurons.** RA exposure (2−/5+ protocol) generated neurons characteristic of the ventral CNS, somatic (Islet+), and cranial (Phox2b+) motoneurons and interneurons (islet−). RT-PCR: upregulated expression of GAD, TH, ChAT, TrkB, TrkC.	Murine CCE	Yes 2−/5+ hanging drops	Somatic MNs and interneurons, nestin+ NPs, Pax6+, few Pax7+, Islet, -1/2+,Phox2b+, Lim $1/2$+, Lim3+ EN1+, peripherin+ GAD, TH, ChAT. RA enhanced expression of these, particularly ChAT.		48
	Demonstration of a strict temporal differentiation profile for neuroglial cells in RA-induced ES cells in culture. The cell types first appeared on the following postplating days: neurons (5), astrocytes and oligodendrocytes (9), and microglia (16). First demonstration of microglial differentiation from whole EBs.	Murine BLC6 (129/Sv Gat mouse blastocyst)	Yes RA as EBs	NSE+ Synaptophysin+ O4+ oligos, C56+ microglia	GFAP+ astrocytes,	38

Therefore, EBs recapitulate the temporal order of neural cell development in the CNS.

Year	Finding	Cell source			Neuronal markers	Glial markers	Ref.
1999	**First demonstration that murine ES cells could myelinate in the immature CNS.** Development of procedure to enrich for GRPs from murine ES cells with potential for forming oligodendrocytes and astrocytes. No RA used for induction. EB growth (4 d), then plated in ITSFn (5 d), then sequential propagation in (a) bFGF, (b) bFGF and EGF, (c) bFGF and PDGF. Demonstrated that the GRPs could myelinate axons in the developing nervous system of myelin-deficient mutant rats (spinal cord, 1 wk old; intraventricular at E17).	Murine J1 Cj7	No	NE		A2B5+ precursors, O4+ CNPase+, GFAP+	59
	Demonstration that BMP-4, a TGF-β superfamily member, inhibited RA-induced neural differentiation and enhanced mesodermal differentiation in murine ES cells. The effect of BMP-4 was restricted from d 5–8 of the 4–/4+ EB aggregation protocol. BMP-4 did not alter cell proliferation or death in EBs. As a baseline, ~25% of EB (4–/4+ stage at 5–6 d) cells were TUNEL+, as in earlier observations (122). Coincubation with the anti-apoptotic molecule BAF (30–50 μM) reduced TUNEL+ by 35% but did not alter neural differentiation. The effect of BMP-4 could be reversed by coapplication of noggin, a BMP-4 antagonist.	Murine D3	Yes 4–/4+	β-tubulin III+ (decreased 5- to 10-fold by BMP-4) NeuN+, HNK–1+		GFAP+ (decreased by BMP-4)	

(continued)

Table 2
Published Studies of ES Cell-Derived Neural Cells

Date	Observation	Cell line	RA induction	Neurons	Glia	Ref.
1999	**Demonstration of murine ES cells as a source of late embryonic neural precursor cells.** NRPs (E-NCAM+) and GRPs (A2B5+/E- NCAM-) can be immuno-derived from ES cells and can differentiate into postmitotic neurons and glia, respectively. ES cells grown as aggregates for 4 d, then plated on fibronectin-coated dishes in NEP basal medium. Upon differentiation, E-NCAM+ cells expressed early neuronal markers (β-tubulin III+, MAP-2+) but not GFAP glial markers. Immunopanned E-NCAM+ ES cell-derived precursor cells expressed markers for neurons upon (glutamate, GAD, glycine+) differentiation. ES-cell-derived A2B5-immunoreactive cells differentiated into oligodendrocytes and two types of astrocytes (type I [A2B5-/GFAP+], II [A2B5+/GFAP+] astrocytes).	Murine D3	Yes used for long-term culture PDGF used for long-term glial induction	E-NAM+ differentiated cell: glutamate+, GAD+, Glycine+ RT-PCR:ChaT, GAD, glutaminase A2B5+E-NCAM- E-differen-tiated cells:(−) for markers of differentiated neurons.	E-NCAM+ differentiated cells: no oligo or astrocytes A2B5+ NCAM- differentiated cells: Gal-C+Olig type I, II	109
	First demonstration that ES-cell transplantation can enhance recovery of lost function. Transplantation of dissociated 4−/4+ RA-stimulated EBs into the spinal cord 9 d after contusion injury improved spontaneous locomotion of hindlimbs. ES cell-derived cells integrated, migrated 1 cm, and differentiated into neurons, oligodendrocytes, and astrocytes.	Murine D3 ROSA26	Yes 4−/4+	NeuN+, EMA+	APC CC-1+ GFAP+	76

Year	Description	Species / Cell line		Markers			Ref.
2000	**Differentiation of neurons from human ES cell-like cells.** Second demonstration of isolation of ES cells from human blastocysts. **First demonstration that neural progenitors could be isolated from human ES cells and induced to generate mature neurons.** Two ES cell lines obtained from human blastocysts with demonstration of neuronal differentiation in vitro. The cell lines could be passaged 64 (HES-1) and 44 (HES-2) times, corresponding to a minimum of approx 384 and 264 population doublings, respectively. Both lines were successfully recovered from cryopreservation. First demonstration of Oct-4 expression in human ES cells. Oct-4 is a POU domain transcription factor whose expression is limited in the mouse to pluripotent cells, and zygotic expression of Oct-4 is essential for establishment of the pluripotent stem population of the inner cell mass (103). When both lines were transplanted into testis of SCID mice, teratomas formed that contained derivatives of all three embryonic germ layers (examined 6–7 wk posttransplantation).	Human HES-1 HES-2	No	NFL+ N-CAM+, Nestin+ RT-PCR: β-actin, nestin, Pax-6	Yes (7–15 d RA exposure in culture), hanging drop	NFH+, MAP-2+, AP20+, synaptophysin+, β-tubulin+, GAD+, glutamate+ RT-PCR: GABA-AReca2	NE	7
	Demonstration of marked changes in calcium-binding proteins and voltage-dependent Ca^{2+} channel subtypes during in vitro differentiation of murine ES cell-derived neurons, using immunocytochemistry, patch-clamp, and	Murine	Parvalbumin+, calretinin+, calbindin+	NE				40

(continued)

Table 2
Published Studies of ES Cell-Derived Neural Cells

Date	Observation	Cell line	RA induction	Neurons	Glia	Ref.
	videomicroscopy time-lapse techniques. Neuronal maturation proceeded from apolar to bipolar to multipolar morphologies. All Ca^{2+} channel subtypes were expressed; apolar cells had mainly N- and L-type channels, in contrast to the P/Q- and R-type channels that predominate in bipolar and multipolar cells. Parvalbumin was present in bipolar cells, whereas calretinin and calbindin were preferentially found in multipolar cells.		Physiology: N-, L-, P/Q-, R-type Ca^{2+}			
2000	**First demonstration that ES cell-derived oligodendrocytes could: (a) myelinate axons in culture, (b) myelinate axons in the injured mature nervous system (spinal cord), (c) myelinate axons in the adult myelin mutant *shiverer* mouse.** A simple and rapid method was developed for isolating and purifying oligodendrocyte precursors. It involved an intermediate "oligosphere" step after dissociation of 4−/4+ RA-treated EBs.	Murine D3 ROSA26	Yes 4−/4+	β-tubulin III+, NF+	NG2+, O4+, O1+, MBP+, CNPase+ APC CC-1+	50

matergic, glycinergic, noradrenergic, and cholinergic neurons (Table 2). Subtypes of cholinergic motoneurons have also been identified in ES cell-derived cultures *(48)*. The majority of these cell-typing studies have relied on immunohistologic methods, however. Recent physiologic studies indicate that ES cell-derived neurons develop functional synapses, exhibit spontaneous activity, and possess electrophysiologic properties remarkably similar to those of neurons in primary culture systems *(35,36,40,43,45,47,56*; Table 2).

METHODS FOR ISOLATING ES CELL-DERIVED NEURAL PRECURSORS

Multiple classes of neural precursor cells have been isolated from murine ES cells, and methods for purifying these populations have been developed. In theory, ES cells are capable of generating all later stage stem-cell types if all necessary factors are present. Derivation of stable, genetically normal,

---

Table 2 (cont.)

a Published studies examining ES cell-derived neural cells. Summary of the use of ES cells in the study of neural systems, both in vitro and in vivo. The columns refer to the major observations of each study, which ES cell lines were used, the evidence supporting differentiation of neurons and glia, and the corresponding reference.

Abbreviations: AChE, acetylcholinesterase; AMPAR, AMPA (α-amino-3-hydroxy-5-methyl-4-isoxazoleproprionic acid) receptor; APC CC-1, adenomatous polyposis coli, CC-1 subtype antibody; B27, defined media supplement (with RA); BAF, boc-aspartyl(OMc)fluoromethylketane; BF-1, brain factor-1; bFGF, basic fibroblast growth factor-2; BMP, bone morphosenetic protein; BNP-4, bone morphogenic protein-4; ChAT, choline acetyltransferase; CNPase, 2',3'-cyclic nucleotide 3'-phosphodiesterase; D, day; DIV, days in vitro; EB, embryoid body; EG, embryonic germ; EGF, epidermal growth factor; EMA, mouse-specific antibody that preferentially recognizes neurons *(123)*; ES, embryonic stem; FGF, fibroblast growth factor; GABA, γ-aminobutyric acid; GAD, glutamic acid decarboxylase; Gal-C, galactocerebroside; GD3S, GD3 synthase; GFAP, glial fibrillary acidic protein; GluR, glutamate receptor; GRP, glial-restricted precursors; ITSFn, chemically defined media *(46)*; MAP, microtubule-associated protein; MBP, myelin basic protein; MN, motor neuron; N/A, not applicable; N-CAM, neural cell adhesion molecule; NE, not examined; NEP basal medium, neuroepithelial precursor basal media *(109)*; NeuN, neuron-specific nuclear protein; NF (L, M, H), neurofilament (low, medium, and heavy forms); NG2, NG2 chondroitin sulfate proteoglycan; NGF, nerve growth factor; NMDA, *N*-methyl-D-aspartate; NP, neural precursor; NRP, neuronal restricted precursor; NSE, neuron-specific enolase; O4, preoligodendrocyte marker; O1, mature oligodendrocyte marker; Oligo, oligodendrocyte; PDGF, platelet-derived growth factor; RA, retinoic acid, RAR, retinoic acid receptor; RT-PCR, reverse transcription polymerase chain reaction; SCID, severe combined immunodeficiency; SQ, subcutaneous; TEM, transmission electron microscopy; T3R-α, T3 receptor-α; TH, tyrosine hydroxylase; Trk, tyrosine kinase; TTX, tetrodotoxin; TUNEL, terminal deoxyneucleotidyltransferase (TdT) mediated deoxyuridine triphosphate-biotin nick end labeling; +, positive reactivity; (–/–), double-allele gene inactivation; 4–/4+ RA, induction protocol that exposes EBs to RA only in the last 4 d of an 8-d protocol.

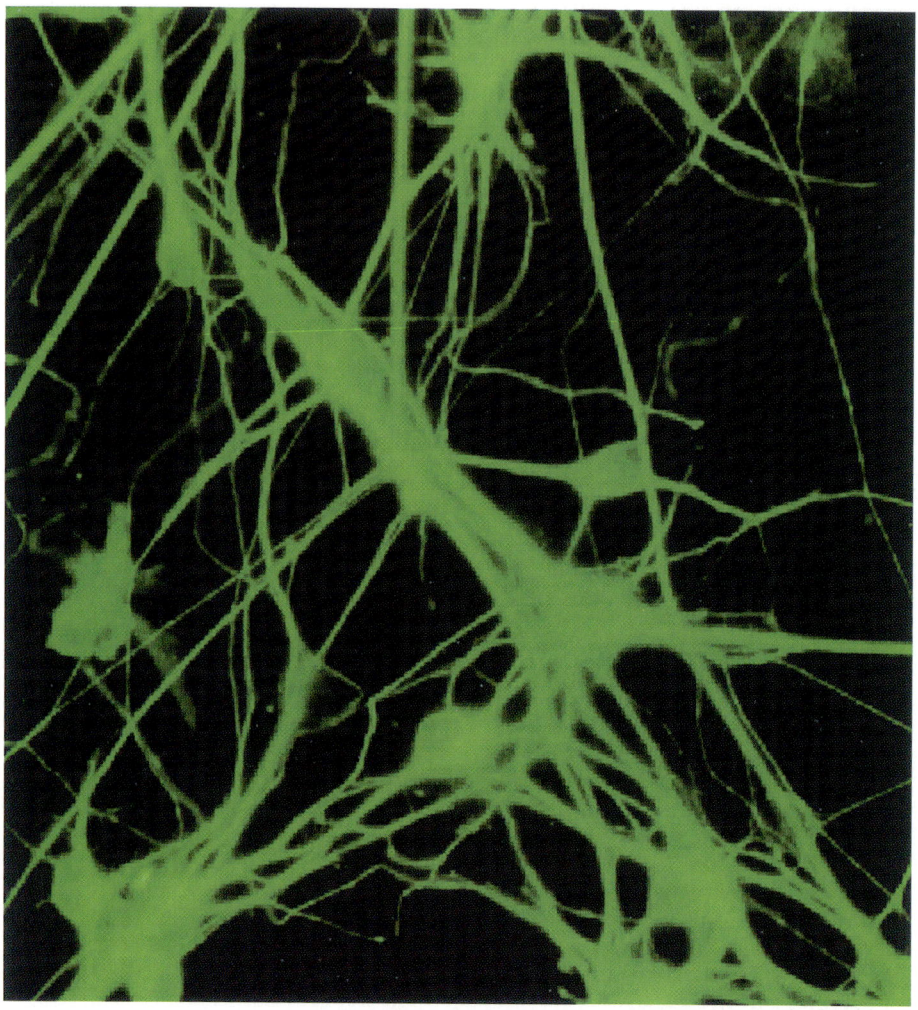

Fig. 4. Embryonic stem cell-derived neural lineage cells rapidly differentiate into the three major types of spinal cord cells when grown under special condition in culture dishes. Cultured ES cell-derived neurons (DIV 9) are shown using epifluorescence to β-tubulin III. 4 – /4 + EBs were dissociated and plated in gelatin-coated culture dishes. Note the many thin line-like axons growing from neurons. Many of these axons are hundreds of times longer than the cell body from which they originate.

later stage, lineage-restricted stem cells will provide powerful scientific tools and consistent sources of cells for transplantation. For example, transplantation of fate-restricted precursor cells would help limit the risk of teratomas. ES cells or RA-induced ES cells, when transplanted (e.g., subcutaneous,

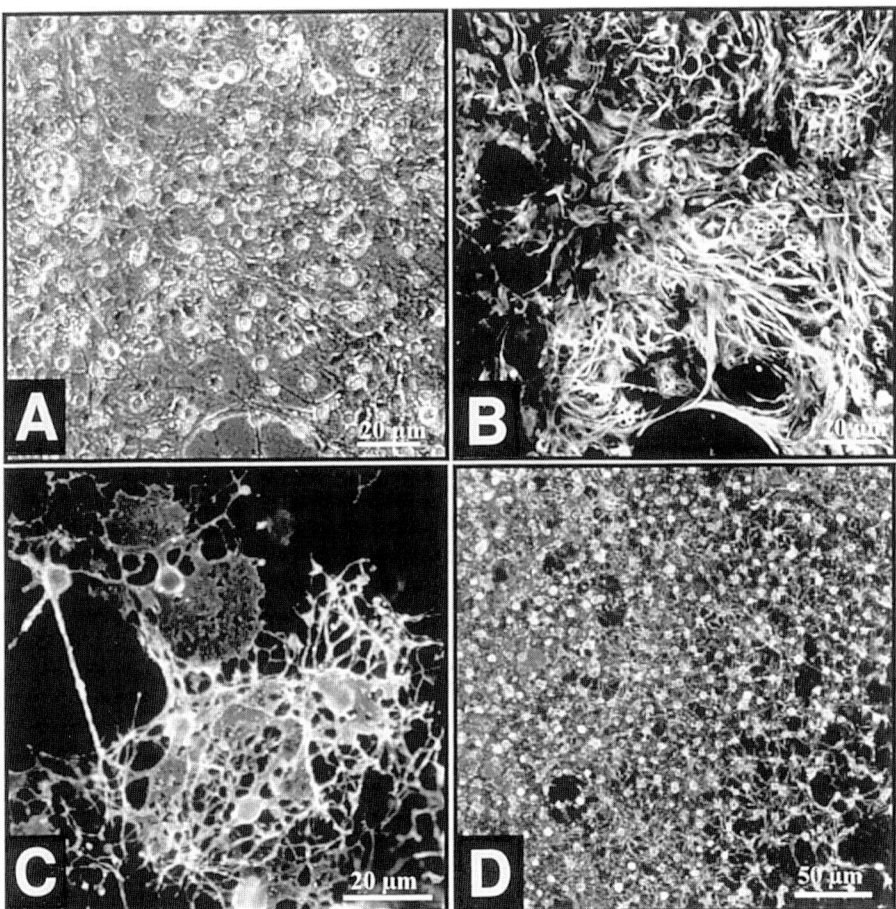

Fig. 5. Mixed cultures of neurons, astrocytes, and oligodendrocytes can be produced by plating dissociated 4 – /4 + stage EBs in SATO defined media *(50,64)* supplemented with 5% fetal calf serum and 5% equine serum. Phase-contrast image of a mixed culture containing primarily oligodendrocytes (phase-bright cells; **A**) on an astrocyte monolayer (shown using anti-GFAP immunoreactivity; **B**). All cultures are DIV 9. Individual mature oligodendrocytes are visualized using anti-O1 immunoreactivity (**C**). Highly enriched cultures of oligodendrocytes can be cultivated through an intermediate "oligosphere" stage. Oligospheres are floating cell clusters produced from EBs through an additional culture step, and they yield highly enriched cultures of oligodendrocytes when plated in oligosphere-conditioned media. Oligosphere-derived cultures are enriched for O4(+) oligodendrocytes (DIV 5; **D**).

into testis) into immunocompromised or syngenic hosts typically form teratomas containing cells derived from each of the three germ layers. In fact,

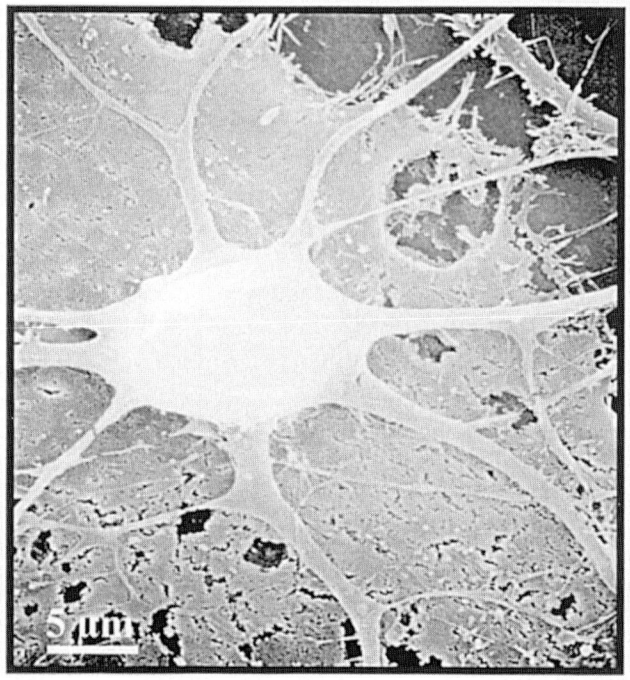

Fig. 6. Scanning EM shows a solitary oligodendrocyte in a mixed culture or ES cell-derived neural cells. Scale bar = 5 μm.

this method is typically used in the characterization of ES-cell lines to demonstrate pluripotency. Any multipotent cell can form normal tissue at an incorrect location (ectopic tissue), provided conditions that favor survival are available. Measures to prevent ectopic tissue formation at the transplantation site include transplanting lineage-restricted neural precursors and engineering cells with fail-safe systems that permit selective removal if conditions become unfavorable. For example, transplantable cells might contain an inducible promoter that activated programmed cell death genes. To date, considerable progress has been made in developing methods for isolating lineage-specific ES cells.

RA Induction Protocols

One of the first strategies for enriching murine neural precursor cell populations involved exposure to RA (*35,39,43,45,47,51,56,57*; see ref. *39* for review of molecular mechanisms of RA induction; Figs. 1 and 2). When EBs are exposed to RA, the overall temporal development of gene and marker expression strongly resembles that of the developing embryo, although differences exist *(39,47,51;* Fig. 2). In general, murine EBs contain a wide variety of cell types but typically include only low numbers of neural cells when induction protocols lack RA *(2,51,57).* In all three of the in vitro neural ES-cell studies published in 1995 (Table 2), more efficient production of neural cells was accomplished by exposure to RA, despite differing exposure periods during the differentiation protocol *(35,43,45).* The RA protocol has also been shown to produce electrophysiologically active neuronal circuits with functional excitatory and inhibitory synapses (see below; *35,43,47).* Subsequent murine ES cell studies demonstrated that the 4 – /4 + RA protocol generates EBs composed primarily of an undifferentiated, nestin+ precursor pool (89% of cells nestin+ = NEPs) capable of generating three of the primary neural lineages: neurons, oligodendrocytes, and astrocytes *(43,50,56).* In fact, less than 5 to 10% of EB cells express markers for differentiated neural cells *(50),* indicating that the vast majority are still in the precursor stage.

Of the three induction protocols mentioned above, the Bain et al. *(43,51)* 4 – / 4+ RA protocol has been studied and used most extensively (Fig. 1). Under this protocol, murine ES cells are cultured as EBs in the absence of LIF for 4 d in uncoated, nonadhesive Petri dishes. EBs are further cultured in suspension for an additional 4 d in the presence of 0.5–1 µM RA (Fig. 3). At the end of the 8 d 4 – /4 + protocol, aggregates are dissociated with trypsin and plated as monolayers on adhesive substrates. Generation of neural subtypes is then largely achieved by providing media known to selectively support the survival of primary neural cultures. Growth in neural basal medium supplemented with serum generates mixed cultures containing mostly neurons and astrocytes with a few oligodendrocytes *(43,44).* Approximately 40% of the cells in these cultures are neurons.

In these original studies, further cellular division was inhibited on days in vitro (DIV) 2–4 by cytosine arabinoside (Ara-C). Ara-C exposure selects against the development of oligodendrocytes, which are highly sensitive to Ara-C-induced death (McDonald et al., unpublished observations). Omitting Ara-C generates cultures containing large numbers of oligodendrocytes *(50;* Fig. 5). We have also observed that 4 – /4 + RA-induced EBs are capable of producing substantial amounts of extracellular matrix when plated. This extracellular matrix markedly reduces the effectiveness of immunocytochemical labeling of cell-surface proteins (McDonald et al., unpublished observations). Most oligodendrocyte lineage-specific antibodies label cell-surface epitopes, making accurate indices of cell labeling in the ES cell-derived cultures possible only after partial permeabilization *(50).* Application of the same types of media used to enrich for oligodendrocytes in primary CNS cultures also allows enrichment of oligodendrocytes in these ES cell-derived mixed cultures, providing useful in vitro models of myelination *(50)*.

Strubing et al. *(35)* have used a similar RA-induction approach, obtaining murine EBs by the "hanging drop" culture method. This method permits differentiation in a minimal volume of medium, which allows effective buildup of EB-conditioned medium. EBs were exposed to RA for the first 2 d and then cultured for an additional 2 d without RA (2 + /2 –). Nondissociated EBs then were cultured on adhesive substrates. All EBs exposed to RA gave rise to neurons, but only 15% of EBs not exposed to RA did so. Alternatively, Fraichard et al. *(45)* exposed murine EBs, generated in mass culture, to RA during the first 2 d. Subsequent plating on adhesive substrate in the absence of RA also generated neurons in 4–5 d.

Renoncourt et al. *(48)* utilized a 2 – /5+ RA induction protocol in murine ES-cell cultures to produce neurons with characteristics of the ventral CNS: somatic (Islet +) and cranial (Phox2b +) motoneurons and interneurons (Islet-; Table 2). Production of ChAT + neurons under this protocol contrasts with the absence of ChAT+ cells in experiments using the 4 – /4 + RA protocol (discussed in ref. *39*). This discrepancy probably results from the use of SATO-defined medium, which is highly supportive of neuron survival *(48)*. These differences may also reflect the effects of the timing of RA exposure. This sensitivity is not entirely surprising, given the sequence of extensive developmental changes that occur over 2 d in the early embryo. For example, a 2 – /2 + EB RA protocol produces cultures largely devoid of neural cells (McDonald et al., unpublished observations).

Li et al. *(58)* used the 4 – /4 + RA protocol to enrich for NEPs. Then they selected genetically for NRPs using *Sox*-2 gene-driven neomycin resistance (see Genetic Stem-Cell Selection, below). Like Brüstle et al. *(59)*, we have observed that subsequent addition of basic fibroblast growth factor (bFGF) to the medium further enriches for glial cells.

bFGF Induction Methods

Additional methods, for enriching ES cell cultures in neural cells do not involve RA. For example, Okabe et al. *(46)* and Brüstle et al. *(59)* made use of bFGF's known ability to propogate neural stem cells *(60,61*; for review, see Chapters 2, 3, and 10) to generate nestin+ NEPs and subsequent mixed neural/glial cultures. Mouse ES cells were cultured as EBs for 4 d and then plated on adhesive substrate in the presence of bFGF in a defined medium lacking serum (Dulbecco's modified Eagle's medium [DMEM]/F-12 supplemented with ITSFn medium -DMEM/F-12 supplemented with insulin, transferrin, progesterone, putrescine, selenium, bFGF, and laminin; see ref. *46* for details). Under these conditions, the majority of cells died, but the bFGF sensitive precursor population survived and continued to replicate in the presence of this potent mitogen. This method selects for nestin + /keratin 8- NEPs *(46)*. Withdrawing bFGF induces spontaneous differentiation into neurons and glia. Although the number of O4+ oligodendrocytes was low in these studies, addition of T3 helped support oligodendrocyte survival. Further studies of CNS-isolated progenitors have even suggested that bFGF induction may be *required* for generation of some neuronal phenotypes (F. H. Gage, personal communication).

Similar methos of bFGF selection and propagation have been used to isolate uniform propulations of ES-derived nestin+ neural precursors (derived from murine TC1 ES cells; *128*) that can then be cryopreserved and later thawed with good viability and then re-expanded with bFGF *(124)*. These methods provide a useful sourse of ES-cell-serived precursor cells, capable of differentiation inot functional neurons. How closely ES-cell-derived neural precursors resemble CNS isolated neural precursors is an important question that is at present unanswered.

In addition to selecting tripotential NEPs, bFGF can be used to enrich glial precursors from murine ES cells *(50,59)*. Brüstle et al. *(59)* produced cultures enriched for neural cells, particularly oligodendrocytes (about 30%), using a temporal exposure to bFGF and platelet-derived growth factor (PDGF). These methods are logical extensions to previous studies of oligodendrocyte precursors (for review, see Chapter 6) and neurospheres/oligospheres *(62,63)*. In the above study by McKay and colleagues *(59)*, no RA induction was used. EBs were grown in suspension for 4 d and then plated for 5 d in a chemically defined medium adapted from SATO medium *(64)*, ITSFn. Cultures were trypsinized and sequentially propagated as neurospheres in suspension cultures in medium supplemented first with bFGF, then with bFGF and EGF, and finally with bFGF and PDGF. The general approach is similar to that applied by Duncan and colleagues to generate oligospheres from CNS-isolated neural progenitors *(62,63)*. These methods enriched for oligodendrocytes when the EBs were dissociated and plated in vitro (about 30% oligodendrocytes). Transplantation of neurosphere-derived cells into the embryonic and postnatal rat CNS resulted in oligodendrocyte differentiation and myelination of host axons.

Subsequently, a simple procedure was developed to highly enrich oligodendrocytes in culture *(50)*. Plating dissociated 4 – /4 + RA-induced murine EBs into SATO-defined medium produced mixed neural cultures that readily supported oligodendrocyte myelination of ES cell-derived neurons. A key point for success was not inhibiting cellular division with Ara-C, which limits oligodendrocyte survival.

Oligodendrocytes can be enriched further by plating dissociated 4 –/4 + RA-induced EBs on nonadhesive plates in a chemically defined medium that lacks serum and is supplemented with bFGF. Under these conditions, aggregates of cells re-form. We term these aggregates oligospheres because subsequent plating produces cultures highly enriched for oligodendrocytes *(50)*. Oligospheres are dissociated after 2 d and replated under similar conditions for another 4 d. Trypsin-dissociated oligospheres can then be plated on adhesive substrates in oligodendrocyte-supporting defined medium to obtain highly enriched cultures of oligodendrocytes. Further enrichment to about 90% can be achieved by plating dissociated oligospheres on adhesive substrate in the presence of oligosphere-conditioned medium *(50;* Figs. 5 and 6).

Studies of adult CNS-isolated NEPs indicate that RA and neurotrophins work synergistically to regulate neurogenesis *(65)*. Exposing bFGF-propagated hippocampal NEPs to RA rapidly induces neural differentiation, with upregulation of neuron-specific genes. RA also upregulates or sustains expression of trkA, trkB, trkC, and p75NGFR receptors. Without RA, cells respond minimally to neurotrophins. Addition of neurotrophins (NGF, BDNF, NT-3) following RA induction promotes neuronal maturation into phenotypes including GABAergic (GABA+), dopaminergic (TH +), and cholinergic (AChE +) cells. BDNF is the most effective agent for increasing the number of AChE+ neurons, whereas a combination of BNDF and NT-3 is most effective treatment condition for increasing GABA+ and TH+ neurons.

ADDITIONAL METHODS FOR SELECTIVELY ENRICHING PRECURSOR POPULATIONS

Immunopanning and FACS

Traditional methods developed for primary CNS cultures can be applied to ES cell-derived cells to enrich or purify different stages of neural precursors. These methods include cell immunopanning (an immunocytochemical sorting technique) and fluorescence-activated cell sorting (FACS). Rao et al. have used immunopanning to derive NEPs *(66,67)* and GRPs *(66,68,69)*, first from embryonic rodent spinal cord and later from murine ES cells. NEPs can be positively immunopanned using E-NCAM + selection. GRPs can be selected by first using negative immunopanning with E-NCAM to remove NEPs and early neurons and then by utilizing positive panning using A2B5 immunoreactivity. The same immunopositive profiles can be used to complete FACS analysis. The latter method allows higher levels of purification at the cost of much lower yields.

Genetic Stem-Cell Selection

In future studies with genetically modified cells, FACS will be particularly efficient for detecting progenitor-specific gene expression that is coupled to fluorescent markers. An elegant, simple, and more cost-effective method for isolating precursor populations takes advantage of the genetic flexibility of ES cells. This method involves genetically engineering resistance genes (e.g., to G418) by linking resistance expression to precursor-specific gene expression. One of the first examples was the isolation of cardiomyocytes from murine ES cells by Klug and co-workers *(30)*, who introduced cardiac-specific neomycin (neo)-resistance constructs into ES cells and isolated nearly pure (99.6%) preparations of cardiomyocytes from differentiating ES cell cultures. Without selection, cultures contained less than 1% cardiomyocytes. Subsequently, a similar approach maintained pure cultures of undifferentiated murine ES cells by killing cells that differentiated *(70)*. The OKO160 ES cell line *(71)*, which contains a targeted insertion of the *lacZ*/neo gene fusion (β-geo; *72*) inserted into the *Oct 3/4* gene, was rederived from OKO160 mice. *Oct 3/4* is an octamer binding transcription factor that is expressed in undifferentiated ES cells, oocytes, pluripotent early embryo cells, and primordial germ cells *(73)*. When the cells differentiate, they lose *Oct 3/4* expression and their resistance to neomycin and therefore die.

Smith and colleagues *(58)* first used this approach to purify neural precursors. They targeted the *SOX-2* gene in a murine cell line. This gene codes for a DNA-binding protein that is selectively expressed in the early neural plate. The targeting event placed a promoterless neomycin cassette in the *SOX-2* gene so that only cells expressing the gene were neomycin resistant. ES cells were induced by the 4 – /4 + RA protocol and selected with neomycin. This selection procedure produced an enriched population of precursor cells that were 94% nestin+ and could be expanded using bFGF for at least 3 wk. Withdrawl of bFGF in serum free media induced rapid neuronal differentiation and cultures contained greater thatn 90% cells expressing neuronal markers by 96 h. GABAergic and glutamatergic neurons were frequent. Nonneuronal cells were rare. This selection procedure produced pure cultures of neurons that expressed numerous regulatory genes known to be associated with neural precursor cells, including *delta1*, *mash1*, *pax3*, *pax6*, and *SOX-1 (58)*. It is clear that gene activation specific to each of the neural precursors described above will soon be used to isolate pure precursor populations. Candidate genes are already available for most of the neural precursors.

TRANSPLANTING ES CELL-DERIVED CELLS INTO THE CNS

Only six published articles have described transplantation of ES cell-derived cells into the CNS (Table 2). In the first and third studies, differentiated ES cell-derived neurons and glia were implanted into the injured adult corpus striatum *(57,74)*. In the second, ES cell-derived cells were implanted into embryonic rat brain *(75)*. In the fourth, ES cell-derived cells were transplanted into the embryonic rat brain and early postnatal mutant rat spinal cord *(59)*. In the fifth, dissociated 4 − /4 + RA-induced EBs were transplanted into the contusion-injured spinal cord of adult rats 9 d after injury *(76)*. In the sixth, dissociated 4 − /4 + RA-induced EBs were transplanted into the spinal cords of adult mutant mice lacking the MBP and into chemically demyelinated adult rat spinal cord *(50)*.

In the study by Dinsmore et al. *(57)*, mouse ES cells (D3 line) were either induced as EBs with RA for 4 d, then trypsinized and transplanted, or induced for 4 d with RA, trypsinized, and plated in Neuro-C medium *(57)* for two more days and then transplanted. Adult Sprague-Dawley rats, with lesions induced in the striatum by injection of quinolinic acid, were used in this study. Dissociated cells (0.1–1 million) in 5 µL of medium were transplanted into the lesion site 7 d after injury. The animals were given a daily cyclosporine injection to prevent transplant rejection. Six weeks after transplantation, immunohistochemistry revealed the presence of ES cell-derived cells exhibiting neuronal markers that included neuron-specific enolase (NSE), Thy 1.2, TuJ1, AChE, and GABA. Transplanting undifferentiated ES cells failed to produce cells expressing neuronal markers.

In the second study *(75)*, 4-d mouse (J1 ES cell line) EBs (without RA) were trypsinized and plated in gelatin dishes containing ITSFn media for 5–12 d. Then the resulting aggregates were trypsinized, and the cells were transplanted into E16 or E18 Sprague-Dawley rat embryos. ITSFn selects for nestin+ neural precursor cells *(46)*. In animals sacrificed between postnatal day (PND) 0–15, immunohistochemistry indentified ES cell-derived cells in most brain regions, including the cortex, striatum, septum, thalamus, hypothalamus, and tectum. The ES cell-derived cells had become integrated into host tissue and showed markers for neurons, oligodendrocytes, and astrocytes. ES cell-derived neurons appeared by PND 0, ES cell-derived astrocytes became prominent in the second postnatal week, but ES cell-derived oligodendrocytes were not observed until PND 15. Immunostaining with mouse-specific M6 antibody *(77)* revealed ES cell-derived cells with highly arborized dendrites and bundles of axons projecting into white-matter tracts. Some nonneural ES cell-derived tissues were also observed developing ectopically in the region of the transplant.

In the third study *(74)*, RA-induced or non-RA-induced 4-d EBs were trypsinized and transplanted into normal adult mouse brain or the striatum of immunosuppressed rats injured by 6-hydroxydopamine injection. Histologic examination of the rats 2–4 wk after transplantation revealed the presence of neurons and astrocytes derived from ES cells. ES cell-derived cells were immunopositive for neuronal markers, including TH, 5-HT, NSE, and AChE. TH+ and 5-HT+ axons derived from ES neurons were observed in the striatum.

In the fourth study *(59)*, ES cell-derived cells were shown to myelinate axons in CNS of the embryonic normal and postnatal myelin-deficient (MD) mutant rat. GRPs were isolated from ES cells through a complicated multiple mitogenic exposure system using bFGF and PDGF. No RA was used for induction. EBs were grown for 4 d and then plated in ITSFn for 5 d. Cells were then sequentially propagated in (a) bFGF, (b) bFGF and EGF, and (c) bFGF and PDGF. Two transplantation paradigms were used. Cells were transplanted into the E17 embryos by intraventricular injection or directly into the spinal cords of 7-d-old mutant rats. The data clearly demonstrate that ES cell-derived glial precursors and immature oligodendrocytes can myelinate axons in the immature normal and mutant CNS.

In the fifth study *(76)*, ES cells were induced with the 4 – /4 + RA protocol. Adult rats received a moderate-to-severe thoracic spinal cord contusion injury from the NYU weight-drop device *(78)*. Dissociated EBs were transplanted 9 d later directly into the cavity that was forming in the center of the spinal cord. Examinations 2 and 5 wk after transplant revealed that ES cell-derived cells had migrated into host tissues up to 1 cm from the injection site. Immunostaining indicated that ES cells had differentiated into neurons (<10% of all identifiable ES cells), astrocytes (~30%), and oligodendrocytes (~60%). Transplants also improved locomotor activity in the hindlimbs as assessed by the Basso, Beattie, Bresnahan (BBB) locomotor rating scale *(79)*. ES cell-transplanted animals, but not control animals transplanted with vehicle medium or adult mouse neocortical cells, recovered the ability to partially support their body weight with their hindlimbs and to make occasional steps (BBB scale 9–10). Controls could not bear weight on their hindlimbs and dragged their hindquarters (BBB score 7–8).

In the sixth study *(50)*, GRPs were derived from ES cells using *oligospheres* as an intermediate stage of differentiation. RA-induced (4 – /4 +) EBs were dissociated and plated on a nonadhesive substrate in defined medium containing bFGF. Spherical cell aggregates formed under these conditions; when dissociated and plated on adhesive substrates in the presence of oligosphere-conditioned medium, they differentiated primarily into oligodendrocytes. ES cell-derived oligodendrocytes were shown to be capable of myelinating multiple axons in vitro. Transplanted ES cell-derived oligodendrocytes were also shown to myelinate axons in the MBP mutant *shiverer* mouse and in the chemically demyelinated spinal cord of adult rats.

In these initial studies, no tumors were observed other than teratoma-like growths in the ventricles of some of the rats that Brüstle et al. transplanted intraventricularly on E17 *(75)*. Together, these studies demonstrate the potential of ES cells for transplantation and for replacing lost neural cells. Genetically engineering ES cells with readily detectable markers will allow for easier identification of transplanted cells and their processes. This will be a substantial advance over past studies in which transplanted cells and their processes were difficult to discern from host cells.

NONMURINE ES CELLS

ES cell-like pluripotent cells have been successfully derived from a growing number of species other than mouse, including fish *(80–82)*, chicken *(83)*, rat *(84,125)*, hamster *(85)*, rabbit *(86–88)*, mink *(89,90)*, pigs *(91–97)*, cows *(92,98,99)*, sheep *(92,100)*, nonhuman primates *(3,4,101,102)*, and most recently, humans *(5,6,40)*. Strict criteria satisfying designation of ES cells are only available in mouse ES cells (for review of these criteria, see ref. *129*). In particular, the clonal demonstration of somatic cell differentiation from all three germ layers is not practical with non-human primate and human cells because of technical limitations and ethical considerations. Evaluation of neural differentiation in these additional species has been limited. However, neural structures are prominent in teratomas derived from transplanted rhesus ES cells *(3,101)*. In particular, ES-derived cells have a propensity to produce organized neuroepithelia in culture or in teratomas *(5)*.

The emphasis on murine ES-cell studies has been based primarily on the genetic flexibility this system, which has fueled the transgenic animal boom over the past 10 yr. Moreover, the mouse embryo develops as a ball that can be cultured easily in nonadhesive plates as floating aggregates of cells. This feature simplifies differentiation procedures because the aggregates can be easily handled. In most other species, the early embryo grows as a disc that tends to adhere to the culture dish and undergo spontaneous differentation. Early receptor/signal transduction systems also vary between species. For example, LIF does not prevent nonhuman primate and human ES cell differentiation. This problem has hindered studies with primate ES cells *(3–5,40,101,102)* so identification of primate LIF analogs would progress.

NONHUMAN PRIMATE ES CELLS, HUMAN ES CELLS, AND HUMAN PGC-DERIVED CELLS

ES cell-like cells have been isolated from excess human blastocysts produced through in vitro fertilization *(5,7)*. Simultaneously, in 1998, Gearhart and colleagues derived human PGCs *(6)*. These scientific strides heightened public and governmental awareness and ethical concerns. Discovery became harnessed to the growing debate over right-to-life issues, which is likely to continue. Such debates have dramatically hindered scientific progress in this arena because federal funding and even private monies have been withheld from work relating to human ES cells. Despite these hindrances, some progress has been made. Thomson and colleagues in Wisconsin, Gearhart and colleagues in Baltimore, and Reubinoff's group in Australia have isolated pluripotent-like human cells *(5–7)*. Benvenisty and colleagues recently replicated the differentiation of human ES cells into embryoid bodies comprising cells from all three embryonic germ layers *(126)*. Other groups in Europe and Asia have also invested heavily in the technology but have yet to report in peer-reviewed journals. The private sector is perhaps the most heavily invested, with many biotechnology companies such as Geron venturing into human ES-cell technology. It is probable that technical advances with human ES cells will parallel those in mouse, and may even may move forward more quickly by benefitting from the advances already made in ES-cell technology from other vertebrate species. The recent approval of federal funding for scientific study of human ES cells will allow much needed progress in the U.S.

Nonhuman Primate ES-Cell Lines

The first nonhuman primate ES cell-like cells were derived from the *Rhesus* monkey, an Old World species, in the laboratory of Thompson and colleagues in Wisconsin *(3)*. They were isolated by culturing and subsequently cloning cells from the inner cell mass of a preimplantation blastocyst (R278.5 cell line). The line remained undifferentiated in continuous passage for more than a year. It also maintained a normal XY karyotype and expressed cell-surface markers characteristic of ES cells and human embryonal carcinoma (EC) cells. These markers included alkaline phosphatase, stage-specific embryonic antigen 3, stage-specific embryonic antigen 4, TRA-1-60, and TRA-1-81. The cells grew well on mouse fibroblast feeder layers, but they differentiated and died in the absence of fibroblasts, despite the presence of LIF, which permits the cultivation of undifferentiated murine ES cells without feeder layers. Differentiation even in the presence of LIF appears to be a feature common to all ES cell-like cells derived from primates to date *(3,4,6,7,101,102)*.

Cultured rhesus ES cells were shown to form derivatives of the trophoectoderm. In additional tests, they were transplanted into leg muscle and testis of severe combined immunodeficiency (SCID) mice. By 8 to 15 wk after transplantation, teratomas had formed that consistently contained derivatives of all three embryonic germ layers: endoderm, mesoderm, and ectoderm. These included gut-like structures containing smooth muscle, ciliated and nonciliated columnar epithelium, mucus-secreting goblet cells (endoderm); bone, cartilage, striated and smooth muscle (mesoderm); and stratified squamous epithelium with hair follicles, ganglia, and stratified layers of neural cells in patterns of developing neural tubes (ectoderm).

Neural differentiation in vitro has not been demonstrated in rhesus ES cells, but a subsequent study by Thomson's group demonstrated expression of markers for differentiated neuronal and glial cells in teratomas formed 5–12 wk after transplantation into SCID mouse muscle *(101)*. All the teratomas exhibited neural differentiation that included neural tube-like structures, embryonic ganglia, dispersed neurons, and brain-like gray matter. The neural tube-like structures were at different developmental stages, but there was a general progression from simple tubular structures at 5–6 wk to organized, highly stratified structures possessing well-defined ventricular and mantle layers at 7 wk. More advanced neural structures were not observed. Ganglion-like structures were always present. Individual neurons were NF+ and seen in teratomas at all stages (between 5–12 wk). Axonal tracts were rare at 5 wk but consistently present at 6 wk and later. Increased astrocytic differentiation (GFAP+) was observed as teratomas aged: not present at 5 wk, rare at 6 wk, and consistently present at 7 wk or later. Therefore, some aspects of normal neural differentiation were recapitulated in teratomas from rhesus ES cells. To date, Thomson's group has isolated a total of seven rhesus ES cell lines *(102)*. These cells offer a more complete model of the differentiation of diverse neuronal cell types, as described above, than human EC cells (NTERA2 cells). NTERA2 cells do not form well-organized ganglia or neural tube-like structures *(18)*.

ES cell-like cells have also been isolated from another nonhuman primate, the common marmoset (*Callithrix jacchus*), a New World monkey *(102)*. Eight pluripotent cell lines were derived from preimplantation blastocysts, all expressing an antigen profile consistent with pluripotent ES cells. Two cell lines were passaged for more than 1 yr and remained undifferentiated and euploid. Although EB formation was demonstrated in culture, evaluation of neural differentiation was incomplete. A recent report by Thomson indicates that ES cells can consistently differentiated into neurons in vitro *(102)*.

There are many developmental differences between mice and primates, so nonhuman primate ES cells will eventually provide important models for human development and human disease. Thus, transplantation studies in nonhuman primates species will be important steps toward potential transplantation of stem cells in humans for neural cell replacement. The need for rapid expansion of nonhuman primate ES cell studies is heightened by the current U.S. limitations on federally funded studies using human ES cells.

Human ES-Cell Lines

The first derivation of human ES cell-like cells was reported in 1998 *(5)*. Human embryos produced by in vitro fertilization were cultured to the bastocyst stage, and cells from the inner cell mass were isolated by immunosurgery. These cells were plated on irradiated mouse embryonic fibroblast feeder layers, and ES cell-like cells were isolated. However, neural differentiation studies were limited to evaluating teratoma formation in five isolated cell lines transplanted into SCID mice. The teratomas contained all three germ layers as well as neuroepithelial rosettes and neural ganglia (after 4 mo).

More recently, Pera and colleagues *(7)* isolated ES cells from human blastocysts and provided evidence of neural cell differentiation and the formation of neural progenitors. They isolated two ES cell lines that could be passaged 64 (HES-1) and 44 (HES-2) times, corresponding to a minimum of approx 384 and 264 population doublings, respectively. Both lines were successfully recovered from cryopreservation. This chapter was also the first to describe Oct-4 expression in human ES cells. Oct-4 is a POU domain transcription factor whose expression is limited to pluripotent cells, at least in mice. Expression of Oct-4 in the zygote is essential for establishing the pluripotent stem-cell population of the inner cell mass *(103)*. When either the HES-1 or -2 lines were transplanted into the testis of SCID mice, teratomas formed. The tumors contained derivatives of all three embryonic germ layers, including ganglionic structures and neural rosettes (examined 6–7 wk after transplantation). Prolonged culture (4–7 wk) of these ES cells at high densities and under conditions that supported differentiation produced multicellular aggregates that contained differentiated neural structures. N-CAM+ and NF+ cells were identified. Typically the N-CAM+ cells grew in small cellular aggregates with short processes. Visually guided isolation of the characteristic aggregates and further culturing in serum-free medium generated spherical structures primarily containing cells immunopositive for N-CAM, the intermediate neurofilament proteins nestin and vimentin, and the transcription factor Pax-6.

These neurospheres adhered to adhesive substrate, and their edges sprouted a monolayer of differentiated cells that displayed morphologic characteristics of postmitotic, differentiated neurons. They expressed NF-H, β-tubulin, NF-M, MAP-2, synaptophysin, glutamate, GAD, and $GABA_A\alpha2$ receptors. It is unclear whether the population of N-CAM+ cells in this study is similar to the N-CAM + neural stem population of NRPs described by Rao and colleagues *(109)* that can give rise to differentiated neurons. The presence or absence of glial cells was not assessed in these studies of human-derived N-CAM+ cells. It is possible that multiple stem cells were present in the visually identified aggregates used to culture N-CAM+ and differentiated neurons. Clonal examination of the putative N-CAM+ population is readily feasible using standard cloning techniques, FACS, and immunopanning. Most recently, Benvensity and colleagues reported successful differentiation of human ES cells into polycystic embryoid bodies that contain cells representative of the three embryoinic germ layers *(126)*. They used the H9 ES cell clone from Thomson and colleagues *(5)*. They demonstrated evidence for differentiation of hematopoetic cells, myocardial cells, and neuronal cells.

Human PGC-Derived Cell Lines

In 1998, isolation of both human PGC cells *(6)* and human ES cells was reported. PGCs were isolated from the gonadal ridges and mesenteries of 5–9 wk postfertilization embryos and cultured on STO fibroblast feeder layers. During 2–3 wk in culture, embryonic germ-cell colonies formed. Normal karyotypes were maintained after prolonged continuous passage. Formation of EBs in culture produced a wide variety of differentiated cell types. Derivatives of all three embryonic germ layers, including neuroepithelia and cells immunopositive for the neuronal marker NF and the glial marker S-100 appeared.

It should be noted that blastocyst-derived human ES cells differ from the stem cell lines derived from PGCs *(6)*. Unlike human ES cells, PGC-derived cell lines do not grow as flat monolayers or express the cell-surface SSEA-1 epitope, and they are at least partially dependent on exogenous LIF and bFGF. However, they are quite similar in most other characteristics.

Relationship of Human ES Cells to Other Human and Nonprimate Pluripotent Cell Lines

The human ES cell lines isolated in the two studies discussed above have similar properties and also share features with pluripotent human EC cells and monkey ES cells. Aside from pluripotency, common features of spontaneously differentiating human EC *(104–105)* and primate ES cells *(3,4,101,102)* include similar morphologies, expression of surface carbohydrate epitopes (e.g., SSEA-3, SSEA-4, TRA 1-60, TRA 1-81), expression of the pericellular matrix proteoglycan detected by GCTM-2, and lack of response to LIF or related members of the LIF cytokine family.

Human and monkey ES and EC cells differ substantially from mouse ES cells in terms of morphology, surface-marker expression, and response to LIF. However, both mice and human ES and EC cells express the transcription factor Oct-4 and downregulate its expression as they differentiate. This shared characteristic may aid in isolating and characterizing human and nonhuman primate ES cells.

At present, spontaneous differentiation hinders the culture of human ES cells; therefore, the culture system cannot support clonal growth. Identification of a factor equivalent to LIF that prevents spontaneous differentiation will dramatically facilitate the study of human ES cells in vitro. Clonal isolation of NPs, NRPs, and GRPs is theoretically possible and also may facilitate simple studies of human neural function in vitro. To date, however, no diploid human pluripotent stem cell has been cloned.

RELATIONSHIP BETWEEN NEURAL STEM CELLS DERIVED FROM ES CELLS AND THOSE DERIVED FROM EMBRYONIC AND ADULT CNS

The relationship between multipotent neural stem cells derived from adult and embryonic CNS has just begun to be evaluated. The relationship between these stem cells and neural stem cells isolated from ES cells is also in its infancy. It is possible that the three types of cells will possess different characteristics and behave in different ways. The characteristics used to select and classify stem cells are based on: (a) a few immunologic markers that are held to be stage-specific, (b) mitogen/growth factor-dependence of stem cell propagation, and (c) the phenotypes of the cells that differentiate. Stem cells isolated from ES cells are influenced neither by the myriad of three-dimensional interactions between the three embryonic germ layers nor by contact with the terminally differentiated cell types to which CNS-derived stem cells are exposed. For example, later staged CNS stem cells isolated by these methods have varying characteristics based on the CNS location from which they were isolated (for review, see refs. *106–108* and other relevant chapters in this volume). Rao and colleagues *(109)*, one of the first groups to compare murine neural stem cells derived from embryos or ES cells, observed numerous differences between the two. Recently, gene chip technology has been applied to ES cell differentiation and such techniques may prove very useful in comparing "genetic fingerprints" of ES-derived stem cells and CNS isolated stem cells *(127)*.

THE ES CELL SYSTEM: ADVANTAGES, LIMITATIONS, AND POTENTIAL

ES-cell culture systems offer several advantages over primary CNS-derived culture systems for neuroscience studies. First, ES cell lines can replicate indefinitely and therefore provide an endless supply of normal cells without further harvesting of tissues *(1,2,10)*. Second, ES cells are totipotent and can give rise to the entire developmental lineage, providing an otherwise unobtainable model for studying early embryonic development. Third, and most important, ES cells are the most genetically flexible stem cell. Recent advances in ES-cell genetic engineering make it possible to generate diploid gene knockouts and knockins at predetermined genetic loci

(110,111). This provides powerful genetic tools for examining mechanisms such as myelination or cell injury and death. Therefore, hypotheses can be tested genetically in culture, avoiding the costs and manpower associated with generating transgenic animals and maintaining animal colonies. The ES-cell system also avoids the lethality issues troubling transgenic animal knockout studies: deletion of important genes often impairs survival.

The ability to insert ES-cell markers that may be expressed in differentiated cells is another important advantage of the ES-cell system. For example, protein expression in specific regions of the CNS can be monitored with markers such as LacZ or GFP, while markers such as tau-LacZ (for axons) or PLP-LacZ (for myelin) can be used in specific cell types or subcellular strucutres. It may also be possible to activate or deactivate the expression of these markers depending on different experimental objectives. A similar tagging system could be used to follow gene activation. These systems can be employed in vivo and in vitro, both in living animals and postmortem. The ability to place genes in known genetic sites helps prevent downregulation during differentiation, a common problem with simple transfection techniques. Genetic expression of marker proteins and/or resistance genes can be used to select, enrich, or purify cell subtypes or later stage stem-cell populations.

Insertion of genetic markers will prove to be a simple and powerful mechanism for obtaining purified populations of selected stem cells or differentiated cells with specific phenotypes. As the number of genetically modified ES cell lines grows, supplies of frozen ES cells will be accessible to researchers for "off-the-shelf" culture studies. As long as the genetic manipulations are completed in germline competent ES cell clones, the possibility of producing corresponding transgenic animals also exists.

The fourth advantage of the ES cell system is the ability to prove that ES cells are genetically normal by producing normal chimeric animals after reimplanting the cells into host blastocysts. This ultimate test is not yet possible for any other stem cell. However, it is not always practical (e.g., in primates or other species with long fetal periods) or ethical (e.g., in humans). Fifth, there are many advantages to studying human ES cells instead of primary human cells.

Although animal models can be used to approximate the human condition, they are only models, and most are not entirely accurate. There are many disorders unique to humans for which animal models do not even exist. For example, hepatitis kills millions of people annually, but there are no suitable animal models of this disease because the viruses that cause human hepatitis infect only human cells. The advantages of establishing in vitro model systems, using human ES cells, to evaluate such human cell-specific diseases, are innumerable. The teratogenic properties and cellular toxicities of many drugs are also species-specific. One further advantage would be the ability to test cultured human cells for toxic effects before conducting Phase I clinical trials.

The greatest impediment to progress in the ES cell field is the federal ban on funding for human ES cell-related work. Although federally funded scientists can now study human ES cells, they cannot create embryos, and the solution to the political and ethical problems involved will probably take time and courage to work out. The absence of an LIF-like molecule to prevent spontaneous cell differentiation of primate ES cells is also a big hurdle. Without it, the process of passaging and maintaining undifferentiated primate ES cells is very difficult.

The risk of generating tumors in transplantation studies deserves specific discussion. All dividing cells, including ES cells, share the theoretical risk of tumor formation. Furthermore, all dividing cells will form teratomas (ectopic normal tissues), if the required combination of factors is present at the site of transplantation. In fact, teratoma formation is used during in ES cell derivation to demonstrate the ability to generate cells from all three embryonic germ-cell layers. Typically, these tests must be conducted in hosts with markedly impaired cellular immune responses (SCID mice, systemic immunosuppression) and the cells must be transplanted into supportive tissues (subcutaneous, intratesticular, intramuscular).

Despite teratoma risks, it is important to note that signs of malignant tumor formation, such as tissue invasion or metastasis, have never been reported in any study of ES cell transplantation. However, more and longer-term studies need to be completed to evaluate this danger fully. Also, several ES cell characteristics may be used to limit the risk or to eliminate transplanted cells if such an aberration occurred. Transplanting isolated, pure populations of lineage-restricted, later staged stem cells would inherently prevent the formation of teratomas or ectopic tissue. It may also be possible to genetically engineer ES cells with a fail-safe mechanism that could selectively eliminate them in a controlled fashion. Cells could, for instance, be equipped with an inducible promoter capable of being driven by a unique, exogenously administered agent such as a pro-apoptotic molecule. The resulting inducible self-destruction program could effectively eliminate unwanted transplanted cells via a mechanism—apoptosis—that does not trigger inflammation.

Ultimately, the ES cell system will provide a powerful set of biologic tools for scientific study. The magnitude of its potential can be predicted from the recent boom in transgenic animals, which are created from ES cells.

This new field has revolutionized neuroscience as will studies of the ES cells themselves. This system offers powerful therapeutic possibilities, particularly in humans, because all disease ultimately involves the death or dysfunction of cells. Replacing specfic types or abnormal cells, such as those involved in Parkinson's disease (dopaminergic neurons), Alzheimer's disease (cholinergic neurons), and juvenile-onset diabetes mellitus (pancreatic β-islet cells), could be potentially lifelong treatments.

Because undifferentiated ES cells can proliferate indefinitely in culture, individual cell lines could be extensively evaluated for pathogens or karyotypic abnormalities before being used for transplantation. Moreover, their pluripotency, infinite expandability, and their ability to be generated and differentiated in vitro make them a potentially limitless source of replacement cells. Multiple human ES cell lines could be typed for major histocompatability complex (MHC) characteristics to closely match patient and donor cells. Their capacity for indefinite proliferation also could permit the use of genetic techniques for changing their surface proteins or other characteristics that might provoke unwanted immune responses. Finally, ES cells could be genetically modified so that their differentiated derivatives could actively combat specific diseases.

ES cells have proved to be powerful allies in our quest for scientific discovery, and their full potential is only now being realized. It will be exciting to look back in the year 2010 to see how far we have come with their aid.

ACKNOWLEDGMENTS

The author would like to thank Drs. David Gottlieb, Terrance Holekamp, and Mike Howard for critical discussions and editorial comments. This work was supported by NIH grants, NS01931, NS39577, and NS37927, as well as bt the W. M. Keck Foundation and National Football League Charities.

REFERENCES

1. Evans, M. J. and Kaufman, M. H. (1981) Establishment in culture of pluripotential cells from mouse embryos. *Nature* **292,** 154–156.
2. Martin, G. R. (1981) Isolation of a pluripotent cell line from early mouse embryos cultured in medium conditioned by teratocarcinoma stem cells. *Proc. Natl. Acad. Sci. USA* **7,** 7634–7638.
3. Thomson, J. A., Kalishman, J., Golos, T. G., Durning, M., Harris, C. P., Becker, R. A., and Hearn, J. P. (1995) Isolation of a primate embryonic stem cell line. *Proc. Natl. Acad. Sci. USA* **92,** 7844–7848.
4. Thomson, J. A., Kalishman, J., Golos, T. G., Durning, M., Harris, C. P., and Hearn, J. P. (1996) Pluripotent cell lines derived from common marmoset (*Callithrix jacchus*) blastocysts. *Biol. Reprod.* **55,** 254–259.

5. Thomson, J. A., Itskovitz-Eldor, J., Shapiro, S. S., Waknitz, M. A., Swiergiel, J. J., Marshall, V. S., and Jones, J. M. (1998) Embryonic stem cell lines derived from human blastocysts. *Science* **282,** 1145–1147.

6. Shamblott, M. J., Axelman, J., Wang, S., Bugg, E. M., Littlefield, J. W., Donovan, P. J., et al. (1998) Derivation of pluripotent stem cells from cultured human primordial germ cells. *Proc. Natl. Acad. Sci. USA* **95,** 13,726–13,731.

7. Reubinoff, B. E., Pera, M. F., Fong, C. Y., Trounson, A., and Bongso, A. (2000) Embryonic stem cell lines from human blastocysts: somatic differentiation in vitro. *Nature Biotechnol.* **18,** 399–404.

8. Bradley, A., Evans, M., Kaufman, M. H., and Robertson, E. (1984) Formation of germ-line chimeras from embryo-derived teratocarcinoma cell lines. *Nature* **309,** 255–256.

9. Nagy, A., Rossant, J., Nagy, R., Abramow-Newerly, W., and Roder, J. C. (1993) Derivation of completely cell-culture-derived mice from early-passage embryonic stem cells. *Proc. Natl. Acad. Sci. USA* **90,** 8424–8428.

10. Suda, Y., Suzuki, M., Ikawa, Y., and Aizawa, S. (1987) Mouse embryonic stem cells exhibit indefinite proliferative potential. *J. Cell. Physiol.* **133,** 197–201.

11. Matsui, Y., Toksoz, D., Nishikawa, S., Nishikawa, S., Williams, D., Zsebo, K., and Hogan, B. L. (1991) Effect of Steel factor and leukaemia inhibitory factor on murine primordial germ cells in culture. *Nature* **353,** 750–752.

12. Resnick, J. L., Bixler, L. S., Cheng, L., and Donovan, P. J. (1992) Long-term proliferation of mouse primordial germ cells in culture. *Nature* **359,** 550–551.

13. Stewart, C., Gadi, I., and Bhatt, H. (1994) Stem cells from primordial germ cells can reenter the germ line. *Dev. Biol.* **161,** 626–628.

14. Labosky, P., Barlow, D., and Hogan, B. (1994) Mouse embryonic germ (EG) cell lines: transmission through the germline and differences in the methylation imprint of insulin-like growth factor 2 receptor *(Igf2r)* gene compared with embryonic stem (ES) cell lines. *Development* **120,** 3197–3204.

15. Solter, D. and Knowles, B. (1978) Monoclonal antibody defining a stage-specific mouse embryonic antigen (SSEA-1). *Proc. Natl. Acad. Sci. USA* **75,** 5565–5569.

16. Kannagi, R., Cochran, N., Ishigami, F., Hakomori, S., Andrews, P., Knowles, B., and Solter, D. (1983) Stage-specific embryonic antigens (SSEA-3 and -4) are epitopes of a unique globo-series ganglioside isolated from human teratocarcinoma cells. *EMBO J.* **2,** 2355–2361.

17. Andrews, P. W., Banting, G., Damjanov, I., Arnaud, D., and Avner, P. (1984) Three monoclonal antibodies defining distinct differentiation antigens associated with different high molecular weight polypeptides on the surface of human embryonal carcinoma cells. *Hybridoma* **3,** 347–361.

18. Andrews, P. W., Damjanov, J., Simon, D., Banting, G., Carlin, C., Dracopoli, N., and Fogh, J. (1984) Pluripotent embryonal carcinoma clones derived from the human teratocarcinoma cell line Tera-2. *Lab. Invest.* **50,** 147–162.

19. Whittemore, S. R. and Snyder, E. Y. (1996) Physiological relevance and functional potential of central nervous system-derived cell lines. *Mol. Neurobiol.* **12,** 13–38.

20. Andrews, P. W. (1998) Teratocarcinomas and human embryology: pluripotent human EC cell lines. Review article. *APMIS* **106,** 158–167.
21. Ourednik, K., Ourednik, J., Park, K. I., and Snyder, E.Y. (1999) Neural stem cells — versatile tool for cell replacement and gene therapy in the central nervous system. *Clin. Genet.* **56,** 267–278.
22. Park, K. I., Liu, S., Flax, J. D., Nissim, S., Stieg, P. E., and Snyder, E. Y. (1999) Transplantation of neural progenitor and stem cells: developmental insights may suggest new therapies for spinal cord and other CNS dysfunction. *J. Neurotrauma* **16,** 675–687.
23. Vescovi, A. L. and Snyder, E. Y. (1999) Establishment and properties of neural stem cell clones: plasticity in vitro and in vivo. *Brain Pathol.* **9,** 569–598.
24. Williams, R. L., Hilton, D. J., Pease, S., Willson, T. A., Stewart, C. L., Genring, D. P., et al. (1988) Myeloid leukemia inhibitory factor maintains the developmental potential of embryonic stem cells. *Nature* **336,** 684–687.
25. Rohwedel, J., Maltsev, V., Bober, E., Arnold, H. H., Hescheler, J., and Wobus, A. M. (1994) Muscle cell differentiation of embryonic stem cells reflects myogenesis in vivo: developmentally regulated expression of myogenic determination genes and functional expression of ionic currents. *Dev. Biol.* **164,** 87–101.
26. Risau, W., Sariola, H., Zerwes, H. G., Sasse, J., Ekblom, P., Kemler, R., and Doetschman, T. (1988) Vasculogenesis and angiogenesis in embryonic-stem-cell-derived embryoid bodies. *Development* **102,** 471–478.
27. Wang, R., Clark, R., and Bautch, V. L. (1992) Embryonic stem cell-derived cystic embryoid bodies form vascular channels: an in vitro model of blood vessel development. *Development* **114,** 303–316.
28. Doetschman, T. C., Eistetter, H., Katz, M., Schmidt, W., and Kemler, R. (1985) The in vitro development of blastocyst-derived embryonic stem cell lines: formation of visceral yolk sac, blood islands and myocardium. *J. Embryol. Exp. Morphol.* **87,** 27–45.
29. Wiles, M. and Keller, G. (1991) Multiple hematopoietic lineages develop from embryonic stem (ES) cells in culture. *Development* **111,** 259–267.
30. Klug, M. G., Soonpaa, M. H., Koh, G. Y., and Field, L. J. (1996) Genetically selected cardiomyocytes from differentiating embryonic stem cells form stable intracardiac grafts. *J. Clin. Invest.* **98,** 216–224.
31. Miller-Hance, W., LaCorbiere, M., Fuller, S., Evans, S., Lyons, G., Schmidt, C., et al. (1993) In vitro chamber specification during embryonic stem cell cardiogenesis. *J. Biol. Chem.* **268,** 25,244–25,252.
32. Rohwedel, J., Sehlmeyer, U., Shan, J., Meister, A.,and Wobus, A. M. (1996) Primordial germ cell-derived mouse embryonic germ (EG) cells in vitro resemble undifferentiated stem cells with respect to differentiation capacity and cell cycle distribution. *Cell Biol. Int.* **20,** 579–587.
33. Pederson, R. A. (1994) Studies of in vitro differentiation with embryonic stem cells. *Reprod. Fertil. Dev* **6,** 543–552.
34. Keller, G. M. (1995) In vitro differentiation of embryonic stem cells. *Curr. Opin. Cell. Biol.* **7,** 862–869.

35. Strubing, C., Ahnert-Hilger, G., Shan, J., Wiedenmann, B., Hescheler, J., and Wobus, A. M. (1995) Differentiation of pluripotent embryonic stem cells into the neuronal lineage in vitro gives rise to mature inhibitory and excitatory neurons. *Mech. Dev.* **53**, 275–287.

36. Strubing, C., Rohwedel, J., Ahnert-Hilger, G., Wiedenmann, B., Hescheler, J., and Wobus, A.M. (1997) Development of G protein-mediated Ca2 + channel regulation in mouse embryonic stem cell-derived neurons. *Eur. J. Neurosci.* **9**, 824–832.

37. Rohwedel, J., Kleppisch, T., Pich, U., Guan, K., Jin, S., Zuschratter, W., et al. (1998) Formation of postsynaptic-like membranes during differentiation of embryonic stem cells in vitro. *Exp. Cell Res.* **239**, 214–225.

38. Angelov, D. N., Arnhold, S., Andressen, C., Grabsch, H., Puschmann, M., Hescheler, J., and Addicks, K. (1998) Temporospatial relationships between macroglia and microglia during in vitro differentiation of murine stem cells. *Dev. Neurosci.* **20**, 42–51.

39. Gottlieb, D. I. and Huettner, J. E. (1999) An in vitro pathway from embryonic stem cells to neurons and glia. *Cells Tissues Organs* **165**, 165–172.

40. Arnhold, S., Andressen, C., Angelov, D. N., Vajna, R., Volsen, S. G., Hescheler, J., and Addicks, K. (2000) Embryonic stem-cell derived neurons express a maturation dependent pattern of voltage-gated calcium channels and calcium-binding proteins. *Int. J. Dev. Neurosci.* **18**, 201–212.

41. Wobus, A. M., Grosse, R., and Schoneich, J. (1988) Specific effects of nerve growth factor on the differentiation pattern of mouse embryonic stem cells in vitro. *Biomed. Biochim. Acta* **47**, 965–973.

42. Wobus, A. M., Holzhausen, H., Jakel, P., and Schoneich, J. (1984) Characterization of a pluripotent stem cell line derived from a mouse embryo. *Exp. Cell Res.* **152**, 212–219.

43. Bain, G., Kitchens, D., Yao, M., and Gottlieb, D. I. (1995) Embryonic stem cells express neuronal properties in vitro. *Dev. Biol.* **168**, 342–357.

44. Bain, G., Yao, M., Huettner, J. E., Finley, M. F. A., and Gottlieb, D. I. (1998) Neuron-like cells derived in culture from P19 embryonal carcinoma and embryonic stem cells, in *Culturing Nerve Cells,* 2nd ed. (Banker, G. and Goslin, K., eds.), MIT Press, Cambridge, MA, pp. 189–212.

45. Fraichard, A., Chassande, O., Bilbaut, G., Dehay, C., Savatier, P., and Samarut, J. (1995) In vitro differentiation of embryonic stem cells into glial cells and functional neurons. *J. Cell Sci.* **108**, 3181–3188.

46. Okabe, S., Forsberg-Nilsson, K., Spiro, A. C., Segal, M., and McKay, R. D. G. (1996) Development of neuronal precursor cells and functional postmitotic neurons from embryonic stem cells in vitro. *Mech. Dev.* **59**, 89–102.

47. Finley, M. F., Kulkarni, N., and Huettner, J. E. (1996) Synapse formation and establishment of neuronal polarity by P19 and embryonic stem cells. *J. Neurosci.* **16**, 1056–1065.

48. Renoncourt, Y., Carroll, P., Filippi, P., Arce, V., and Alonso, S. (1998) Neurons derived in vitro from ES cells express homeoproteins characteristic of motoneurons and interneurons. *Mech. Dev.* **79**, 185–197.

49. Rohwedel, J., Guan, K., Zuschratter, S., Jin, S., Ahnert-Hilger, G., Furst, D., et al. (1998) Loss of beta1 integrin function results in a retardation of myogenic, but an acceleration of neuronal, differentiation of embryonic stem cells in vitro. *Dev. Biol.* **201,** 167–184.

50. Liu, S., Qu, Y., Stewart, T., Howard, M., Chakrabortty, S., Holekamp, T., and McDonald, J. W. (2000) Embryonic stem cells differentiate into oligodendrocytes and myelinate in culture and after spinal cord transplantation. *Proc. Natl. Acad. Sci. USA* **97,** 6126–6131.

51. Bain, G., Ray, W. J., Yao, M., and Gottlieb, D. I. (1996) Retinoic acid promotes neural and represses mesodermal gene expression in mouse embryonic stem cells in culture. *Biochem. Biophys. Res. Commun.* **223,** 691–694.

52. Tanabe, Y. and Jessell, T. M. (1996) Diversity and pattern in the developing spinal cord. *Science* **274,** 1115–1123.

53. Ericson, J., Morton, S., Kawakami, A., Roelink, H., and Jessell, T. M. (1996) Two critical periods of Sonic Hedgehog signaling required for the specification of motor neuron identity. *Cell* **87,** 661–673.

54. Ericson, J., Briscoe, J., Rashbass, P., van Heyningen, V., and Jessell, T. M. (1997) Graded sonic hedgehog signaling and the specification of cell fate in the ventral neural tube. *Cold Spring Harbor Symp. Quant. Biol.* **62,** 451–466.

55. Tao, W. and Lai, E. (1992) Telencephalon-restricted expression of BF-1, a new member of the HNF-3/fork head gene family, in the developing rat brain. *Neuron* **8,** 957–966.

56. Finley, M. F., Devata, S., and Huettner, J. E. (1999) BMP-4 inhibits neural differentiation of murine embryonic stem cells. *J. Neurobiol.* **40,** 271–287.

57. Dinsmore, J., Ratliff, J., Deacon, T., Pakzaban, P., Jacoby, D., Galpern, W., and Isacson, O. (1996) Embryonic stem cells differentiated in vitro as a novel source of cells for transplantation. *Cell Transplant.* **5,** 131–143.

58. Li, M., Pevny, L., Lovell-Badge, R., and Smith, A. (1998) Generation of purified neural precursors from embryonic stem cells by lineage selection. *Curr. Biol.* **8,** 971–974.

59. Brüstle, O., Jones, K. N., Learish, R. D., Karram, K., Choudhary, K., Wiestler, O. D., et al. (1999) Embryonic stem cell-derived glial precursors: a source of myelinating transplants. *Science* **285,** 754–756.

60. Richards, L. J., Kilpatrick, T. J., and Bartlett, P. F. (1992) De novo generation of neuronal cells from the adult mouse brain. *Proc. Natl. Acad. Sci. USA* **89,** 8591–8595.

61. Gritti, A., Parati, E. A., Cova, L., Frolichsthal, P., Galli, R., Wanke, E., et al. (1996) Multipotential stem cells from the adult mouse brain proliferate and self-renew in response to basic fibroblast growth factor. *J. Neurosci.* **16,** 1091–1100.

62. Zhang, S. C., Lipsitz, D., and Duncan, I. D. (1998) Self-renewing canine oligodendroglial progenitor expanded as oligospheres. *J. Neurosci. Res.* **54,** 181–190.

63. Zhang, S. C., Ge, B., and Duncan, I. D. (1999) Adult brain retains the potential to generate oligodendroglial progenitors with extensive myelination capacity. *Proc. Natl. Acad. Sci. USA* **96,** 4089–4094.

64. Bottenstein, J. E. and Sato, G. H. (1979) Growth of a rat neuroblastoma cell line in serum-free supplemented medium. *Proc. Natl. Acad. Sci. USA* **76,** 514–517.

65. Takahashi, J., Palmer, T. D., and Gage F. H. (1999) Retinoic acid and neurotrophins collaborate to regulate neurogenesis in adult-derived neural stem cell cultures. *J. Neurobiol.* **38,** 65–81.

66. Mayer-Proschel, M., Kalyani, A. J., Mujtaba, T., and Rao, M. S. (1997) Isolation of lineage-restricted precursors from multipotent neuroepithelial stem cells. *Neuron* **19,** 773–785.

67. Kalyani, A., Piper, D., Mujtaba, T., Lucero, M. T., and Rao, M. S. (1998) Spinal cord neuronal precursors generate multiple neuronal phenotypes in culture. *J. Neurosci.* **18,** 7856–7868.

68. Rao, M. S. and Mayer-Proschel, M. (1997) Glial-restricted precursors are derived from multipotent neuroepithelial stem cells. *Dev. Biol.* **188,** 48–63.

69. Rao, M. S., Noble, M., and Mayer-Proschel, M. (1998) A tripotential glial precursor is present in the developing spinal cord. *Proc. Natl. Acad. Sci. USA* **95,** 3996–4001.

70. Mountford, P., Nichols, J., Zevnik, B., O'Brien, C., and Smith, A. (1998) Maintenance of pluripotential embryonic stem cells by stem cell selection. *Reprod. Fertil. Dev.* **10,** 527–533.

71. Mountford, P. S., Zevnik, B., Duwel, A., Nichols, J., Li, M., Dani, C., et al. (1994) Dicistronic targeting vectors: reporters and modifiers of mammalian gene expression. *Proc. Natl. Acad. Sci. USA* **91,** 4303–4307.

72. Friedrich, G. and Soriano, P. (1991) Promotor traps in embryonic stem cells: a genetic screen to identify and mutate developmental genes in mice. *Genes Dev.* **5,** 1513–1523.

73. Scholer, H. R. (1991) Octomania: the POU factors in murine development. *Trends Genet.* **7,** 323–329.

74. Deacon, T., Dinsmore, J., Costantini, L. C., Ratliff, J., and Isacson, O. (1998) Blastula-stage stem cells can differentiate into dopaminergic and serotonergic neurons after transplantation. *Exp. Neurol.* **149,** 28–41.

75. Brüstle, O. et al. (1997) In vitro-generated neural precursors participate in mammalian brain development. *Proc. Natl. Acad. Sci. USA* **94,** 14,809–14,814.

76. McDonald, J. W., Liu, X.-Z., Qu, Y., Liu, S., Turetsky, D., Mickey, S. K., et al. (1999) Transplanted embryonic stem cells survive, differentiate, and promote recovery in injured rat spinal cord. *Nature Med.* **5,** 1410–1412.

77. Lund, R. D., Chang, F. L., Hankin, M. H., and Lagenaur, C. F. (1985) Use of a species-specific antibody for demonstrating mouse neurons transplanted to rat brains. *Neurosci. Lett.* **61,** 221–226.

78. Gruner, J. A. (1992) A monitored contusion model of spinal cord injury in the rat. *J. Neurotrauma* **9,** 123–128.

79. Basso, D. M., Beattie, M. S., and Bresnahan, J. C. (1995) A sensitive and reliable locomotor rating scale for open field testing in rats. *J. Neurotrauma* **12,** 1–21.

80. Sun, L., Bradford, C. S., Ghosh, C., Collodi, P., and Barnes, D. W. (1995) ES-like cell cultures derived from early zebrafish embryos. *Mol. Mar. Biol. Biotechnol.* **4,** 193–199.

81. Hong, Y., Winkler, C., and Schartl, M. (1996) Pluripotency and differentiation of embryonic stem cell lines from the medakafish (*Oryzias latipes*). *Mech. Dev.* **60,** 33–44.

82. Hong, Y., Winkler, C., and Scharti, M. (1998) Production of medakafish chimeras from a stable embryonic stem cell line. *Proc. Natl. Acad. Sci. USA* **95,** 3679–3684.

83. Pain, B., Clark, M. E., Shen, M., Nakazawa, H., Sakurai, M., Samarut, J., and Teches, R. J. (1996) Long-term in vitro culture and characterization of avian embryonic stem cells with multiple morphogenetic potentialities. *Development* **122,** 2339–2348.

84. Iannaccone, P. M., Taborn, G. U., Garton, R. L., Caplice, M. D., and Brenin, D. R. (1994) Pluripotent embryonic stem cells from the rat are capable of producing chimeras. *Dev. Biol.* **163,** 288–292.

85. Doetschman, T., Williams, P., and Maeda, N. (1988) Establishment of hamster blastocyst-derived embryonic stem (ES) cells. *Dev. Biol.* **127,** 224–227.

86. Giles, J. R., Yang, X., Mark, W., and Foote, R. H. (1993) Pluripotency of cultured rabbit inner cell mass cells detected by isozyme analysis and eye pigmentation of fetuses following injection into blastocysts or morulae. *Mol. Reprod. Dev.* **36,** 130–138.

87. Graves, K. H. and Moreadith, R. W. (1993) Derivation and characterization of putative pluripotential embryonic stem cells from preimplantation rabbit embryos. *Mol. Reprod. Dev.* **36,** 424–433.

88. Schoonjans, L., Albright, G. M., Li, J., Collen, D., and Moreadith, R. W. (1996) Pluripotential rabbit embryonic stem (ES) cells are capable of forming overt coat color chimeras following injection into blastocysts. *Mol. Reprod. Dev.* **45,** 439–443.

89. Sukoyan, M. A., Golubitsa, A. N., Zhelezova, A. L., Shilov, A. G., Vatolin, S. Y., Maximovsky, L. P., et al. (1992) Isolation and cultivation of blastocyst-derived stem cell lines from American mink (*Mustela vision*). *Mol. Reprod. Dev.* **33,** 418–431.

90. Sukoyan, M. A., Vatolin, S. Y., Golubitsa, A. N., Zhelezova, A. L., Semenova, L. A., and Serov, O. L. (1993) Embryonic stem cells derived from morulae, inner cell mass, and blastocysts of mink: comparisons of their pluripotencies. *Mol. Reprod. Dev.* **36,** 148–158.

91. Notarianni, E., Laurie, S., Moor, R. M., and Evans, M. J. (1990) Maintenance and differentiation in culture of pluripotential embryonic cell lines from pig blastocysts. *J. Reprod. Fertil.* **Suppl 41,** 51–56.

92. Evans, M. J., Notarianni, E., Laurie, S., and Moor, R. M. (1990) Derivation and preliminary characterization of pluripotent cell lines from porcine and bovine blastocysts. *Theriogenology* **33,** 125–128.

93. Piedrahita, J. A., Anderson, G. B., and Bondurant, R. H. (1990) On the isolation of embryonic stem cells: comparative behavior of murine, porcine, and ovine embryos. *Theriogenology* **34,** 879–901.

94. Piedrahita, J. A., Anderson, G. B., and Bondurant, R. H. (1990) Influence of feeder layer type on the efficiency of isolation of porcine embryo-derived cell lines. *Theriogenology* **34,** 865–867.

95. Strojek, R. M., Reed, M. A., Hoover, J. L., and Wagner, T. E. (1990) A method for cultivating morphologically undifferentiated embryonic stem cells from porcine blastocysts. *Theriogenology* **33,** 901–913.

96. Wheeler, M. B. (1994) Development and validation of swine embryonic stem cells: a review. *Reprod. Fertil. Dev.* **6,** 563–568.
97. Shim, H., Gutierrez-Adan, A., Chen, L., BonDurant, R., Behboodi, E., and Anderson, G. (1997) Isolation of pluripotent stem cells from cultured porcine primordial germ cells. *Biol. Reprod.* **57,** 1089–1095.
98. Saito, S., Strelchenko, N., and Niemann, H. (1992) Bovine embryonic stem cell-like cell lines cultured over several passages. *Roux Arch. Dev. Biol.* **201,** 134–141.
99. Cherny, R. A., Stokes, T. M., Merei, J., Lom, L., Brandon, M. R., and Williams, L. (1994) Strategies for the isolation and characterization of bovine embryonic stem cells. *Reprod. Fertil. Dev.* **6,** 569–575.
100. Handyside, A., Hooper, M. L., Kaufman, M. H., and Wilmut, I. (1987) Towards the isolation of embryonal stem cell lines from sheep. *Roux's Arch. Dev. Biol.* **196,** 185–190.
101. Thomson, J. A., Marshall, V. S., and Trojanowski, J. Q. (1998) Neural differentiation of rhesus embryonic stem cells. *APMIS* **106,** 149–156, 156–157.
102. Thomson, J. A. and Marshall, V. S. (1998) Primate embryonic stem cells. *Curr. Top. Dev. Biol.* **38,** 133–165.
103. Nichols, J., Zevnik, B., Anastassiadis, K., Niwa, H., Klewe-Nebenius, D., Chambers, I., et al. (1998) Formation of pluripotent stem cells in the mammalian embryo depends of the POU transcription factor Oct4. *Cell* **95,** 379–391.
104. Pera, M. F., Cooper, S., Mills, J., and Parrington, J.M. (1989) Isolation and characterization of a multipotent clone of human embryonal carcinoma-cells. *Differentiation* **42,** 10–23.
105. Roach, S., Cooper, S., Bennett, W., and Pera, M.F. (1993) Cultured cell lines from human germ cell tumours: windows into tumour growth and differentiation and early human development. *Eur. Urol.* **23,** 82–88.
106. McKay, R. (1997) Stem cells in the central nervous system. *Science* **276,** 66–71.
107. Kalyani, A. and Rao, M. S. (1999) Lineage determination in the developing spinal cord. *Cell Biol. Biochem.* **76,** 1–17.
108. Shihabuddin, L. S., Ray, J., and Gage, F. H. (1999) Stem cell technology for basic science and clinical applications. *Arch. Neurol.* **56,** 29–32.
109. Mujtaba, T., Piper, D. R., Kalyani, A., Groves, A. K., Lucero, M. T., and Rao, M. S. (1999) Lineage-restricted neural precursors can be isolated from both the mouse neural tube and cultured ES cells. *Dev. Biol.* **214,** 113–127.
110. Wilder, P. J., Kelly, D., Brigman, K., Peterson, C. L., Nowling, T., Gao, Q. S., et al. (1997) Inactivation of the FGF-4 gene in embryonic stem cells alters the growth and/or the survival of their early differentiated progeny. *Dev. Biol.* **192,** 614–629.
111. Hakem, R., Hakem, A., Duncan, G. S., Henderson, J. T., Woo, M., Soengas, M. S., et al. (1998) Differential requirement for caspase 9 in apoptotic pathways in vivo. *Cell* **94,** 339–352.
112. Matsui, Y., Zsebo, K., and Hogan, B. L. M. (1992) Derivation of pluripotential embryonic stem cells from murine primordial germ cells in culture. *Cell* **70,** 841–847.

113. Weiss, M., Keller, G., and Orkin, S. (1994) Novel insights into erythroid development revealed through in vitro differentiation of GATA-1-embryonic stem cells. *Genes Dev.* **8,** 1184–1197.

114. Lee, L. R., Mortensen, R. M., Larson, C. A., and Brent, G. A. (1994) Thyroid hormone receptor-α inhibits retinoic acid-responsive gene expression and modulates retinoic acid-stimulated neural differentiation in mouse embryonic stem cells. *Mol. Endocrinol.* **8,** 746–756.

115. Kawai, H., Sango, K., Mullin, K. A., and Proia, R. L. (1998) Embryonic stem cells with a disrupted GD3 synthase gene undergo neuronal differentiation in the absence of b-series gangliosides. *J. Biol. Chem.* **273,** 19,634–19,638.

116. Sirard, C., Kim, S., Mirtsos, C., Tadich, P., Hoodless, P. A., Itie, A., et al. (2000) Targeted disruption in murine cells reveals variable requirement for Smad4 in transforming growth factor beta-related signaling. *J. Biol. Chem.* **275,** 2063–2070.

117. Talbot, N. C., Rexroad, C. E. Jr., Pursel, V. G., and Powell, A. M. (1993) Alkaline phosphatase staining of pig and sheep epiblast cells in culture. *Mol. Reprod. Dev.* **36,** 139–147.

118. Talbot, N. C., Powell, A. M., Nel, N. D., Pursel, V. G., and Rexrod, C. E. Jr. (1993b) Culturing the epiblast cells of the pig blastocyst. *In Vitro Cell Dev. Biol.* **29,** 543–554.

119. Gajovic, S., St-Onge, L., Yokota, Y., and Gruss, P. (1997) Retinoic acid mediates Pax6 expression during in vitro differentiation of embryonic stem cells. *Differentiation* **62,** 187–192.

120. Grantyn, R., Perouansky, M., Rodriguez-Tebar, A., and Lux, H. D. (1989) Expression of depolarizing voltage- and transmitter-activated currents in neuronal precursor cells from the rat brain is preceded by a proton-activated sodium current. *Dev. Brain Res.* **49,** 150–155.

121. Petitte, J. N., Clark, M. E., Verrinder Gibbons, A. M., and Etches, R. J. (1990) Production of somatic and germline chimeras in the chicken by transfer of early blastodermal cells. *Development* **108,** 185–189.

122. Coucouvanis, E. and Martin, G. R. (1995) Signals for death and survival: a two-step mechanism for cavitation in the vertebrate embryo. *Cell* **83,** 279–287.

123. Baumrind, N. L., Parkinson, D., Wayne, D. B., Heuser, J. E., and Pearlman, A. L. (1992) EMA: a developmentally regulated cell-surface glycoprotein of CNS neurons that is concentrated at the leading edge of growth cones. *Dev. Dyn.* **194,** 311–325.

124. Hancock, C. R., Wetherington, J. P., Lambert, N. A., and Condie, B. G. (2000) Neuronal differentiation of cyropreserved neural progenitor cells derived from mouse embryonic stem cells. *Biochem. Biophys. Res. Comm.* **271,** 418–421.

125. Vassilieva, S., Guan, K., Pich, U., and Wobus, A. M. (2000) Establishment of SSEA-1 and Oct-4-expressing rat embryonic stem-like cell lines and effects of cytokines of the IL-6 family on clonal growth. *Exp. Cell. Res.* **258,** 361–373.

126. Itskovitz-Eldor, J. Schuldiner, M., Karsenti, D., Eden, A., Yanuka, O., Amit, M., Soreq, H., and Benvenisty, N (2000) Differentiation of human embryonic stem cells into embryoid bodies comprising the three embryonic germ layers. *Mol. Med.* **6,** 88–95.

127. Kelly, D. L. and Rizzinio, A. (2000) DNA microarray analyses of genes regulated during the differentiation of embryonic stem cells. *Mol. Reprod. Dev.* **56,** 113–123.

128. Deng, C., Zhang, P., Harper, J. W., Elledge, S., and Leder, P. (1995) Mice lacking p21CIP1/WAF1 undergo normal development, but are defective in G1 checkpoint control. *Cell* **82,** 675–684.

129. Pera, M. F., Reubinoff, B., and Trounson, A. (2000) Human embryonic stem cells. *J. Cell Sci.* **113,** 5–10.

130. Kleinsmith, L. J. and Pierce, G. B. (1964) Multipotentiality of single embryonal carcinoma cells. *Cancer Res.* **24,** 1544–1549.

131. Martin, G. R. (1980) Teratocarcinoma and mammalian embryogenesis. *Science* **209,** 768–776.

132. Stevens, L. C. (1980) Testicular, ovarian, and embryo-derived teratomas. *Cancer Surveys* **2,** 75–91.

10

Mobilizing Endogenous Stem Cells

Theo D. Palmer, Sophia Colamarino, and Fred H. Gage

INTRODUCTION

Over the last 50 yr, the long-held dogma that neurogenesis stops at birth has been gradually modified to allow specific exceptions to the rule. In many mammalian species, new neurons are continually added to the adult olfactory bulb *(1)* and hippocampus *(2–5)*. As more information is gathered, additional sites may be found that are unique to individual species *(6)*. Although natural neurogenic processes respond in a limited manner to damage in the CNS *(7–9)* the inability of the adult brain to regenerate fully following disease or injury still provides a concrete example of how the dogma holds true from a clinical point of view. In this context, the insight gained in defining neurogenic mechanisms in the adult may be relevant to central nervous system (CNS) repair since the successful therapeutic approaches will ultimately depend on the precise modulation of an extensive regulatory network.

THE DOGMA OF "NO-NEW-NEURONS": ACCUMULATED EVIDENCE FOR ANATOMIC EXCEPTIONS TO THE RULE

In spite of the recent attention to the phenomenon of adult neurogenesis, Ramón y Cajal's dogma of no new neurons is still largely correct when viewed in the context of developmental neurogenesis. *In utero*, neurogenesis operates with extraordinary dynamics to generate all the basic brain structures and circuitry that will be used for the remainder of an individual's adult life. At birth, all areas of the brain downregulate neurogenesis, and most areas stop producing neurons altogether (at least within the current limits of detection). The last 50 yr, however, have provided increasingly robust evidence for persistent neurogenesis in several anatomically restricted areas in the adult brain. The unambiguous examples are the hippocampal granule cell layer and olfactory bulb. Ongoing studies may show that this process extends to other areas of the neocortex as well *(6)*.

From: *Stem Cells and CNS Development*
Edited by: M. S. Rao © Humana Press Inc., Totowa, NJ

Monitoring Mitosis In Vivo

In the late 1950s, the earliest evidence for neurogenesis in the adult was observed following the systemic injection of radiolabeled thymidine ([³H]Tdr). Endogenous nucleoside pools become substituted with the labeled nucleoside, and newly synthesized DNA incorporates analog in proportion to its relative systemic concentration. The α-particle emissions can be visualized by autoradiography *(10–13)*. Since the number of silver grains developed in the photographic emulsion is dependent on the amount of [³H]Tdr incorporated, grain counts associated with each nucleus can provide reliable quantitative information on whether a cell has completed an entire S-phase or how many times a cell transits S-phase during a labeling period *(14)*.

More recently, antibodies that recognize nonradioactive thymidine analogs such as bromodeoxyuridine (BrdU) have been used in lieu of autoradiography *(2,3)*. In contrast to [³H]Tdr, nonradioactive analogs can be immunologically detected in any thickness of tissue that can be penetrated by the detecting antibody. Most studies now take further advantage of the exquisite sensitivity of immunodetection by combining it with avidin-biotin amplification and enzyme-linked detection. With enough amplification, even trace amounts of nucleoside analog can be detected and, by combining antibodies to BrdU with those that recognize lineage-specific epitopes, it is possible to identify the phenotype of labeled cells unambiguously.

Technical Caveats

Unfortunately, the remarkable sensitivity of immunodetection comes at a price. In the context of neurogenesis in the adult, it is possible that the small amounts of nucleoside analog incorporated into neuronal nuclei during DNA repair may introduce false positives. In addition, thymidine analogs can significantly influence cell behavior. At moderate substitution levels, BrdU can be mutagenic *(15,16)* and is known to alter gene expression patterns *(17–20)*. Subsequent changes in physiology may directly impact proliferative activity. For example, BrdU is known to enhance adrenal glucocorticoid levels in rats *(21,22)*, and adrenal steroids are known to suppress the proliferative activity of neural precursors in the adult hippocampus *(23)*. At higher concentrations, both [3H]Tdr and BrdU become directly cytotoxic, an effect that can potentially lead to the ablation of the population being evaluated *(24)*. Even now, it is possible to argue that labeling a cell may change its fate or make a nonneuronal cell inappropriately express phenotypic attributes of neurons. At the very least, the labeling paradigm may affect proliferative activity itself.

Fortunately, evidence for the addition of functional neurons in the adult is not entirely based on nucleoside incorporation data, and most studies include supporting evidence for the generation of new neurons. For example, in areas suspected of neurogenesis, the number of neurons actually increases over the life of the organism *(25)*. In addition, these newborn cells send long projections to appropriate target fields *(6,26,27)*. This is clearly a neuronal attribute that is unlikely to be displayed by abnormal glia. Ultrastructural analysis adds confidence that these new projecting cells display the intracellular architecture of neurons *(28)*. Furthermore, viral vectors that only infect dividing cells have unambiguously shown that precursors in the ventricular zone migrate to the olfactory bulb, where they differentiate into new periglomerular and granule cell interneurons *(29,30)*.

The State of the Art

Most recently, studies of adult neurogenesis utilize continuing advances in confocal imaging and multiple-immunolabeling methods to elegantly demonstrate co-localization of numerous neuron-specific markers with the labeled nucleus *(3)*. For example, more than two-thirds of the BrdU-labeled cells in the granule cell layer also express markers such as calbindin, neuron-specific enolase, and neuronal nuclear antigen (NeuN) *(3,31)* and are surrounded by synaptophysin, a vesicular protein localized to the synapse *(27)*. In the olfactory bulb, newborn cells express NeuN along with tyrosine hydroxylase or calbindin and are indistinguishable from surrounding periglomerular and granule cell interneurons that are not labeled with BrdU *(32,33)*.

THE ANATOMY OF NEUROGENESIS: THE VENTRICULAR AND HIPPOCAMPAL SUBGRANULE ZONES

The Ventricular Zone

The anatomy of neurogenesis provides the first clues as to how the adult may regulate the production of new neurons. As elegantly described by Alvarez-Buylla in Chapter 4, the dividing cells are not neurons but immature precursors similar to those seen in the developing brain. Cells destined to become neurons in the olfactory bulb originate within a residual ventricular zone (VZ) of the lateral ventricle. In rodents, the zone of highest neurogenic activity is in the rostral portion of the lateral ventricle overlying the striatum *(1,30)*. Within the VZ, which includes the ependyma and the adjacent subventricular zone (SVZ), precursors of several types form a contiguous lamina of dividing cells that are readily visualized by incorporation of nucleoside analogs (Fig. 1).

Both glia and neurons are generated by the precursors in the VZ *(1,30,34,35)*. Glial precursors (primarily of oligodendrocyte lineage) migrate throughout the brain *(34)*, whereas neuroblasts organize into chains of

B
HC
Ventricular
A
Zone
RMS
OB
C
A
SVZ
LV
Str
B
DG
B1
GCL
SGZ
C
OB
RMS

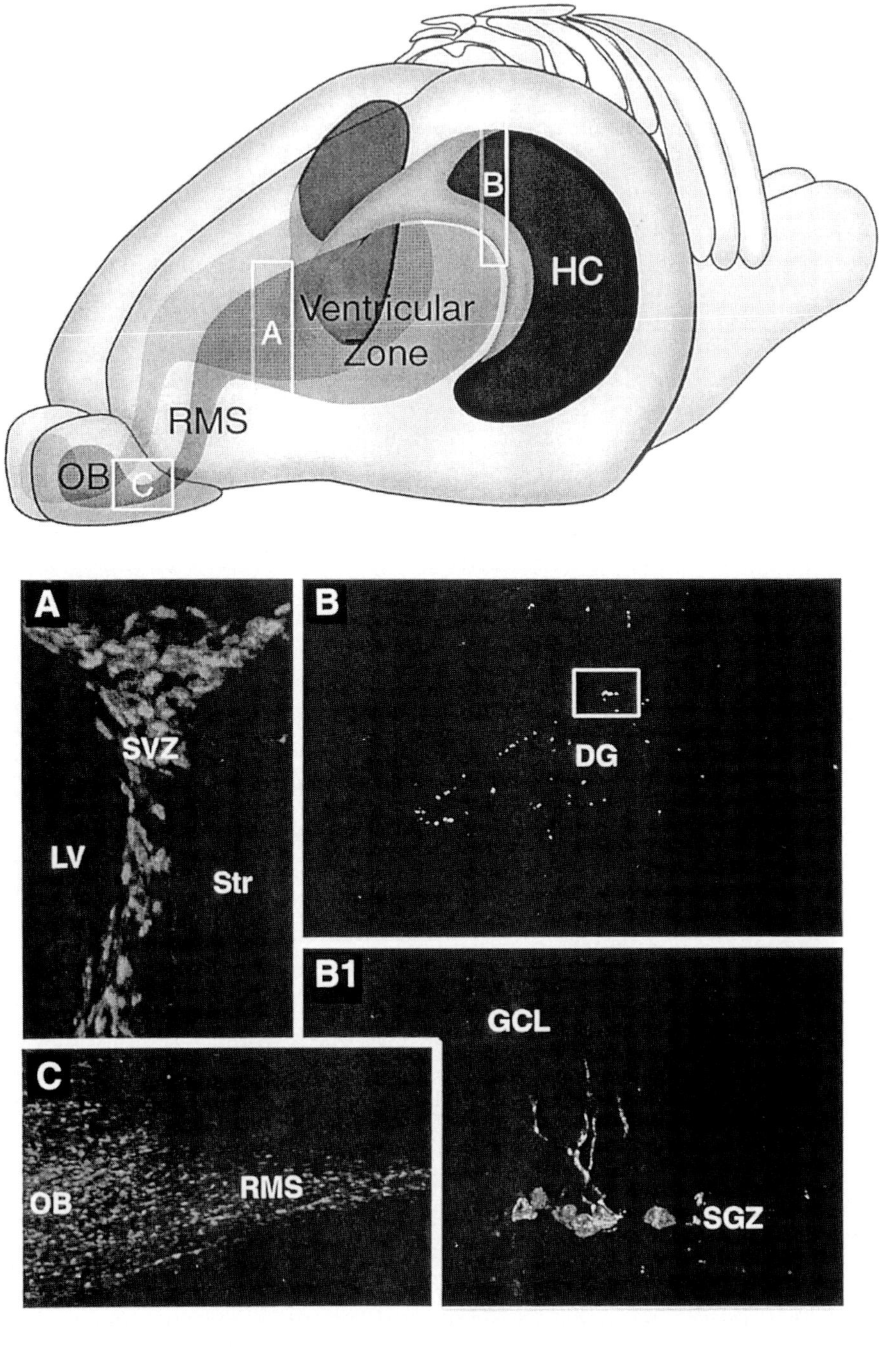

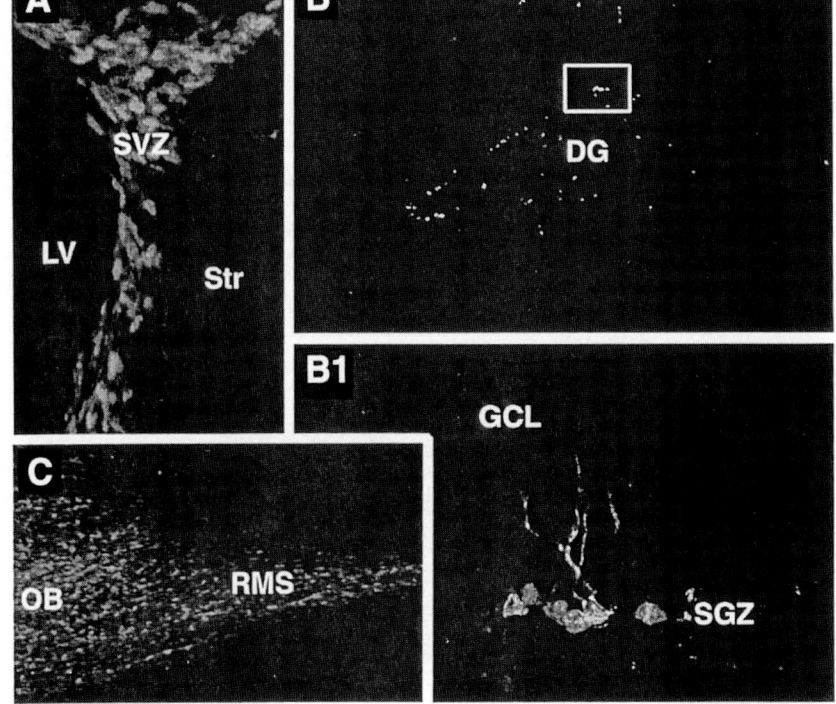

migrating cells that follow a rostral migratory path to the olfactory bulb *(1,30,36)*. Although both precursor types divide primarily within the SVZ proper, the migrating cells may complete one or more divisions en route to their final locations. Once neuroblasts arrive within the olfactory bulb, they complete the differentiation process and become integrated within the periglomerular and granule cell interneuron pools. A recent study in primates also suggests that some of the migrating cells may exit tangentially from the rostral migratory stream, where they cross the corpus callosum and finally come to rest within the prefrontal cortex and become interneurons appropriate for their location *(6)*.

A considerable amount of indirect evidence suggests that the primary proliferative pools within the VZ are derived from a relatively quiescent multipotent precursor. Ablation of the proliferative cells shows that the entire proliferative zone can be repopulated from the progeny of a small number of relatively quiescent precursors *(24,37)*. If these precursors are marked with an inheritable genetic marker (i.e., a retroviral vector), it can be shown that both neurons and glia are generated from single cells *(37)*.

The nature of the stem-like cells that reside within the VZ is still under debate. Careful ultrastructural evaluation of the cells that repopulate the SVZ shows striking similarities to astrocytes *(37)*; however, others have provided equally convincing data showing that cells within the ependymal layer proper also exhibit the stem cell-like properties of proliferative self-renewal and multilineage potential when isolated in culture *(38)*. It may be possible that cells displaying a number of similar, yet distinct phenotypes are all able to display stem-like properties under appropriate conditions. Regardless of the exact identity(s) of the stem cell, the presence of an immature stem-like cell in the adult provides the underpinnings for most working hypotheses regarding adult neurogenesis.

Fig. 1. After 1 wk of treatment with BrdU, dividing neural progenitors are readily observed within the adult rat anterior subventricular zone (SVZ, **A**) and subgranule zone (SGZ, **B**) of the hippocampal dentate gyrus (DG). Cells that proliferate in the SGZ adjacent to the lateral ventricle (LV) migrate along the surface of the striatum (Str) and converge in a rostral migratory stream (RMS) leading to the olfactory bulb (OB; **C**). Within the SGZ, cells proliferate in clusters and then migrate short distances to become disseminated throughout the GCL. Type III β-tubulin is an early intermediate filament marker for the neuronal lineage that is expressed by migrating cells within the RMS and the SGZ. Immunoreactivity for type III β-tubulin and BrdU have been combined in **B1** to show an individual neuroblast within a small cluster of BrdU-labeled precursors.

The Hippocampal Subgranule Zone

Proliferative cells within the hippocampus are found within a distinct zone of proliferation at the margin between the hilus and granule cell layer proper, or subgranule zone (SGZ) *(2,3)*. Precursors in this zone cycle relatively rapidly and then exit the cell cycle to differentiate into immature neurons (a complete cell cycle is roughly 16 h in the p20 mouse; *14*). Although the dividing cells do not themselves express neuronal markers, their progeny begin to express several markers of immature neuroblasts within days of cell division. These markers include turned-on-after-division-64 (TOAD-64), type III β-tubulin and polysialated neuronal cell adhesion molecule (PSA-NCAM) *(2,7,8,39,40)*. With time, the young neurons are found progressively deeper within the granule cell layer (GCL), and marker profiles switch from immature markers to those typically expressed by mature neurons of the GCL. Some of the markers used in this context include NeuN, an epitope of unknown function expressed by nearly all postmitotic neurons, and calbindin, a calcium-binding protein abundantly expressed by granule cell neurons *(2,3,41)*.

Although most newborn neurons become phenotypically indistinguishable from the preexisting GCL neurons, their functional role in the hippocampus is not yet known. The inability to identify living newborn cells has made it difficult to determine whether neurogenesis contributes neurons with known properties of granule cell neurons, i.e., neuronal membrane potentials, action potentials, or indications of plasticity such as that exhibited in the paradigm of long-term potentiation (LTP). However, there is considerable correlative evidence for functional integration. Retrograde tracing shows that most newborn neurons project appropriately to the CA3 region of the hippocampus *(26,27,42)*, and recordings from the inner layer, where TOAD-64-positive cells are distributed, show that the putative immature neurons do display LTP, albeit with a somewhat altered physiology *(43)*. Interestingly, stimulation of LTP in these cells does not require γ-aminobutyric-acid (GABA)a-blockade, suggesting that eliciting LTP is easiest in the youngest population of neurons within the GCL. The implication is that new neurons may provide a naive and easily imprinted circuitry that facilitates the acquisition and processing of new information.

The concept that new granule cell neurons are important for processing novel information has gained support from several studies showing that the number of new neurons generated correlates well with performance in learning- and memory-related tasks *(4,44–47)*. This correlation extends well beyond any potential variation between species or even differences between individuals since neurogenesis in any given group of individuals can be dramatically influenced by environmental, physical, psychological, and cognitive processes. This intrinsic modulation and its apparent correlation with learning and memory makes hippocampal neurogenesis an excellent platform for unraveling the molecular basis of neurogenesis and, eventually, understanding how stimulating ectopic neurogenesis might positively or negatively effect cognition following attempts to repair the CNS.

THE HIPPOCAMPUS AS AN EXPERIMENTAL MODEL FOR THE REGULATION OF ADULT NEUROGENESIS

Although very little is known about the specific molecules that regulate neural precursor activity and fate, neurogenesis in the adult dentate gyrus is a dynamic process that responds to numerous intrinsic and extrinsic influences (Table 1).

Genetic Factors

We all may have wondered, when faced with the dizzying array of intellect in a typical kindergarten class or recent faculty meeting, how the relative contributions of genetics vs environment dictate cognitive ability. In the context of hippocampal neurogenesis, it seems clear that both genetics and experience have a significant impact on the number of new neurons generated. For example, different strains of the common laboratory mouse show striking differences in baseline neurogenesis when housed under identical conditions *(48)*. Differences can be seen in both the size of the proliferative precursor pool and the fraction of newborn cells that survive and differentiate into neurons. Although the allelic variations that control this modulation are not yet known, classical genetics along with transgenic methods should identify genetic loci responsible for individual aspects of the adult neurogenic process.

Environmental Enrichment

For each genetic makeup, neurogenesis is further modulated by numerous extrinsic factors. For example, animals housed under standard laboratory conditions can be compared with those placed in large population cages containing toys, edible treats, and numerous social cohorts *(4,41)*. The differences in hippocampal neurogenesis are striking. Those in the "enriched" environment (see also refs. *49–51*) generate roughly two times more neurons than their underprivileged compatriots. This increase is generated in the absence of any increases in proliferation, suggesting that enrichment induces more of the newborn cells to differentiate and survive as neurons. Although this enrichment yields robust experience-related changes in neurogenesis, the complexity of the provided environment made it difficult to determine which elements had the most influence.

Table 1
Regulators of Neural Precursor Activity in the Hippocampal SGZ

	Effects	Context	References
Genetic	Changes in proliferation	Unknown loci	48
	Number of new neurons	Unknown loci	48
Environmental enrichment	Increased number of new neurons	Population cage, food treats, toys, running wheel.	4, 41, 44; see also 49–51
Physical exercise	Increases in proliferation and number of new neurons	Running wheel	46, 47
Learning	Increased number of new neurons	Water maze eye-blink conditioning	45
Stress and other Hormones	Increased proliferation	Adrenalectomy	23
		Estrogen	148
	Reduced proliferation	Glucocorticoids Predator/psychosocial stress	23, 31,52
Neurotransmitters	Increased proliferation	nmda antagonists, serotonin, norepinephrine, and dopamine signaling	55,57,59, 60,149,150
	Reduced proliferation	NMDA agonists Serotonin receptor antagonists, depletion of 5-HT	57–60, 149
Growth factors/ mitogens	Increased proliferation and number of neurons	FGF-2 EGF IGF-1	89,90,92–94

Physical Exercise

One element of the enriched environment is access to running wheels, and subsequent studies have shown that physical exercise alone can increase both the size of the proliferative precursor pool and the number of new neurons that survive and become integrated into the GCL *(46,47)*. When given the opportunity to run, individual animals will run more than 10,000 revolutions per night on a running wheel (3–5 miles). One would anticipate numerous physiologic changes in both the CNS and periphery. These would include increased blood flow and changes in oxygen and glucose metabolism and, within the CNS alone, the propagation of numerous motor and cognitive patterns, some of which may feed back into the neurogenic regulatory cascade via specific neural transmitters (see below).

Learning

Even the act of learning a spatial task appears to trigger an increase in the number of newborn precursors that survive and differentiate into neurons *(45)*. When animals housed under standard laboratory conditions are placed in a water maze, those animals that have the opportunity to learn the position of a submerged platform retain more newborn neurons than those that are simply asked to swim for the same amount of time. The extent to which purely cerebral activities influence neurogenesis may be subtly embedded within other regulatory influences. For example, similar learning paradigms do not seem to have a measurable effect, and the absence of increased neurogenesis following swimming (a form of physical exercise) seems to contradict the running data *(46,47)*. However, the very brief swimming periods may not be equivalent to the extended periods of exercise provided by a wheel. In addition, swimming is not an activity a rat would normally choose, and the psychological stress of swimming may actually counteract any neurogenic stimuli.

Stress, Glucocorticoids, and Neural Transmitters

Stress induced in a number of paradigms can rapidly influence neural precursor proliferation in the hippocampus. Within 24 h of being placed in an environment of psychosocial stress, e.g., the odor of a predator, rodents show a significant decrease in the number of dividing cells in the subgranule zone *(31,52)*. Artificial modulation of stress-related hormones by adrenalectomy or exogenous administration of adrenal steroids shows that at least some of this neurogenic suppression is moderated via circulating corticosteroids *(23,52,53)*. However, steroids may in part impact neurogenesis via changes in neurotransmitter signaling since blockade of *N*-methyl-D-aspartate (NMDA) receptor activation with MK-801 can counteract the effects of stress on proliferation, and amplification of glutamate signaling via NMDA receptor agonists mimics the suppression seen in the stress *(23,54)*.

In addition to NMDA receptors, perturbations in several other transmitter systems also influence proliferation in the SGZ. Reduction in serotonin is accompanied by decreased neurogenesis, whereas augmentation increases the number of new neurons *(55,56)*. Similar changes are seen following manipulation of norepinephrine and dopamine systems, each of which are also modulated following physical exercise. The extensive correlations seem to indicate a additional role for monoamine signaling in modulating neurogenesis *(57–60)*.

When taken together, these environmental and CNS-intrinsic modulatory mechanisms indicate that neurogenesis in the hippocampus responds dynamically to a complex set of overlapping cues. Perturbation of these cues within no-neurogenic regions might provide the means for mobilizing stem cells in repair.

REGULATORY MECHANISMS IDENTIFIED IN PRIMARY CULTURE MODELS

The endogenous process of neurogenesis shows that all the key elements for repair are present in the adult. The long-distance migration from the VZ to the olfactory bulb shows that precursor migration is possible within the adult parenchyma, and the fact that new neurons within the GCL successfully extend processes to CA3 confirms that newborn neurons can establish projections in the context of preexisting neural networks. Finally, the dynamic modulation of neurogenesis shows that the process can be extrinsically modulated. A working hierarchy of mechanisms that are important to understand for repair might include information on the nature and location of cells competent to make neurons, the presence or absence of signals that recruit precursors into cycle, the region-specific cues that dictate neuronal fate and phenotype, and finally attractive or repulsive factors that direct migration and connectivity (Fig. 2).

Although it is convenient to think of neurogenic signals as stimulating a homogeneous population of immature neural precursors, the mixture of proliferating cell types found within the neurogenic zones *(35,37,38)* suggests that each signaling cascade may impinge on only a subset of cells. Although the signaling complexity is daunting when contemplating the carefully controlled mobilization of cells, the manipulation of these immature precursors in culture is beginning to provide considerable insight into the phenotypes and regulatory mechanisms available to neural precursors from the adult (Table 2).

Cellular Substrates for Repair: Evidence for Neural Stem Cells

In development, stem cells are defined as an undifferentiated precursor that can divide to give rise to an identical precursor (self-renewal) as well as

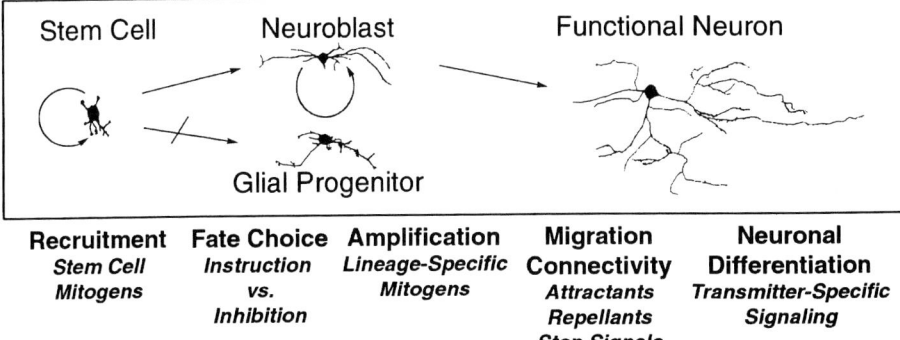

Recruitment	Fate Choice	Amplification	Migration	Neuronal
Stem Cell Mitogens	*Instruction vs. Inhibition*	*Lineage-Specific Mitogens*	**Connectivity** *Attractants Repellants Stop Signals*	**Differentiation** *Transmitter-Specific Signaling*

Fig. 2. Mobilization of neural progenitor for neuronal replacement or augmentation may require the manipulation of precursors at numerous points along a complex regulatory pathway. First, stem cells must be recruited into the cycle and then influenced to adopt a neuronal fate in regions that may normally only produce glia. The neuroblasts must then be amplified and instructed to migrate along routes that may not naturally exist within the adult brain. The newborn neurons may then require additional instructions that direct the appropriate connectivity and transmitter phenotype.

progeny that go on to differentiate into one or more terminal phenotypes (as reviewed in refs. *33, 61,* and *62*). In the adult CNS, the strongest evidence for stem-like cells has been generated in primary cultures isolated from the VZ or hippocampus. Precursors from the adult germinal zones can be stimulated to proliferate in vitro using epidermal growth factor (EGF) and/or basic fibroblast growth factor-2 (FGF-2) The cultures contain a mixture of cell phenotypes similar to that seen in vivo, with only a minor population of cells displaying definitive markers for mature neurons or glia *(33,34,63–66)*. By isolating or marking single cells, it has been shown that the entire array of cell phenotypes originates from a small population of multipotent stem-like cells within the relatively heterogeneous population of dividing precursors *(24,66–69)*.

Although one might expect to isolate stem-like cells from neurogenic zones, other areas also seem to contain cells that acquire stem cell-like characteristics in vitro. In vivo, white matter contains an abundant population of dividing precursors, and although they differentiate exclusively into glia in vivo *(70–72)*, cells isolated from spinal cord or optic nerve readily generate neurons when cultured in vitro *(73–77)*. The fact that neurons can be generated in vitro from areas that produce only glial precursors suggests that some of the regional restriction noted for neurogenesis may be the result of the presence or absence of neurogenic signals rather than the absence of competent precursors.

Table 2
Regulators of Stem Cell Activity In Vitro

	Effectors	Effect	Res.
Proliferation	FGF-2	Mitogenic, recruits stem cells into cycle and may reprogram cells to have broader fate potential than displayed in vivo	*33,63 –65,78, 84,85, 151,152*
	FGF-1, -4, -7, -8	Mitogenic, can substitute for FGF-2 to maintain rat neural precursors in cycle	*86,87*
	EGF	Mitogenic, recruits mouse stem cells but also stimulates increased gliogenesis	*34,65, 69,82, 84*
	PDGF	Mitogen for glial precursors	*98,99*
Differentiation	Retinoic acid	Increase in neuronal differentiation Upregulation of Trk receptors	*77,79, 153*
	Cyclic AMP	Increase in neuronal differentiation	*77,79, 153*
	Neurotrophins: NGF, BDNF, NT-3	Increase in neuronal differentiation Neuronal types include those that express TH, GABA, acetyl cholinesterase, NP-Y, and calbindin	*64,66, 67,69, 78,79, 84*
	CNTF	Glial differentiation	*154*
Transmitter phenotype	Nurr1	Direct activation of tyrosine hydroxylase expression	*153*
	Nurr1 + glial co-factors	Increased differentiation of dopaminergic neurons	*147*
	Neurotrophins: NGF, BDNF, NT-3	Pleiotrophic effects: promote maturation of several neuronal phenotypes including TH, GABA, cholinesterase, NP-Y, and calbindin-expressing neurons.	*66,79, 82–84*

Stem Cell Plasticity

In vitro, the neuronal progeny from stem-like cells display an unexpected phenotypic plasticity. Several groups have demonstrated that stem cells isolated from the SVZ or hippocampus develop in vitro into GABAergic, dopaminergic, cholinergic, or neuropeptide-Y expressing neurons

(67,78,79). This plasticity does not appear to be owing to in vitro mutation since transplant of hippocampal stem-like precursors back into the hippocampus shows that they revert back to a program of generating only granule cell neurons *(33)*. The concept that in vitro plasticity may directly translate to similar potentials in vivo is demonstrated when stem cells from the hippocampus are introduced into the rostral migratory stream. The same cells that generate only granule cell neurons in the hippocampus generate tyrosine hydroxylase-positive interneurons in the olfactory bulb, a pheno-type never seen in the hippocampal GCL *(33)*.

The plasticity of precursors in the adult brain may actually extend considerably beyond that of a neural fate. Recent work by Vescovi and colleagues *(80)* indicates that some of the precursors isolated from the adult brain have the potential to change their lineage program entirely. When clonally derived populations of neural precursors were peripherally injected following partial bone marrow ablation, the injected cells were able to participate in reconstituting the hematopoetic system. In reciprocal experiments, Eglitis and colleagues *(81)* had earlier shown that astrocyte-like cells are generated in the CNS following the reconstitution of the hematopoetic system with whole bone marrow. These data challenge the concept that stem cells found within a particular location are inherently restricted to local cell fates. In the context of how precursors resident in the brain are directed to generate neurons or glia, this surprising stem cell plasticity suggests that fate may be controlled primarily by the local microenvironment rather than by cell-intrinsic programming. Perhaps the first element of this hypothetical neurogenic microenvironment is the presence of specific mitogens that influence fate and/or lineage potential of the responding stem cell.

Mitogenic Recruitment

The nature of the mitogen(s) that regulates neural stem cell proliferation in vivo is unknown, but in vitro studies have implicated several potential candidates. The two most widely used mitogens are EGF *(65,82–84)* and FGF-2 *(33,63,78,85)*. However, numerous other growth factors are also mitogenic *(86,87)*. To a certain extent, the species being studied (mice, rats, or humans) dictates which mitogenic cocktail is preferred in culture *(34,65,84,88)*. Precursors from mouse or human tissues proliferate readily in a mixture of EGF and FGF-2, whereas rat precursor cells can be maintained in FGF-2 alone. In most cases, proliferation also depends on the presence of high levels of insulin, which could act either through the insulin receptor or the insulin-like growth factor-1 (IGF-1) receptor.

In vivo, EGF, FGF-2, and IGF-1 also induce proliferation within the VZ and/or SGZ. When injected into the lateral ventricle, EGF stimulates a dramatic proliferation of the VZ precursors with smaller, yet measurable effects in the SGZ *(89,90)*. FGF-2 administered in the ventricle shows similar effects on SGZ precursors but is not able to diffuse into the parenchyma and has little effect on hippocampal neurogenesis *(90,91)*. Peripheral injection of FGF-2 in neonatal animals does, however, have striking effects on hippocampal neurogenesis *(92,93)*, but only during the first few postnatal weeks, suggesting that FGF-2 can access precursors in the SGZ only until the blood-brain barrier becomes complete. In contrast, peripheral injections of IGF-1 induce a twofold increase in the number of dividing cells within the adult SGZ *(94)*. Presumably, this activity is owing to delivery via the vascular system, indicating that some circulating factors may have considerable influences on precursors resident within the parenchyma. When the growth factor data are pooled with earlier data on glucocorticoids, glutamate, and serotonin, it becomes increasingly evident that even the first step in neurogenesis (recruitment of precursors into cycle) involves the integration of a complex array of signals that could originate from either the CNS or the periphery.

Fate Choice

Although several signaling pathways are known to influence the neuron/glia choice in the embryo, the exact cues that function in the adult are unknown. The type of cell generated in vivo is probably regulated by both instructive and selective cues. For example, mitogens may themselves influence the fate potential of neural stem cells. In the embryo, FGF-2 can trigger a multilineage differentiation program (both neurons and glia) at a time when precursors normally generate only neurons *(95)*. In the adult, precursors that generate only glia can also be switched to a multilineage fate under the influence of FGF-2 *(76)*. Other growth factors appear to be more selective in their action. Intraventricular injection of brain-derived neurotrophic factor (BDNF) increases the number of neurons produced by SVZ precursors *(96,97)*. Precursors that select a glial fate can be amplified by platelet-derived growth factor (PDGF) *(98–100)*, and serum appears to favor the accumulation of astrocytes in cultures initially established from multipotent precursors *(66)*.

The cues that modulate the fate-choice outcome of mitogenic amplification are not known in the adult, but it seems likely that elements of developmentally relevant signaling pathways may be retained. Early in development, factors such as the bone morphogen proteins are instrumental in determining peripheral vs central fates *(101–103)*, and within the CNS further choices to adopt neuronal or glial fates may be regulated via a balance of instructive, selective, and inhibitory cues, the later being typified by the notch-delta complex *(104–107)*. Since precursors from nonneurogenic areas do generate neurons after they are removed from their local environment, it is possible that precursors may simply be prevented from differentiating into neurons in areas where neurogenesis does not occur. Antibodies to several members of the notch family do recognize related epitopes in the postnatal brain, but the role of notch signaling in the adult remains to be determined *(108–110)*.

Migration

In the adult SVZ and SGZ, newborn neurons migrate from their site of proliferation to a final destination and then send projections into the surrounding parenchyma. As in development, getting to the right place and connecting to the appropriate targets must involve attractive and/or repulsive signals. To a large extent, these signals are unknown in the adult, but some insights are being gained within the SVZ. Precursors within the SVZ migrate in unique self-assembling chains along tracts rich in PSA-NCAM *(35,36)*. It is unlikely that chemoattractants are produced by the olfactory bulb since removal of the olfactory bulb itself has little effect on migration *(111)*. Instead, repulsive proteins related to the Slit family members may drive rostral migration. The evidence for Slit involvement is still correlative, but two reports show that the septum and choroid plexus produce repulsive factors that act on precursors in the anterior SVZ *(112)*, and both tissues express Slit. In vitro, Slit can repel SVZ precursors, and migration within the rostral migratory stream can be inhibited by a soluble form of Robo, the receptor for Slit proteins *(113)*.

Slit family members are also expressed in the developing and postnatal hippocampus, and expression patterns are consistent with a role in guiding migration by repulsion. However, most studies in the hippocampus have focused on Slit effects on axonal extension rather than cell migration *(114)*. Deficits in the migration of granule cells in development may provide some insight into the molecules active in the adult. For example, Reelin, a large extracellular matrix protein highly expressed by Cajal-Retzius cells *(115–119)*, may provide cues defining where specific neuronal lamina should form by inducing migrating cells to stop *(119–121)*. The combination of repulsion, attraction, and stop signals provided by the Slits and Reelins probably act in concert with other extracellular matrix proteins expressed in the dentate gyrus, such as F-spondin *(122)*, Mindin *(122)*, and PSA-NCAM *(123–125)* to establish the precisely defined migration patterns of precursors in the adult brain. Although the guidance signals that normally target cells to a particular location are likely to be somewhat complex, it may not be necessary to perturb each individual element in turn if more global effectors could be identified. For example, ventricular infusion of BDNF amplifies neurogenesis in the SVZ and induces some newborn neuron to migrate tangentially into the overlying striatum *(96)*. These relatively nonspecific stimuli might eventually be refined as the relevance of individual guidance proteins becomes more apparent in the adult.

Axonal Pathfinding

Newborn neurons appear to elaborate axons and dendrites quite rapidly after their last division. In the adult hippocampus, newborn granule cells project to CA3 within a few days of incorporating BrdU *(42)* and receive afferent connections sometime in the following weeks *(27)*. In development, neurites are directed to their final targets by a variety of cues found in the environment through which they are navigating. Within the postnatal hippocampus, Slit proteins may provide some of the repulsive cues that initiate projection away from the granule cell layer. Slit-2 is expressed by cells within the developing and postnatal dentate and may be one of the signals that tell the growth cone to migrate away from the GCL. The fact that exogenously applied Slit-2 is able to repel the axons emanating from dentate explants *(114)* seems to support this possibility; however, Slit-1 and Slit-2 are also expressed in the CA3 target field. If the Slit-responsive axons emanating from dentate explants are in fact those that normally project to CA3 in vivo, then Slits or Slit-like repellants may both initiate extension and help refine the topography of synapse formation by acting at short distances within the target field.

Although we have used the Slit interactions as examples above, connectivity is ultimately modulated by multiple arrays of repulsive, attractive, and stop signals. In addition to Slit, semaphorin-neuropilin interactions provide repellent cues within the developing hippocampus, and genetic removal of the neuropilin-2 receptor results in aberrant mossy fiber targeting *(126–131)*. Additional modulation through Eph receptor signaling *(132–136)* provides yet further patterning that may be relevant to establishing connections for the newborn dentate granule cells in the adult. Reeler mice also display subtle defects in axon targeting and synaptogenesis within the dentate gyrus, which may suggest a role for Reelin, which is independent of its participation in guidance of cell body migrations. Even the cues that stimulate axonal fasciculation may influence patterns of connectivity since disruption of NCAM *(124)* or limbic system-associated membrane protein (LAMP)-mediated signaling *(137)* results in improper pathfinding of the mossy fiber axons within the pyramidal layers.

Neuronal Differentiation and Transmitter Phenotype

In addition to directing cell fate, final location, and connectivity, local cues also instruct precursors to consolidate location-specific transmitter phenotypes. The numerous transmitter phenotypes generated by adult-derived stem cells in vitro suggest that it may be possible to trigger specific transcriptional programs to produce a wider range of neuronal types than naturally generated in vivo. For example, the simple act of removing mitogen and stimulating cultured stem cells with retinoic acid dramatically upregulates Trk receptor expression. Subsequent application of neurotrophic factors can promote cells to acquire attributes of dopaminergic, cholinergic, or GABAergic neurons *(79)*. If relatively generic "differentiate" signals such as this are combined with the specific manipulation of key transcription factors, it may be possible to direct a specific neuronal fate precisely. An example of how this might be done can be seen in the experimental manipulation of stem cells to generate dopaminergic neurons, the cells at risk in Parkinson's disease.

Dopaminergic neurons are developmentally generated under the influence of a partially defined signaling cascade. Sonic hedgehog protein (Shh) and FGF-8 expression are known to intersect in regions of the developing CNS that become induction sites for dopaminergic neurons in the midbrain and forebrain *(138–140)*. Downstream and independent of Shh signaling are additional factors, such as Nurr1, that act in concert to implement the gene expression patterns of a midbrain dopaminergic neuron *(141–145)*. In culture, the combination of genetic manipulation (ectopic expression of Nurr1) and exposure to Shh or glial feeder layers can induce cells to acquire many of the dopaminergic cells' phenotypes, the most important of which is the ability to generate dopamine *(146,147)*. Since the adult brain does generate new tyrosine-hydroxylase positive neurons within the olfactory bulb, the ability to recapitulate dopamine production in vitro might not be unexpected. However, it seems possible that the ultimate range of neuronal phenotypes that can be generated by stem cells in the adult will be expanded as similar cascades are identified for other neuronal populations.

CONCLUSIONS

Native neurogenesis appears to involve the local control of neural precursors that are widely distributed throughout the adult brain. Within each anatomic context, local microenvironments dictate the fate of these precursors. Additional local cues control migration, connectivity, and ultimate neuronal phenotype. Although a number of candidate molecules have been identified for some of these steps, there are still a considerable number of unknowns. It is not known how the decision to generate neurons vs glia is made in the adult. With few exceptions, the cues that direct migration or the projection of neurites in the adult are entirely unknown, and, for the vast majority of neuronal types, the transcriptional regulators that control transmitter phenotype remain anonymous.

Perhaps the single largest uncertainty relates to the long-standing fact that evolutionary pressures have unambiguously selected for the absence of global reconstruction in the mammalian brain. It seems likely that the advantages provided by the absence of large-scale regeneration may ultimately provide one of the more difficult obstacles in repair if the cognitive repercussions of generating new neurons outweigh the anticipated advantages. Perhaps the first emphasis in exploring the therapeutic potential of mobilizing stem cells should be a careful evaluation of how ectopic neurogenesis in the intact CNS affects behavior and cognition.

REFERENCES

1. Luskin, M. B. and Boone, M. S. (1994) Rate and pattern of migration of lineally related olfactory bulb interneurons generated postnatally in the subventricular zone of the rat. *Chem. Senses* **19,** 695–714.
2. Gould, E., Reeves, A. J., Fallah, M., Tanapat, P., Gross, C. G., and Fuchs, E. (1999) Hippocampal neurogenesis in adult Old World primates. *Proc. Natl. Acad.Sci. USA* **96,** 5263–5267.
3. Kuhn, H. G., Dickinson-Anson, H., and Gage, F. H. (1996) Neurogenesis in the dentate gyrus of the adult rat: age-related decrease of neuronal progenitor proliferation. *J. Neurosci.* **16,** 2027–2033.
4. Kempermann, G., Kuhn, H. G., and Gage, F. H. (1997) More hippocampal neurons in adult mice living in an enriched environment. *Nature* **386,** 493–495.
5. Cameron, H. A. and McKay, R. (1998) Stem cells and neurogenesis in the adult brain. *Curr. Opin. Neurobiol.* **8,** 677–680.
6. Gould, E., Reeves, A. J., Graziano, M. S., and Gross, C. G. (1999) Neurogenesis in the neocortex of adult primates. *Science* **286,** 548–552.
7. Parent, J. M., Janumpalli, S., McNamara, J. O., and Lowenstein, D. H. (1998) Increased dentate granule cell neurogenesis following amygdala kindling in the adult rat. *Neurosci. Lett.* **247,** 9–12.
8. Parent, J. M., Yu, T. W., Leibowitz, R. T., Geschwind, D. H., Sloviter, R. S., and Lowenstein, D. H. (1997) Dentate granule cell neurogenesis is increased by seizures and contributes to aberrant network reorganization in the adult rat hippocampus. *J. Neurosci.* **17,** 3727–3738.
9. Liu, J., Solway, K., Messing, R. O., and Sharp, F. R. (1998) Increased neurogenesis in the dentate gyrus after transient global ischemia in gerbils. *J. Neurosci.* **18,** 7768–7778.
10. Altman, J. and Das, G. D. (1965) Post-natal origin of microneurones in the rat brain. *Nature* **207,** 953–956.
11. Altman, J. and Das, G. D. (1965) Autoradiographic and histological evidence of postnatal hippocampal neurogenesis in rats. *J. Comp. Neurol.* **124,** 319–335.
12. Altman, J. and Bayer, S. A. (1990) Migration and distribution of two populations of hippocampal granule cell precursors during the perinatal and postnatal periods. *J. Comp. Neurol.* **301,** 365–381.
13. Altman, J. and Bayer, S. A. (1990) Mosaic organization of the hippocampal neuroepithelium and the multiple germinal sources of dentate granule cells. *J. Comp. Neurol.* **301,** 325–342.

14. Nowakowski, R. S. and Lewin, S. B. (1989) Bromodeoxyuridine immunohistochemical determination of the lengths of the cell cycle and the DNA-synthetic phase for an anatomically defined population. *J. Neurocytol.* **18,** 311–318.
15. Anisimov, V. N. (1995) Carcinogenesis induced by neonatal exposure to various doses of 5-bromo-2'-deoxyuridine in rats. *Cancer Lett.* **91,** 63–71.
16. Ashman, C. R., Reddy, G. P., and Davidson, R. L. (1981) Bromodeoxyuridine mutagenesis, ribonucleotide reductase activity, and deoxyribonucleotide pools in hydroxyurea-resistant mutants. *Somatic. Cell Genet.* **7,** 751–768.
17. Comi, P., Ottolenghi, S., Giglioni, B., Migliaccio, G., Migliaccio, A. R., Bassano, E., et al. (1986) Bromodeoxyuridine treatment of normal adult erythroid colonies: an in vitro model for reactivation of human fetal globin genes. *Blood* **68,** 1036–1041.
18. Keoffler, H. P., Yen, J., and Carlson, J. (1983) The study of human myeloid differentiation using bromodeoxyuridine (BrdU). *J. Cell. Physiol.* **116,** 111–117.
19. Kinoshita, Y., Makita, A., and Takeuchi, T. (1982) Bromodeoxyuridine-induced molecular species conversion of sialic acids of gangliosides and the alteration of cellular phenotypic expression in B16 mouse melanoma cells. *J. Biochem. (Tokyo)* **92,** 801–808.
20. Ashman, C. R. and Davidson, R. L. (1980) Inhibition of Friend erythroleukemic cell differentiation by bromodeoxyuridine: correlation with the amount of bromodeoxyuridine in DNA. *J. Cell. Physiol.* **102,** 45–50.
21. Malendowicz, L. K., Rebuffat, P., Andreis, P. G., Nussdorfer, G. G., and Nowak, M. (1997) Different mechanisms mediate the in vivo aldosterone and corticosterone responses to 5-bromo-2'-deoxyuridine in rats. *Exp. Clin. Endocrinol. Diabetes* **105,** 277–281.
22. Malendowicz, L. K. and Nussdorfer, G. G. (1996) 5-Bromo-2'-deoxyuridine stimulates the pituitary-adrenal axis in the rat: an effect blocked partially by endothelin-receptor antagonists. *J. Int. Med. Res.* **24,** 363–368.
23. Cameron, H. A., Tanapat, P., and Gould, E. (1998) Adrenal steroids and N-methyl-D-aspartate receptor activation regulate neurogenesis in the dentate gyrus of adult rats through a common pathway. *Neuroscience* **82,** 349–354.
24. Morshead, C. M., Reynolds, B. A., Craig, C. G., McBurney, M. W., Staines, W. A., Morassutti, D., et al. (1994) Neural stem cells in the adult mammalian forebrain: a relatively quiescent subpopulation of subependymal cells. *Neuron* **13,** 1071–1082.
25. Bayer, S. A. (1985) Neuron production in the hippocampus and olfactory bulb of the adult rat brain: addition or replacement? *Ann. N. Y. Acad. Sci.* **457,** 163–172.
26. Stanfield, B. B. and Trice, J. E. (1988) Evidence that granule cells generated in the dentate gyrus of adult rats extend axonal projections. *Exp. Brain Res.* **72,** 399–406.
27. Markakis, E. A. and Gage, F. H. (1999) Adult-generated neurons in the dentate gyrus send axonal projections to field CA3 and are surrounded by synaptic vesicles. *J. Comp. Neurol.* **406,** 449–460.
28. Kaplan, M. S. and Hinds, J. W. (1977) Neurogenesis in the adult rat: electron microscopic analysis of light radioautographs. *Science* **197,** 1092–1094.

29. Betarbet, R., Zigova, T., Bakay, R. A., and Luskin, M. B. (1996) Dopaminergic and GABAergic interneurons of the olfactory bulb are derived from the neonatal subventricular zone. *Int. J. Dev. Neurosci.* **14,** 921–930.

30. Craig, C. G., D'sa, R., Morshead, C. M., Roach, A., and van der, Kooy D. (1999) Migrational analysis of the constitutively proliferating subependyma population in adult mouse forebrain. *Neuroscience* **93,** 1197–1206.

31. Gould, E., McEwen, B. S., Tanapat, P., Galea, L. A., and Fuchs, E. (1997) Neurogenesis in the dentate gyrus of the adult tree shrew is regulated by psychosocial stress and NMDA receptor activation. *J. Neurosci.* **17,** 2492–2498.

32. Luskin, M. B. (1998) Neuroblasts of the postnatal mammalian forebrain: their phenotype and fate. *J. Neurobiol.* **36,** 221–233.

33. Gage, F. H., Coates, P. W., Palmer, T. D., Kuhn, H. G., Fisher, L. J., Suhonen, J. O., et al. (1995) Survival and differentiation of adult neuronal progenitor cells transplanted to the adult brain. *Proc. Natl. Acad. Sci. USA* **92,** 11,879–11,883.

34. Pincus, D. W., Harrison-Restelli, C., Barry, J., Goodman, R. R., Fraser, R. A., et al. (1997) In vitro neurogenesis by adult human epileptic temporal neocortex. *Clin. Neurosurg.* **44,** 17–25.

35. Doetsch, F., Garcia-Verdugo, J. M., and Alvarez-Buylla, A. (1997) Cellular composition and three-dimensional organization of the subventricular germinal zone in the adult mammalian brain. *J. Neurosci.* **17,** 5046–5061.

36. Lois, C., Garcia-Verdugo, J. M., and Alvarez-Buylla, A. (1996) Chain migration of neuronal precursors. *Science* **271,** 978–981.

37. Doetsch, F., Caille, I., Lim, D. A., Garcia-Verdugo, J. M., and Alvarez-Buylla, A. (1999) Subventricular zone astrocytes are neural stem cells in the adult mammalian brain. *Cell* **97,** 703–716.

38. Johansson, C. B., Momma, S., Clarke, D. L., Risling, M., Lendahl, U., and Frisen, J. (1999) Identification of a neural stem cell in the adult mammalian central nervous system. *Cell* **96,** 25–34.

39. Minturn, J. E., Fryer, H. J., Geschwind, D. H., and Hockfield, S. (1995) TOAD-64, a gene expressed early in neuronal differentiation in the rat, is related to unc-33, a C. elegans gene involved in axon outgrowth. *J. Neurosci.* **15,** 6757–6766.

40. Minturn, J. E., Geschwind, D. H., Fryer, H. J., and Hockfield, S. (1995) Early postmitotic neurons transiently express TOAD-64, a neural specific protein. *J. Comp. Neurol.* **355,** 369–379.

41. Eriksson, P. S., Perfilieva, E., Bjork-Eriksson, T., Alborn, A. M., Nordborg, C., Peterson, D. A., and Gage, F. H. (1998) Neurogenesis in the adult human hippocampus [see comments]. *Nature Med.* **4,** 1313–1317.

42. Hastings, N. B. and Gould, E. (1999) Rapid extension of axons into the CA3 region by adult-generated granule cells [published erratum appears in J. Comp. Neurol. 1999; 415:144]. *J. Comp. Neurol.* **413,** 146–154.

43. Wang, S., Scott, B. W., and Wojtowicz, J. M. (2000) Heterogenous properties of dentate granule neurons in the adult rat. *J. Neurobiol. 2000* **42,** 248–251.

44. Kempermann, G., Kuhn, H. G., and Gage, F. H. (1998) Experience-induced neurogenesis in the senescent dentate gyrus. *J. Neurosci.* **18,** 3206–3212.

45. Gould, E., Beylin, A., Tanapat, P., Reeves, A., and Shors, T. J. (1999) Learning enhances adult neurogenesis in the hippocampal formation [see comments]. *Nature Neurosci.* **2,** 260–265.
46. van Praag, H., Kempermann, G., and Gage, F. H. (1999) Running increases cell proliferation and neurogenesis in the adult mouse dentate gyrus [see comments]. *Nature Neurosci.* **2,** 266–270.
47. van Praag, H., Christie, B. R., Sejnowski, T. J., and Gage, F. H. (1999) Running enhances neurogenesis, learning, and long-term potentiation in mice. *Proc. Natl. Acad. Sci. USA* **96,** 13,427–13,431.
48. Kempermann, G., Kuhn, H. G., and Gage, F. H. (1997) Genetic influence on neurogenesis in the dentate gyrus of adult mice. *Proc. Natl. Acad. Sci. USA* **94,** 10,409–10,414.
49. Greenough, W. T., Cohen, N. J., and Juraska, J. M. (1999) New neurons in old brains: learning to survive? [news; comment]. *Nature Neurosci.* **2,** 203–205.
50. Diamond, M. C., Rosenzweig, M. R., Bennett, E. L., Lindner, B., and Lyon, L. (1972) Effects of environmental enrichment and impoverishment on rat cerebral cortex. *J. Neurobiol.* **3,** 47–64.
51. Diamond, M. C., Ingham, C. A., Johnson, R. E., Bennett, E. L., and Rosenzweig, M. R. (1976) Effects of environment on morphology of rat cerebral cortex and hippocampus. *J. Neurobiol.* **7,** 75–85.
52. Tanapat, P., Galea, L. A., and Gould, E. (1998) Stress inhibits the proliferation of granule cell precursors in the developing dentate gyrus. *Int. J. Dev. Neurosci.* **16,** 235–239.
53. McEwen, B. S. (1999) Stress and hippocampal plasticity. *Annu. Rev. Neurosci.* **22,** 105–122.
54. Cameron, H. A., McEwen, B. S., and Gould, E. (1995) Regulation of adult neurogenesis by excitatory input and NMDA receptor activation in the dentate gyrus. *J. Neurosci.* **15,** 4687–4692.
55. Brezun, J. M. and Daszuta, A. (2000) Serotonin may stimulate granule cell proliferation in the adult hippocampus, as observed in rats grafted with foetal raphe neurons. *Eur. J. Neurosci. 2000.* **12,** 391–396.
56. Brezun, J. M. and Daszuta, A. (1999) Depletion in serotonin decreases neurogenesis in the dentate gyrus and the subventricular zone of adult rats. *Neuroscience* **89,** 999–1002.
57. Dawirs, R. R., Hildebrandt, K., and Teuchert-Noodt, G. (1998) Adult treatment with haloperidol increases dentate granule cell proliferation in the gerbil hippocampus. *J. Neural Transm.* 105, 317–327.
58. Hildebrandt, K., Teuchert-Noodt, G., and Dawirs, R. R. (1999) A single neonatal dose of methamphetamine suppresses dentate granule cell proliferation in adult gerbils which is restored to control values by acute doses of haloperidol. *J. Neural Transm.* **106,** 549–558.
59. Jacobs, B. L. and Fornal, C. A. (1999) Activity of serotonergic neurons in behaving animals. *Neuropsychopharmacology* **21,** 9S–15S.
60. Duman, R. S., Malberg, J., and Thome, J. (1999) Neural plasticity to stress and antidepressant treatment. *Biol. Psychiatry* **46,** 1181–1191.
61. Morrison, S. J., Shah, N. M., and Anderson, D. J. (1997) Regulatory mechanisms in stem cell biology. *Cell* **88,** 287–298.

62. Temple, S. and Alvarez-Buylla, A. (1999) Stem cells in the adult mammalian central nervous system. *Curr. Opin. Neurobiol.* **9,** 135–141.

63. Kilpatrick, T. J. and Bartlett, P. F. (1995) Cloned multipotential precursors from the mouse cerebrum require FGF-2, whereas glial restricted precursors are stimulated with either FGF-2 or EGF. *J. Neurosci.* **15,** 3653–3661.

64. Gritti, A., Frolichsthal-Schoeller, P., Galli, R., Parati, E. A., Cova, L., Pagano, S. F., et al. (1999) Epidermal and fibroblast growth factors behave as mitogenic regulators for a single multipotent stem cell-like population from the subventricular region of the adult mouse forebrain. *J. Neurosci.* **19,** 3287–3297.

65. Ciccolini, F. and Svendsen, C. N. (10-1-1998) Fibroblast growth factor 2 (FGF-2) promotes acquisition of epidermal growth factor (EGF) responsiveness in mouse striatal precursor cells: identification of neural precursors responding to both EGF and FGF-2. *J. Neurosci.* **18,** 7869–7880.

66. Palmer, T. D., Takahashi, J., and Gage, F. H. (1997) The adult rat hippocampus contains primordial neural stem cells. *Mol. Cell. Neurosci.* **8,** 389–404.

67. Gritti, A., Parati, E. A., Cova, L., Frolichsthal, P., Galli, R., Wanke, E., et al. (1996) Multipotential stem cells from the adult mouse brain proliferate and self-renew in response to basic fibroblast growth factor. *J. Neurosci.* **16,** 1091–1100.

68. Chiasson, B. J., Tropepe, V., Morshead, C. M., and van der, Kooy D. (1999) Adult mammalian forebrain ependymal and subependymal cells demonstrate proliferative potential, but only subependymal cells have neural stem cell characteristics. *J. Neurosci.* **19,** 4462–4471.

69. Tropepe, V., Sibilia, M., Ciruna, B. G., Rossant, J., Wagner, E. F., and van der, Kooy D. (1999) Distinct neural stem cells proliferate in response to EGF and FGF in the developing mouse telencephalon. *Dev. Biol.* **208,** 166–188.

70. Goldman, J. E. (2000) Glial differentiation and lineages. *J. Neurosci. Res.* **259,** 410–412.

71. Levison, S. W. and Goldman, J. E. (1997) Multipotential and lineage restricted precursors coexist in the mammalian perinatal subventricular zone. *J. Neurosci. Res.* **48,** 83–94.

72. Horner, P. J., Power, A. E., Kempermann, G., Kuhn, H. G., Palmer, T. D., Winkler, J., et al. (2000) Proliferation and differentiation of progenitor cells throughout the intact adult rat spinal cord [In process citation]. *J. Neurosci. 2000.* **20,** 2218–2228.

73. Omlin, F. X. and Riederer, B. M. (1991) Optic nerve explant cultures of newborn rats: evidence for a neuron or common neuron-glia progenitor. *Schweiz. Arch. Neurol. Psychiatr.* **142,** 104–107.

74. Omlin, F. X. and Waldmeyer, J. (1989) Differentiation of neuron-like cells in cultured rat optic nerves: a neuron or common neuron-glia progenitor? *Dev. Biol.* **133,** 247–253.

75. Shihabuddin, L. S., Ray, J., and Gage, F. H. (1997) FGF-2 is sufficient to isolate progenitors found in the adult mammalian spinal cord. *Exp. Neurol.* **148,** 577–586.

76. Palmer, T. D., Markakis, E. A., Willhoite, A. R., Safar, F., and Gage, F. H. (1999) Fibroblast growth factor-2 activates a latent neurogenic program in neural stem cells from diverse regions of the adult CNS. *J. Neurosci.* **19,** 8487–8497.

77. Palmer, T. D., Ray, J., and Gage, F. H. (1995) FGF-2-responsive neuronal progenitors reside in proliferative and quiescent regions of the adult rodent brain. *Mol. Cell. Neurosci.* **6,** 474–486.

78. Cameron, H. A., Hazel, T. G., and McKay, R. D. (1998) Regulation of neurogenesis by growth factors and neurotransmitters. *J. Neurobiol.* **36,** 287–306.

79. Takahashi, J., Palmer, T. D., and Gage, F. H. (1999) Retinoic acid and neurotrophins collaborate to regulate neurogenesis in adult-derived neural stem cell cultures. *J. Neurobiol.* **38,** 65–81.

80. Bjornson, C. R., Rietze, R. L., Reynolds, B. A., Magli, M. C., and Vescovi, A. L. (1999) Turning brain into blood: a hematopoietic fate adopted by adult neural stem cells in vivo [see comments]. *Science* **283,** 534–537.

81. Eglitis, M. A. and Mezey, E. (1997) Hematopoietic cells differentiate into both microglia and macroglia in the brains of adult mice. *Proc. Natl. Acad. Sci. USA* **94,** 4080–4085.

82. Reynolds, B. A., Tetzlaff, W., and Weiss, S. (1992) A multipotent EGF-responsive striatal embryonic progenitor cell produces neurons and astrocytes. *J. Neurosci.* **12,** 4565–4574.

83. Reynolds, B. A. and Weiss, S. (1992) Generation of neurons and astrocytes from isolated cells of the adult mammalian central nervous system [see comments]. *Science* **255,** 1707–1710.

84. Vescovi, A. L., Reynolds, B. A., Fraser, D. D., and Weiss, S. (1993) bFGF regulates the proliferative fate of unipotent (neuronal) and bipotent (neuronal/astroglial) EGF-generated CNS progenitor cells. *Neuron* **11,** 951–966.

85. Bartlett, P. F., Dutton, R., Likiardopoulos, V., and Brooker, G. (1994) Regulation of neurogenesis in the embryonic and adult brain by fibroblast growth factors. *Alcohol Alcohol* **Suppl 2,** 387–394.

86. Ray, J., Baird, A., and Gage, F. H. (1997) A 10-amino acid sequence of fibroblast growth factor 2 is sufficient for its mitogenic activity on neural progenitor cells. *Proc. Natl. Acad. Sci. USA* **94,** 7047–7052.

87. DeHamer, M. K., Guevara, J. L., Hannon, K., Olwin, B. B., and Calof, A. L. (1994) Genesis of olfactory receptor neurons in vitro: regulation of progenitor cell divisions by fibroblast growth factors. *Neuron* **13,** 1083–1097.

88. Pincus, D. W., Keyoung, H. M., Harrison-Restelli, C., Goodman, R. R., Fraser, R. A., Edgar, M., et al. (1998) Fibroblast growth factor-2/brain-derived neurotrophic factor-associated maturation of new neurons generated from adult human subependymal cells. *Ann. Neurol.* **43,** 576–585.

89. Craig, C. G., Tropepe, V., Morshead, C. M., Reynolds, B. A., Weiss, S., and van der, Kooy D. (1996) In vivo growth factor expansion of endogenous subependymal neural precursor cell populations in the adult mouse brain. *J. Neurosci.* **16,** 2649–2658.

90. Kuhn, H. G., Winkler, J., Kempermann, G., Thal, L. J., and Gage, F. H. (1997) Epidermal growth factor and fibroblast growth factor-2 have different effects on neural progenitors in the adult rat brain. *J. Neurosci.* **17,** 5820–5829.

91. Gonzalez, A. M., Carman, L. S., Ong, M., Ray, J., Gage, F. H., Shults, C. W., and Baird, A. (1994) Storage, metabolism, and processing of 125I-fibroblast growth factor-2 after intracerebral injection. *Brain Res.* **665,** 285–292.

92. Tao, Y., Black, I. B., and DiCicco-Bloom, E. (1996) Neurogenesis in neonatal rat brain is regulated by peripheral injection of basic fibroblast growth factor (bFGF). *J. Comp. Neurol.* **376,** 653–663.

93. Wagner, J. P., Black, I. B., and DiCicco-Bloom, E. (1999) Stimulation of neonatal and adult brain neurogenesis by subcutaneous injection of basic fibroblast growth factor. *J. Neurosci.* **19,** 6006–6016.

94. Aberg, M. A. I., Aberg, D. N., Hedbacker, H., Oscarsson, J., Hagberg, H., and Eriksson, P. S. (2000) Peripheral infusion of IGF-1 selectively induces neurogenesis in the adult rat hippocampus. *J. Neurosci.* **20,** 2896–2903.

95. Stemple, D. L. and Mahanthappa, N. K. (1997) Neural stem cells are blasting off. *Neuron* **18,** 1–4.

96. Zigova, T., Pencea, V., Wiegand, S. J., and Luskin, M. B. (1998) Intraventricular administration of BDNF increases the number of newly generated neurons in the adult olfactory bulb. *Mol. Cell. Neurosci.* **11,** 234–245.

97. Leventhal, C., Rafii, S., Rafii, D., Shahar, A., and Goldman, S. A. (1999) Endothelial trophic support of neuronal production and recruitment from the adult mammalian subependyma. *Mol. Cell. Neurosci.* **13,** 450–464.

98. Barres, B. A., Schmid, R., Sendtner, M., and Raff, M. C. (1993) Multiple extracellular signals are required for long-term oligodendrocyte survival. *Development* **118,** 283–295.

99. Wolswijk, G. and Noble, M. (1992) Cooperation between PDGF and FGF converts slowly dividing O-2Aadult progenitor cells to rapidly dividing cells with characteristics of O-2Aperinatal progenitor cells. *J. Cell. Biol.* **118,** 889–900.

100. Wolswijk, G., Riddle, P. N., and Noble, M. (1991) Platelet-derived growth factor is mitogenic for O-2Aadult progenitor cells. *Glia* **4,** 495–503.

101. Mujtaba, T., Mayer-Proschel, M., and Rao, M. S. (1998) A common neural progenitor for the CNS and PNS. *Dev. Biol.* **200,** 1–15.

102. Morrison, S. J., Shah, N. M., and Anderson, D. J. (1997) Regulatory mechanisms in stem cell biology. *Cell* **88,** 287–298.

103. Kalyani, A. J. and Rao, M. S. (1998) Cell lineage in the developing neural tube. *Biochem. Cell Biol.* **76,** 1051–1068.

104. Lindsell, C. E., Boulter, J., diSibio, G., Gossler, A., and Weinmaster, G. (1996) Expression patterns of Jagged, Delta1, Notch1, Notch2, and Notch3 genes identify ligand-receptor pairs that may function in neural development. *Mol. Cell Neurosci.* **8,** 14–27.

105. Zhong, W., Jiang, M. M., Weinmaster, G., Jan, L. Y., and Jan, Y. N. (1997) Differential expression of mammalian Numb, Numblike and Notch1 suggests distinct roles during mouse cortical neurogenesis. *Development* **124,** 1887–1897.

106. Chenn, A. and McConnell, S. K. (1995) Cleavage orientation and the asymmetric inheritance of Notch1 immunoreactivity in mammalian neurogenesis. *Cell* **82,** 631–641.

107. Temple, S. and Qian, X. (1996) Vertebrate neural progenitor cells: subtypes and regulation. *Curr. Opin. Neurobiol.* **6,** 11–17.

108. Allen, T. and Lobe, C. G. (1999) A comparison of Notch, Hes and Grg expression during murine embryonic and post-natal development. *Cell Mol.Biol.* **45,** 687–708.

109. Tanaka, M., Kadokawa, Y., Hamada, Y., and Marunouchi, T. (1999) Notch2 expression negatively correlates with glial differentiation in the postnatal mouse brain. *J. Neurobiol.* **41,** 524–539.

110. Higuchi, M., Kiyama, H., Hayakawa, T., Hamada, Y., and Tsujimoto, Y. (1995) Differential expression of Notch1 and Notch2 in developing and adult mouse brain. *Brain Res. Mol. Brain Res.* **29,** 263–272.

111. Kirschenbaum, B., Doetsch, F., Lois, C., and Alvarez-Buylla, A. (1999) Adult subventricular zone neuronal precursors continue to proliferate and migrate in the absence of the olfactory bulb. *J. Neurosci.* **19,** 2171–2180.

112. Hu, H. (1999) Chemorepulsion of neuronal migration by Slit2 in the developing mammalian forebrain. *Neuron* **23,** 703–711.

113. Wu, W., Wong, K., Chen, J., Jiang, Z., Dupuis, S., Wu, J. Y., and Rao, Y. (1999) Directional guidance of neuronal migration in the olfactory system by the protein Slit [see comments]. *Nature* **400,** 331–336.

114. Nguyen Ba-Charvet, K. T., Brose, K., Marillat, V., Kidd, T., Goodman, C. S., Tessier-Lavigne, M., et al. (1999) Slit2-Mediated chemorepulsion and collapse of developing forebrain axons. *Neuron* **22,** 463–473.

115. Borrell, V., Ruiz, M., Del Rio, J. A., and Soriano, E. (1999) Development of commissural connections in the hippocampus of reeler mice: evidence of an inhibitory influence of Cajal-Retzius cells. *Exp. Neurol.* **156,** 268–282.

116. Borrell, V., Del Rio, J. A., Alcantara, S., Derer, M., Martinez, A., D'Arcangelo, G., et al. (1999) Reelin regulates the development and synaptogenesis of the layer-specific entorhino-hippocampal connections. *J. Neurosci.* **19,** 1345–1358.

117. Del Rio, J. A., Heimrich, B., Borrell, V., Forster, E., Drakew, A., Alcantara, S., et al. (1997) A role for Cajal-Retzius cells and reelin in the development of hippocampal connections [see comments]. *Nature* **385,** 70–74.

118. Deller, T., Drakew, A., and Frotscher, M. (1999) Different primary target cells are important for fiber lamination in the fascia dentata: a lesson from reeler mutant mice. *Exp. Neurol.* **156,** 239–253.

119. Frotscher, M. (1998) Cajal-Retzius cells, Reelin, and the formation of layers. *Curr. Opin. Neurobiol.* **8,** 570–575.

120. Nakajima, K., Mikoshiba, K., Miyata, T., Kudo, C., and Ogawa, M. (1997) Disruption of hippocampal development in vivo by CR-50 mAb against reelin. *Proc. Natl. Acad. Sci. USA* **94,** 8196–8201.

121. Frotscher, M. (1997) Dual role of Cajal-Retzius cells and reelin in cortical development. *Cell Tissue Res.* **290,** 315–322.

122. Feinstein, Y., Borrell, V., Garcia, C., Burstyn-Cohen, T., Tzarfaty, V., Frumkin, A., et al. (1999) F-spondin and mindin: two structurally and functionally related genes expressed in the hippocampus that promote outgrowth of embryonic hippocampal neurons. *Development* **126,** 3637–3648.

123. Aubert, I., Ridet, J. L., Schachner, M., Rougon, G., and Gage, F. H. (1998) Expression of L1 and PSA during sprouting and regeneration in the adult hippocampal formation. *J. Comp Neurol.* **399,** 1–19.

124. Cremer, H., Chazal, G., Goridis, C., and Represa, A. (1997) NCAM is essential for axonal growth and fasciculation in the hippocampus. *Mol. Cell Neurosci.* **8,** 323–335.

125. Chazal, G., Durbec, P., Jankovski, A., Rougon, G., and Cremer, H. (2000) Consequences of neural cell adhesion molecule deficiency on cell migration in the rostral migratory stream of the mouse. *J. Neurosci. 2000.* **20,** 1446–1457.

126. Chedotal, A., Del Rio, J. A., Ruiz, M., He, Z., Borrell, V., de Castro, F., et al. (1998) Semaphorins III and IV repel hippocampal axons via two distinct receptors. *Development* **125,** 4313–4323.

127. Steup, A., Lohrum, M., Hamscho, N., Savaskan, N. E., Ninnemann, O., Nitsch, R., et al. (2000) Sema3C and netrin-1 differentially affect axon growth in the hippocampal formation. *Mol. Cell Neurosci. 2000.* **15,** 141–155.

128. Steup, A., Ninnemann, O., Savaskan, N. E., Nitsch, R., Puschel, A. W., and Skutella, T. (1999) Semaphorin D acts as a repulsive factor for entorhinal and hippocampal neurons. *Eur. J. Neurosci.* **11,** 729–734.

129. Chen, H., Bagri, A., Zupicich, J. A., Zou, Y., Stoeckli, E., Pleasure, S. J., et al. (2000) Neuropilin-2 regulates the development of selective cranial and sensory nerves and hippocampal mossy fiber projections. *Neuron 2000.* **25,** 43–56.

130. Chen, H., He, Z., and Tessier-Lavigne, M. (1998) Axon guidance mechanisms: semaphorins as simultaneous repellents and anti-repellents [news; comment]. *Nature Neurosci.* **1,** 436–439.

131. Giger, R. J., Cloutier, J. F., Sahay, A., Prinjha, R. K., Levengood, D. V., Moore, S. E., et al. (2000) Neuropilin-2 is required in vivo for selective axon guidance responses to secreted semaphorins. *Neuron 2000.* **25,** 29–41.

132. Holder, N. and Klein, R. (1999) Eph receptors and ephrins: effectors of morphogenesis. *Development* **126,** 2033–2044.

133. Zhang, J. H., Cerretti, D. P., Yu, T., Flanagan, J. G., and Zhou, R. (1996) Detection of ligands in regions anatomically connected to neurons expressing the Eph receptor Bsk: potential roles in neuron-target interaction. *J. Neurosci.* **16,** 7182–7192.

134. Gao, P. P., Zhang, J. H., Yokoyama, M., Racey, B., Dreyfus, C. F., Black, I. B., and Zhou, R. (1996) Regulation of topographic projection in the brain: Elf-1 in the hippocamposeptal system. *Proc. Natl. Acad. Sci. USA* **93,** 11,161–11,166.

135. Gao, P. P., Yue, Y., Cerretti, D. P., Dreyfus, C., and Zhou, R. (1999) Ephrin-dependent growth and pruning of hippocampal axons. *Proc. Natl. Acad. Sci. USA* **96,** 4073–4077.

136. Stein, E., Savaskan, N. E., Ninnemann, O., Nitsch, R., Zhou, R., and Skutella, T. (1999) A role for the Eph ligand ephrin-A3 in entorhino-hippocampal axon targeting. *J. Neurosci.* **19,** 8885–8893.

137. Pimenta, A. F., Zhukareva, V., Barbe, M. F., Reinoso, B. S., Grimley, C., Henzel, W., et al. (1995) The limbic system-associated membrane protein is an Ig superfamily member that mediates selective neuronal growth and axon targeting. *Neuron* **15,** 287–297.

138. Hynes, M., Porter, J. A., Chiang, C., Chang, D., Tessier-Lavigne, M., Beachy, P. A., and Rosenthal, A. (1995) Induction of midbrain dopaminergic neurons by Sonic hedgehog. *Neuron* **15,** 35–44.

139. Murone, M., Rosenthal, A., and de Sauvage, F. J. (1999) Sonic hedgehog signaling by the patched smoothened receptor complex. *Curr. Biol.* **9,** 76–84.

140. Ye, W., Shimamura, K., Rubenstein, J. L., Hynes, M. A., and Rosenthal, A. (1998) FGF and Shh signals control dopaminergic and serotonergic cell fate in the anterior neural plate. *Cell* **93,** 755–766.

141. Law, S. W., Conneely, O. M., DeMayo, F. J., and O'Malley, B. W. (1992) Identification of a new brain-specific transcription factor, NURR1. *Mol. Endocrinol.* **6,** 2129–2135.

142. Zetterstrom, R. H., Williams, R., Perlmann, T., and Olson, L. (1996) Cellular expression of the immediate early transcription factors Nurr1 and NGFI-B suggests a gene regulatory role in several brain regions including the nigrostriatal dopamine system. *Brain Res. Mol. Brain Res.* **41,** 111–120.

143. Zetterstrom, R. H., Solomin, L., Jansson, L., Hoffer, B. J., Olson, L., and Perlmann, T. (1997) Dopamine neuron agenesis in Nurr1-deficient mice [see comments]. *Science* **276,** 248–250.

144. Saucedo-Cardenas, O., Quintana-Hau, J. D., Le, W. D., Smidt, M. P., Cox, J. J., De Mayo, F., et al. (1998) Nurr1 is essential for the induction of the dopaminergic phenotype and the survival of ventral mesencephalic late dopaminergic precursor neurons. *Proc. Natl. Acad. Sci. USA* **95,** 4013–4018.

145. Castillo, S. O., Baffi, J. S., Palkovits, M., Goldstein, D. S., Kopin, I. J., Witta, J., et al. (1998) Dopamine biosynthesis is selectively abolished in substantia nigra/ventral tegmental area but not in hypothalamic neurons in mice with targeted disruption of the Nurr1 gene. *Mol. Cell Neurosci.* **11,** 36–46.

146. Sakurada, K., Ohshima-Sakurada, M., Palmer, T. D., and Gage, F. H. (1999) Nurr1, an orphan nuclear receptor, is a transcriptional activator of endogenous tyrosine hydroxylase in neural progenitor cells derived from the adult brain. *Development* **126,** 4017–4026.

147. Wagner, J., Akerud, P., Castro, D. S., Holm, P. C., Canals, J. M., Snyder, E. Y., et al. (1999) Induction of a midbrain dopaminergic phenotype in Nurr1-overexpressing neural stem cells by type 1 astrocytes [see comments]. *Nature Biotechnol.* **17,** 653–659.

148. Tanapat, P., Hastings, N. B., Reeves, A. J., and Gould, E. (1999) Estrogen stimulates a transient increase in the number of new neurons in the dentate gyrus of the adult female rat. *J. Neurosci.* **19,** 5792–5801.

149. Brezun, J. M. and Daszuta, A. (2000) Serotonergic reinnervation reverses lesion-induced decreases in PSA-NCAM labeling and proliferation of hippocampal cells in adult rats [In process citation]. *Hippocampus 2000.* **10,** 37–46.

150. Brezun, J. M. and Daszuta, A. (1999) Depletion in serotonin decreases neurogenesis in the dentate gyrus and the subventricular zone of adult rats. *Neuroscience* **89,** 999–1002.

151. Bartlett, P. F., Richards, L. R., Kilpatrick, T. J., Talman, P. T., Bailey, K. A., Brooker, G. J., et al. (1995) Factors regulating the differentiation of neural precursors in the forebrain. *Ciba Found.Symp.* **193,** 85–99.

152. Svendsen, C. N., Caldwell, M. A., and Ostenfeld, T. (1999) Human neural stem cells: isolation, expansion and transplantation. *Brain Pathol.* **9,** 499–513.

153. Sakurada, K., Ohshima-Sakurada, M., Palmer, T. D., and Gage, F. H. (1999) Nurr1, an orphan nuclear receptor, is a transcriptional activator of endogenous tyrosine hydroxylase in neural progenitor cells derived from the adult brain. *Development* **126,** 4017–4026.

154. Bonni, A., Sun, Y., Nadal-Vicens, M., Bhatt, A., Frank, D. A., Rozovsky, I., et al. (1997) Regulation of gliogenesis in the central nervous system by the JAK-STAT signaling pathway. *Science* **278,** 477–483.

11

Transplant Therapy

Barbara A. Tate, Kate A. Bower, and Evan Y. Snyder

INTRODUCTION

Stem cells have been the focus of intense research in the last several years. Their potential as a therapeutic agent in a wide range of central (CNS) and peripheral nervous system (PNS) disorders is beginning to be understood. In this chapter we review data on the transplantation of stem cells into a variety of inherited and acquired disease models. Stem cells are undifferentiated, dividing cells isolated from germinal tissue, which can be maintained in culture. The types of immature cells include stem cells; progenitor cells, those restricted to one or more lineages; and precursor cells, any type of dividing cell (1). The work reviewed in this chapter is limited to cells derived from nervous tissue or from tissue that will give rise to nervous tissue in the adult. The definition of neural stem cells has become more restrictive in an effort to separate these cells from other types of immature cells.

Definition and Potential

Neural stem cells (NSCs) are immature cells originally isolated from nervous tissue. NSCs self-renew; they will divide in culture. A single cell has the potential to differentiate into all types of nervous system cells, including neurons, oligodendrocytes, and astrocytes. Finally, to meet the definition of an NSC, the monoclonal origin of a particular stem cell line must be established by clonal analysis, i.e., it must be demonstrated that a single cell can give rise to all neural cell types (for example, see work from our laboratory in Fig. 1). In addition to the above criteria, NSCs also have the capacity to replace damaged tissue in both the CNS and PNS (2). NSCs will restore functional neurons and glia and regenerate injured tissue. It is this characteristic of neural stem cells that makes them a valuable transplantation material in a host of disorders.

From: *Stem Cells and CNS Development*
Edited by: M. S. Rao © Humana Press Inc., Totowa, NJ

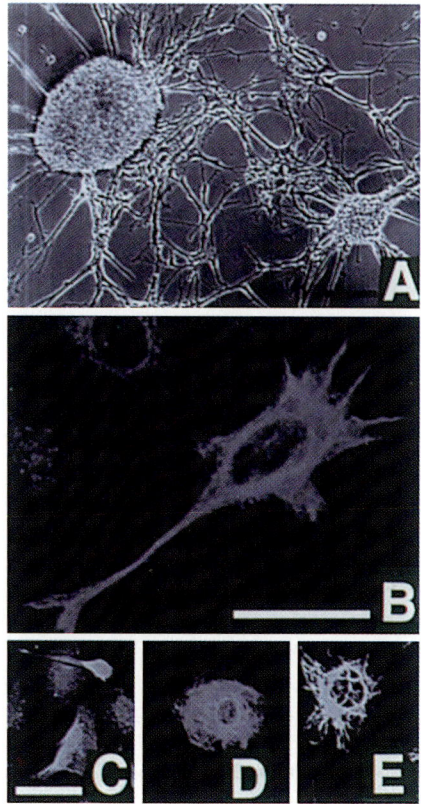

Fig. 1. Human NSCs can differentiate into all neural cell lineages. **(a)** Human
NSCs in vitro. These cells give rise to both neurons and glia. The cells grown in
serum-free media were immunostained for the neuronal marker neurofilament **(b)**
and for the oligodendroglia marker CNPase in **(c)**. The cells showed positive
immunostaining for the astrocyte marker human GFAP upon co-culture with
primary murine CNS cultures **(d)**. Upon transfer to serum-containing medium, the
cells were immunopositive for the immature neural marker vimentin **(e)**. (Modified
with permission from ref. *27*.)

The potential therapeutic value of NSCs is greatly enhanced by the basic
biology of the cells. They are, by definition, clonal, and therefore, homogeneous. The homogeneity of NSCs guarantees the reproducibility of the
results of grafting. The ability of NSCs to self-renew eliminates the need to
harvest tissue repeatedly. NSCs are multipotent, and a single cell can therefore replace both neurons and glia. NSCs are migratory. Transplanted NSCs
will migrate from the site of delivery to distant areas of active neurogenesis
or to damaged tissue, where they reconstruct both neural networks and glial

support. The results of one of the initial studies examining the potency of NSCs to engraft are shown in Fig. 2. The photomicrograph shows a section of a normal mouse brain engrafted with human NSCs transduced with the lacZ gene. The cells were grafted into the subventricular zone (SVZ) of a newborn mouse; the animal was euthanized 3 wk post transplantation, and the brain was sectioned and stained with Xgal and anti-β-galactosidase. The blue Xgal product of the engrafted donor cells appears dark against the lighter host background. Figure 2 demonstrates the stable engraftment of NSCs in the brains of mice. Migration to a similar extent occurs in diseased brains (as will be described throughout the chapter). Following migration to their destination, transplanted NSCs differentiate in host tissue and cease dividing.

The migratory nature of NSCs makes them potent vehicles for gene therapy. NSCs can carry copies of normal genes into tissue with missing or dysfunctional genes. NSCs will migrate, integrate, and produce normal gene product in normal amounts. Because they will differentiate into both neurons and glia, they can carry either neuronal or glial genes into host tissue. In addition, NSCs can be engineered to express a variety of other genes from all organisms including humans, at a range of levels of expression. Table 1 summarizes the characteristics of NSCs that enhance their role as therapeutic agents in CNS and PNS diseases.

We review some of the work in models of diseases of the nervous system for which transplanted stem cells have been demonstrated to have positive effects. The disorders include genetic or inheritable metabolic and deficiency diseases, infectious or inflammatory diseases, and age-related degeneration. These diseases have in common wide spread neural cell loss and/or dysfunction. The widespread nature of these diseases is not easily treated by conventional transplantation approaches, in which a limited number of cells can be delivered to a restricted area. Because NSCs will migrate to distant, multiple, and extensive regions of the nervous system, they can be useful in "global" degeneration, in which all or most of the nervous system is affected. Examples of such disorders include myelin deficiencies, leukodystrophies, and Alzheimer's disease. NSCs are also valuable therapeutic agents in the face of more limited damage to the CNS or PNS. In diseases such as Parkinson's, Huntington's, or specific degeneration of cerebellar neurons, transplantation may be restricted to a region of the CNS. Stroke and CNS or spinal injury also represent situations of a restricted area of damage. However, in these types of disorders, multiple neural cells types must be replaced. The multipotency of NSCs provides the mechanism by which multiple cell types can be regenerated. Finally, much of the biology of regeneration of nervous system tissue has yet to be discovered. To date, many of the factors that contribute to regeneration, or that prevent it, are unknown. NSCs may provide as yet unidentified factors to host tissue, enabling or enhancing regeneration of damaged tissue.

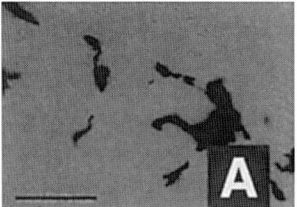

Fig. 2. Human NSCs in vivo following transplantation into the SVZ of neonatal mice. Engrafted cells within the periventricular and subcortical white matter regions (**a**) and the OB granule layer (**b**). The lacZ-expressing human NSCs were detected using Xgal. Human NSCs were also detected in the OB granule layer using anti-β-galactosidase (**c**). Scale bar= 100 μm. (Modified with permission from ref. 27.)

Table 1
Characteristics of Neural Stem Cells

Why NSCs are appealing as therapeutic transplantation material

1. SCs can be grown in culture indefinitely, are easily maintained, and produce large numbers of progeny.
2. NSCs are multipotent and can become both neurons and glia.
3. The clonal NSC can be screened for viruses/diseases that could be transferred to host.
4. NSCs assume the phenotype of surrounding cells into which the cells were engrafted.
5. NSC produce a variety of natural factors and products, including products that contribute to brain repair but have not yet been identified.
6. NSCs have the ability to migrate throughout the CNS.

Why NSCs are excellent gene therapy vehicles

1. NSCs can be implanted directly into brain or into the ventricles, as they circumvent the blood-brain barrier.
2. NSCs are easily transfected/transduced with therapeutic gene, and they can express more than one transgene.
3. NSCs migrate to damaged areas and therefore "deliver" genes to remote areas in the host brain.
4. NSCs show robust engraftment and delivery of transduced genes.

SIGNS OF SUCCESS IN DISEASE MODELS

Table 2 is a noncomprehensive list of some of the work that has demonstrated the success of stem cells in transplantation therapy in models

Table 2
Some Models of Successful Transplantation

Disease/injury	Animal model	Cells used	Gene expression	Outcome	References
Myelin deficiency	Mouse, rat	Murine NSCs	Not engineered	Engrafted NSCs differentiated into mature neurons replacing the dysfunctional host oligodendrocytes	Mouse: 5, 8 Rat: 6
Cerebellar defects	Mouse	Murine NSCs	Not engineered	Cells differentiated into granule cells of anterior lobe	Meandertail: 9
Lysosomal storage disease	Mouse	Murine NSCs Human NSCs	Engineered to express β-glucuronidase Engineered to express β-galactosidase	NSCs cross-corrected the enzyme deficiency	Tay-sachs disease: 11, 27 Sly's disease: 7,10
Stroke injury	Rat	Rat NSCs	Not engineered	NSCs differentiated after engraftment into mature CNS tissue	15
				NSCs prevented the loss of some neurons when implanted before injury	16
Spinal cord injury	Rat	Murine NSCs	Engineered to overproduce NT-3	Engrafted NSCs differentiated into neurons and glia	17
Neurodegenerative Diseases	Rat, mouse	Murine NSCs Human NSCs	Not engineered	Engrafted NSCs differentiated and replaced damaged or deficient areas and	13,24,30
			Engineered to express NGF	NSCs prevented some neuronal atrophy during aging	26
			EGF-responsive NSCs, direct expression of NGF	NSCs delivering NGF prevented degeneration of striatal neurons destined to die in Huntington's disease model rats	25
			Not engineered	Graft restored some complex sensorimotor behavior	18

of disease. The variety of diseases is large and includes genetic disorders, injury, and adult-onset disorders. All work reviewed here has been carried out in animal models; no human data have been reviewed. The types of cells used include embryonic stem cells, progenitor cells derived from fetal and adult tissue, and clonal lines of immortalized neural stem cells.

Demyelination Disorders

Stem cells appear to have the potential to participate in the treatment of a multitude of CNS disorders, including central and peripheral diseases in which host myelination fails. Myelination disorders are mainly of two classes: disorders in which endogenous myelinating cells are genetically impaired and fail to myelinate during development, and disorders that result from loss/failure of remyelination in the adult (for example, multiple sclerosis, allergic encephalomyelitis, and injury). Stem cells have been tested in animal models of several myelination disorders. Early work demonstrated that oligodendroglial progenitor cells were able to myelinate small areas of the brains of myelin-deficient hosts (3–5). The remyelination was limited to focal areas, and the source of the cells was fetal tissue. Later work (6) demonstrated that adult tissue could also provide a source of oligodendroglial progenitors that could myelinate the spinal cord of the 6–8-d-old myelin-deficient (md) rats. In a similar manner, embryonic stem (ES) cells were shown to myelinate the spinal cord of md rats following transplantation into 1-wk-old pups (7). Myelin-forming cells were found at some distance from the implantation site, suggesting that migration of the ES cells had occurred. Finally, work from our laboratory (8) has demonstrated that implantation of NSCs at birth results in widespread replacement of myelin basic deficient cells in the mouse model of dysmyelination, the shiverer (shi). The NSCs migrate throughout the brain, differentiate into oligodendroglia, express myelin basic protein, generate myelin sheaths, and cause phenotypic rescue of the shi mutant (Fig. 3).

Together these studies demonstrate that a number of types of immature cells can replace missing oligodendroglia, with resultant myelination in several different animal models of dysmyelination. Differences in the outcomes of these studies suggest that the degree of replacement and phenotypic rescue of the model appears to depends on the types of cells transplanted and the timing of the transplant.

Dysmyelination diseases represent one category of disease in which transplantation of stem cells may be an effective therapeutic approach. This group of diseases results from the dysfunction of one neural cell type, the oligodendroglia. Therefore, repair requires transplanted cells to assume only one phenotype. Additionally, extensive migration of transplanted cells was not a common out-

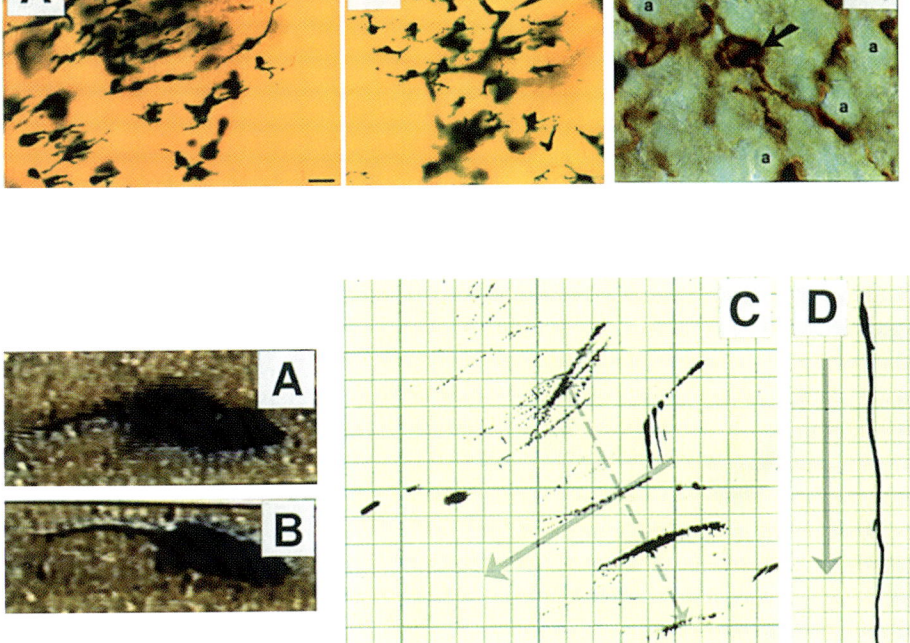

Fig. 3. Top: Murine NSCs transplanted into *shiverer (shi)* mutant mice differentiate into oligodrendrocytes. Representative sections from the corpus callosum (**A** and **B**) show donor-derived Xgal-positive cells with oligodrendrocyte morphology. (**C**) An anti-β-galactosidase-immunoreactive oligodendrocyte with multiple processes extending toward, and beginning to surround, adjacent axonal bundles ("a") viewed on end. Bottom: The degree of motor dysfunction in engrafted and unengrafted *shi* mutant mice with murine NSCs was assessed by blindly scoring videotaped cage behavior and by measuring the amplitude of tail displacement from the body's rostral-caudal axis. (**A**) Video freeze-frame of a representative unengrafted *shi* mutant mouse whose whole-body tremor causes the frame to be blurred. In contrast, the video freeze-frame of an engrafted, now asymptomatic *shi* mouse is focused (**B**). (**C** and **D**) The amplitude of tail displacement in an unsuccessfully engrafted *shi* mouse and a successfully engrafted *shi* mouse, respectively. These measurements were made by dipping the tail of the mouse in India ink and letting the mouse walk across a piece of graph paper. (**C**) The solid arrow indicates the body's long axis, whereas the dotted arrow indicates the points of maximal tail displacement. The displacements were averaged between transplanted and untransplanted animals to determine the effectiveness of the treatment. (**D**) A successfully engrafted, asymptomatic animal shows no tail displacement, allowing the tail to make an uninterrupted ink line on the paper. (Modified with permission from ref. *8*).

come of these studies. Mitigation, but not elimination, of symptoms in all animals in the study from our laboratory *(8)* of the *shiverer* mice following lateral ventricular transplantation may suggest that transplantation into both peripheral and central sites may result in more successful treatment.

Cerebellar Degeneration

Work from our laboratory suggested that NSCs will assume a neuronal phenotype when transplanted into brains with a specific loss of neurons. Rosario et al. *(9)* transplanted NSCs into newborn meander tail mice (*mea*), which are characterized by the lack of development of granule cells in the anterior cerebellar lobe. Transplanted cells preferentially differentiated into neurons (Fig. 4), suggesting that, once transplanted into hosts animals, NSCs will compensate for the specific cell loss characteristic of a disorder rather than being "tracked" to assume a particular cell type. These findings provide further support for stem cell approaches to treatment of disorders that involve the loss of neurons as well as glia.

Metabolic Diseases

A number of metabolic diseases affect the CNS. These diseases represent a second class of disorders for which stem cells offer a powerful approach to therapy. Stem cells can provide therapeutic molecules whose entry into the brain may normally be restricted by the blood-brain barrier. Stem cells can provide these molecules either because they normally produce them or because they have been genetically engineered to do so. Our laboratory demonstrated the successful use of NSCs in a metabolic disorder by cross-correcting a mouse model of the lysosomal storage disease, mucopolysaccharidosis VII (Sly disease; *10*). MPS VII mice have a deficiency of the enzyme β-glucuronidase (GUSB), which results in a lysosomal accumulation of glycosaminoglycans in the brain and other tissues. The result is a progressive, degenerative disorder, with mental retardation and eventually death. Transplantation of NSCs engineered to express the human GUSB enzyme into the ventricles of neonatal MPS VII mice resulted in expression of GUSB throughout the brain and widespread correction of lysosomal storage in neurons and glia. Grafted animals had significant enzymatic activity up to 8 mo post transplantation.

Another study from our laboratory that illustrates a second example of the use of NSCs in a model of a lysosomal storage disease is the work of Lacorazza et al. *(11)*. In this study, NSCs were engineered to express the human form of the β hexosaminidase α-subunit. This enzyme is defective in Tay-Sachs disease, a severe neurodegenerative disorder that results from the accumulation of G_{m2} ganglioside in the CNS. The engineered cells were

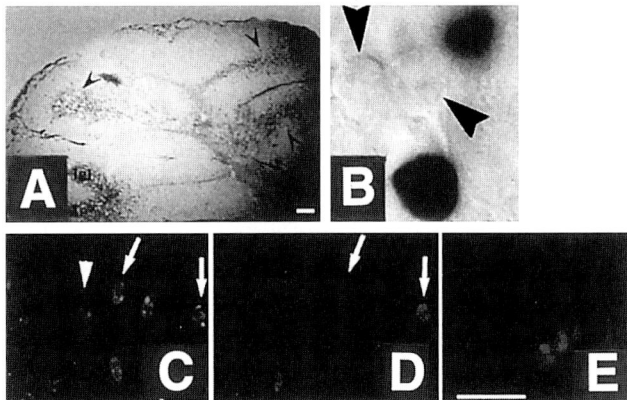

Fig. 4. Human NSCs transplanted into the cerebella of newborn *meandertail* mutant mice. Implanted NSCs in the external germinal layer (EGL) identified in the mature cerebellum by anti-BrDU immunoperoxidase histocytochemistry (brown nuclei in a and b). Cells were found both in the EGL and in sites distant to the injection site. **(a)** The internal granule cell layer (IGL and arrowheads) within the parasagittal section of the cerebellum. **(b)** BrDU-positive donor-derived neuron, indicated by the arrow, adjacent to residual host granule neurons indicated by arrowheads. Transplanted cells co-labeled with anti-BrdU **(c)** and NeuN **(d)**, indicated with arrows. The arrowhead in c indicates a BrdU +/NeuN – cell. Donor cells were located within the IGL using a human-specific probe in a fluorescent *in situ* hybridization procedure **(e)**. Scale bars = 100 μm in a; 10 μm in e. (Modified with permission from ref. *27*.)

transplanted into fetal and neonatal normal mouse brain. Engrafted brains produced high levels of the human protein. This work demonstrated that engrafted cells can migrate throughout the brain and, once integrated into host tissue, continue to produce transgenes. Torchiana et al. *(12)* presented additional evidence for the use of stem cells in the treatment of a lysosomal storage disease, Krabbe disease, by demonstrating that retroviral packing cell lines can be used to transduce immortalized neural progenitor cells to produce the lysosomal enzyme galactocerebrosidase (Galc). These studies provide evidence of the feasibility of stem cell transplantation to aid in the treatment of lysosomal storage diseases, an example of the power of this therapeutic strategy in the treatment of disorders characterized by dysfunction or absence of specific gene products. Stem cells can be both a producer and a delivery vehicle of missing and/or abnormal gene products.

Acute Degeneration

Stroke, CNS injury, and spinal injury represent additional disorders that may be successfully treated by stem cell transplantation. Here, damage is restricted to a circumscribed region (both in space and in time), but multiple neural cells types must be replaced. The multipotency of NSCs provides the mechanism to reconstruct damaged neural tissue. In early work from our group *(13)*, an artificial model of neurodegeneration was produced, and the potency of a clonal line of stem cells to repopulate the brain was examined. Apoptosis was induced in adult mouse brain, resulting in degeneration of cortical neurons. NSCs implanted into the brains engrafted and differentiated not only into glia but also into neurons, although differentiation of engrafted cells in normal adult brain is generally limited to glia phenotypes. Further work in our laboratory extended the investigation to include an examination of the fate of NSCs transplanted into brains that were subjected to ischemic and hypoxic injury. In a study by Park et al. *(14)*, the right common carotid artery of a 1-wk-old mouse was ligated, and the animal was exposed to 8% ambient oxygen. The combination of ischemia and hypoxia resulted in extensive injury to the ipsilateral hemisphere. Following transplantation into the ventricles or directly into the infarct cavity, NSCs migrated to and throughout the infarct cavity and engrafted and differentiated into all major neural cell types *(14)*.

Another study examining functional and structural repair of adult ischemic brain with stem cells is that by Fukunaga et al. *(15)*. Adult male spontaneous hypertensive rats were given middle cerebral artery occlusion (MCAO) followed by reperfusion. Embryonic tissue was implanted into the ischemic striatum. Several weeks later the grafted animals were found to have improved performance in the Morris water maze task, and neurons were seen within the graft.

Stem cells may also be used in the treatment of spinal cord injuries. In animal models of lower motor neuron degeneration, NSCs engrafted and a percentage of cells differentiated into cells morphologically consistent with lower motor neurons *(14)*. Therefore, it appears that NSCs are capable of repopulating and reconstituting areas of injury in the CNS. In addition, it appears that degenerating brain tissue may elicit elaborate signals that draw stem cells to the area and alter the fate of the donor cell.

Ischemia models present another example of the potential of stem cells to be agents of gene delivery, as seen in the work of Andsberg et al. *(16)*. Rat NSCs engineered to overexpress nerve growth factor (NGF) were transplanted into the brains of rats that had been subjected to MCAO. MCAO results in ischemia-induced cell death in the rat striatum. Forty-eight hours after the insult, NGF-secreting NSCs had migrated through the area of infarct, and the loss of striatal neurons was less in animals that had received transplants.

Genetically modified NSCs were also transplanted into spinal cord *(17)*. NSCs engineered to express neurotrophin-3 (NT-3) were transplanted into the spinal cords of immunosuppressed adult rats. The cells migrated long distances, differentiated into neurons and glia, and continued to express NT-3. Thus, stem cells have been used in models of acute neurodegeneration, in some cases with the application of gene therapy. The task in acute injury is to replace all types of neural cells over a discrete but possibly large area of the CNS. A combination of stem cell transplantation and gene therapy may be the most successful approach to the treatment of acute injury.

Chronic and Age-Related Degeneration

The final area we review is the use of NSCs in the treatment of chronic or long-term neurodegeneration, as seen in progressive neurodegenerative diseases such as Parkinson's, Huntington's, and Alzheimer's. In addition, normal aging may be associated with progressive cell loss in the CNS. These diseases, as well as aging, represent perhaps the biggest challenge to stem cell therapy. They require the replacement of all neural cell types in large brain regions (in the case of Alzheimer's disease and aging, virtually the entire brain), in the face of ongoing degeneration and possibly the presence of toxic molecules. Thus, the multipotency of stem cells, their ability to migrate over long distances, and their ability to express foreign genes are all aspects of the cells that must be strongly expressed for transplantation therapy to be of value in the face of chronic neurodegeneration.

Parkinson's disease (PD) represents the CNS disorder in which transplantation therapy has a long history. Fetal tissue grafts have been used in rodent and primate models of PD (*18*; for review see ref. *19*), as well as in clinical trials (*20,21*). However, short graft survival and limited integration of the grafts appeared to reduce the usefulness of fetal tissue. More recently, animal models of PD were given grafts of stem cells, and the behavioral and neurochemical outcomes were assessed. An immortalized rodent neuroprogenitor cell line was transplanted into primate and rodent PD models, with resultant increases in tyrosine hydroxylase (TH) in the brains of transplanted animals (*22*). Lundberg et al. (*23*) transduced astrocytes to produce TH and thus to secrete DOPA. Transplantation of these cells into the striatum of rats with unilateral 6-hydroxydopamine (6-OHDA) lesions resulted in a reduction in behavioral symptoms and the presence of TH-positive cells in the lesioned area. Zigova et al. (*24*) transplanted neuronal progenitor cells derived from neonatal rat SVZ into the striatum of adult rats 1 mo after unilateral 6-OHDA lesions. Animals were examined 1 wk, 2–3 wk, and 5 mo post implant. The donors cells survived and migrated, and many differentiated into neurons. Cells were still present 5 mo after transplantation. The stem cells, therefore, appear to be far superior to fetal tissue grafts and represent new hope for the treatment of PD.

Stem cells were also shown to be of value following transplantation into a rodent model of Huntington's disease *(25)*. Rats given intrastriatal injections of quinolinic acid show large areas of degeneration in the striatum. Transplantation of stem cells derived from transgenic animals overexpressing hNGF reduced the lesion size, with sparing of host striatal neurons. Delivery of hNGF by transplanted stem cells appeared to prevent degeneration of host striatal neurons. These data demonstrate an important aspect of stem cell transplantation therapy, that donor cells may not only replace host cells but may help to prevent loss of host tissue, by secretion of either endogenous factors or transgene products.

Aged rats demonstrate cognitive deficits, which are associated with atrophy of forebrain cholinergic neurons. In a study by Martinez-Serrano and Bjorklund *(26)* middle-aged rats with no cognitive deficits were given grafts of neural progenitor cells engineered to express NGF. The animals were examined during the next 9 mo for cognitive decline. At the end of the 9-mo period, age-matched control rats showned significant behavioral impairment when compared with grafted animals. The transplanted animals maintained cognitive performance at a level similar to that of younger adult animals. In addition, age-related atrophy of cholinergic neurons was not present in transplanted animals. Transplanted cells were integrated into the host tissue and continued to produce the transgene product up to 9 mo post grafting. These results support the use of genetically engineered stem cells in transplantation therapy of age-related neurodegeneration.

FUTURE DIRECTIONS

The work discussed above was generally limited to animal disease models as well as stem cells derived from animal tissue. An exciting area of research is the harvesting, culturing, and characterization of human stem cells. There is an increasing amount of evidence that stem cells can be isolated from embryonic and adult human tissue (for review, see ref. *1*). Work from our laboratory demonstrated that neural stem cells isolated from human fetal telencephalon would engraft in neonatal mouse brain, migrate, and differentiate into all neural lineages, including neurons *(27)*. In addition, these cells were genetically engineered to express foreign transgenes. The cells continued to express these genes following transplantation and engraftment into mouse brain. Finally, these hNSCs were able to cross-correct a genetic defect in a mouse model of Tay-Sachs disease.

Some of the early work on the transplantation of human cells into adult brain demonstrated that cells derived from human embryonic tissue could be transplanted into the normal adult rat CNS *(28)*. However, survival of the cells was limited when they were transplanted into adult rat striatum. Transplantation of growth factor-expanded human neural progenitor cells into neurogenic regions of the adult rat brain resulted in better survival of the cells, with migration and differentiation into both neurons and glia *(29)*. Progenitor cells derived from human tissue and transplanted into rats with unilateral 6-OHDA lesions also showed long-term survival *(30)*.

These data support the notion that human neural stem cells will emulate the attractive features of their rodent counterparts, including the ability to differentiate into all neural lineages, to engraft and migrate through brain tissue, to express foreign genes, to cross-correct genetic defects, and to replace missing cells in both developing and adult tissue. Transplantation therapy using stem cells appears to represent a powerful and versatile tool in the treatment of the most devastating CNS disorders.

ACKNOWLEDGMENTS

The work from our laboratory reviewed here was supported in part by grants from the American Paralysis Association; the Canavan's Research Fund; the Charles H. Hood Foundation; the Late Onset Tay-Sachs Foundation; a Mental Retardation Research Center grant from NIH (HD18655); the National Institute of Neurological Disorders and Stroke (NS07264, NS24707, NS34247, and NS33852); the Paralyzed Veterans of America; Project ALS; the Research Service of the Department of Veterans Affairs; and the William Randolph Hearst Foundation.

REFERENCES

1. Svendsen, C. N., Caldwell, M. A., and Ostenfeld, T. (1999) Human neural stem cells: isolation, expansion and transplantation. *Brain Pathol.* **9,** 499–513.
2. Billinghurst, L. L., Taylor, R. M., and Snyder, E. Y. (1998) Remyelination: Cellular and gene therapy. *Semin. Pediatr. Neurol.* **5,** 211–228.
3. Warrington, A. E., Barbarese, E., and Pfeiffer, S. E. (1993) Differential myelinogenic capacity of specific developmental stages of the oligodendrocyte lineage upon transplantation into hypomyelinating hosts. *J. Neurosci. Res.* **34,** 1–13.
4. Archer, D. R., Cuddon, P. A., Lipsitz, D., and Duncan, L. D. (1997) Myelination of the canine central nervous system by glial cell transplantation: a model for repair of human myelin disease. *Nature Med.* **3,** 54–59.
5. Learish, R. D., Brustle, O., Zhang, S., and Duncan, I. (1999) Intraventricular transplantation of oligodendrocyte progenitors into a fetal myelin mutant results in widespread formation of myelin. *Ann. Neurol.* **46,** 716–722.
6. Zhang, S., Ge, B., and Duncan, I. (1999) Adult brain retains the potential to generate oligodendroglial progenitors with extensive myelination capacity. *Proc. Natl. Acad. Sci. USA* **96,** 4089–4094.

7. Brustle, O., Jones, K. N., Learish, R. D., Karram, K., Choudhary, K., Wiestler, O. D., et al. (1999) Embryonic stem cell-derived glial precursors: a source of myelinating transplants. *Science* **285,** 754–756.

8. Yandava, B. D., Billinghurst, L. L., and Snyder, E. Y. (1999) "Global" cell replacement is feasible via neural cell transplantation: evidence from the dysmyelinated *shiverer* mouse brain. *Proc. Natl. Acad. Sci. USA* **96,** 7029–7034.

9. Rosario, C. M., Yandava, B. D., Kosaras, B., Zurakowski, D., Sidman, R. L., and Snyder, E. Y. (1997) Differentiation of engrafted multipotentneural progenitors towards replacement of missing granule neurons in meander tail cerebellum may help determine the locus of mutant gene action. *Development* **124,** 4213–4224.

10. Snyder, E. Y., Taylor, R. M., and Wolfe, J.H. (1995) Neural progenitor cell engraftment corrects lysosomal storage throughout the MPS VII mouse brain. *Nature* **374,** 367–370.

11. Lacorazza, H. D., Flax, J. D., Snyder, E. Y., and Jendoubi, M. (1996) Expression of human β-galactosidase α-subunit gene (the gene defect of Tay-Sachs disease) in mouse brains upon engraftment of transduced progenitor cells. *Nature Med.* **2,** 424–429.

12. Torchiana, E., Lulli, L., Cattaneo, E., Invernizzi, F., Orefice, R., Bertagnolio, B., et al. (1998) Retroviral-mediated transfer of the galactocerebrosidase gene in neural progenitor cells. *Neuroreport* **9,** 3823–3827.

13. Snyder, E. Y., Yoon, C., Flax, J. D., and Macklis, J. D. (1997) Multipotent neural precursors can differentiate toward replacement of neurons undergoing targeted apoptosic degeneration in adult mouse neocortex. *Proc. Natl. Acad. Sci. USA* **94,** 11,663–11,668.

14. Park, K. I., Liu, S., Flax, J. D., Nissim, S., Stieg, P., and Snyder, E. Y. (1999) Transplantation of neural progenitor and stem cells: Developmental insights may suggest new therapies for spinal cord and other CNS dysfunction. *J. Neurotransm.* **16,** 675–687.

15. Fukunaga, A., Uchida, K., Hara, K., Kuroshima, Y., and Kawase, T. (1999) Differentiation and angiogenesis of central nervous system stem cells implanted with mesenchyme into ischemic rat brain. *Cell Transplant.* **8,** 435–441.

16. Andsberg, G., Kokaia, Z., Bjorklund, A., Lindvall, O., and Martinez-Serrano, A. (1998) Amelioration of ischaemic-induced neuronal death in the rat striatum by NGF-secreting neural stem cells. *Eur. J. Neurosci.* **10,** 2026–2036.

17. Liu, Y., Himes, B. T., Solowska, J., Moul, J., Chow, S. Y., Park, K. I., et al. (1999) Intraspinal delivery of neurotrophin-3 using neural stem cells genetically modified by recombinant retrovirus. *Exp. Neurol.* **158,** 9–26.

18. Mehta, V., Hong, M., Spears, J., and Mendez, I. (1998) Enhancement of graft survival and sensorimotor behavioral recovery in rats undergoing transplantation with dopaminergic cells exposed to glial cell line-derived neurotrophic factor. *J. Neurosurg.* **88,** 1088–1095.

19. Dunnett, S. B. (1999) Repair of the damaged brain. *Neuropathol. Appl. Neurobiol.* **25,** 351–362

20. Lindvall, O., Brundin, P., Widner, H., Rehncrona, S., Gustavii, B., Frackowiak, R., et al. (1990) Grafts of fetal dopamine neurons survive and improve motor function in Parkinson's disease. *Science* **247,** 574–577.

21. Lindvall, O., Rehncrona, S., Brundin, P., Gustavii, B., Astedt, B., Widner, H., et al. (1990) Neural transplantation in Parkinson's disease: the Swedish experience. *Prog. Brain Res.* **82,** 729–734.
22. Anton, R., Kordower, J. H., Maidment, N. T., Manaster, J. S., Kane, D. J., Rabizadeh, S., et al. (1994) Neural-targeted gene therapy for rodent and primate hemiparkinsonism. *Exp. Neurol.* **127,** 207–218.
23. Lundberg, C., Field, P. M., Ajayi, Y. O., Raisman, G., and Bjorklund, A. (1996) Conditionally immortalized neural progenitor cell lines integrate and differentiate after grafting to the adult rat striatum. A combined autoradiographic and electron microscopic study. *Brain Res.* **737,** 295–300.
24. Zigova, T., Pencea, V., Betarbet, R., Wiegand, S. J., Alexander, C., Bakay, R. A. E., and Luskin, M. B. (1998) Neuronal progenitor cells of the neonatal subventricular zone differentiate and disperse following transplantation into the adult rat striatum. *Cell Transplant.* **7,** 137–156.
25. Kordower, J. H., Chen, E., Winkler, C., Fricker, R., Charles, V., Messing, A., et al. (1997) Grafts of EGF-responsive neural stem cells derived from GFAP-hNGF transgenic mice: trophic and tropic effects in a rodent model of Huntington's disease. *J. Comp. Neurol.* **387,** 96–113.
26. Martinez-Serrano, A. and Bjorkland, A. (1998) *Ex-vivo* nerve growth factor gene transfer to the basal forebrain in presymptomatic middle-aged rats prevents the development of cholinergic neuron atrophy and cognitive impairment during aging. *Proc. Natl. Acad. Sci. USA* **95,** 1858–1863.
27. Flax, J. D., Aurora, S., Yang, C., Simonin, C., Wills, A. M., Billinghurst, L. L., et al. (1998) Engraftable human neural stem cells respond to developmental cues, replace neurons, and express foreign genes. *Nature Biotech.* **16,** 1033–1039.
28. Sabate, O., Horellou, P., Vigne, E., Colin, P., Perricaudet, M., Buc-Caron, M., and Mallet, J. (1995) Transplantation to the rat brain of human neural progenitors that were genetically modified using adenovirus. *Nature Genet.* **9,** 256–260.
29. Fricker, R. A., Carpenter, M. K., Winkler, C., Greco, C., Gates, M. A., and Bjorkland, A. (1999) Site-specific migration and neuronal differentiation of human neural progenitor cells after transplantation in the adult rat brain. *J. Neurosci.* **19,** 5990–6005.
30. Svendsen, C. N., Caldwell, M. A., Shen, J., ter Borg, M. G., Rosser, A. E., and Tyers, P. (1997) Long-term survival of human central nervous system progenitor cells transplanted into a rat model of Parkinson's disease. *Exp. Neurol.* **148,** 135–146.

12

Drug Discovery and Gene Discovery

Alexander Kamb and Mahendra S. Rao

INTRODUCTION

Traditional medical therapy relies on small-molecule drugs as therapeutic agents. These compounds are obtained either by serendipity, or by screens of natural products such as those produced by plants and fungi, or, more recently, by screens of synthetic combinatorial chemical libraries. Whereas most early efforts utilized cumbersome screens and bioassays, the modern version of drug discovery depends on high-throughput screens using in vitro assays with purified targets. Often used in combination with structure-based methods and quantitative structure-activity relationship (QSAR) techniques, this current preferred approach to drug discovery has been called *rational design*. Rational design demands known targets, and thus, there is a premium on good, validated targets to begin the drug development process.

In the past 20 yr, recombinant DNA technology has empowered two new classes of therapeutics: protein and gene therapies. Of these, protein therapy has been the more successful, principally because the technical barriers are less severe. Although porcine insulin has been used to treat diabetes for decades, only recently has a more general approach to protein therapy been possible. Human recombinant proteins are now a multibillion-dollar industry. Gene cloning and overexpression provide large quantities of therapeutic human proteins at affordable prices. Gene therapy has been less successful, with no products on the market to date. On the other hand, progress has been made and in some cases preliminary clinical data are encouraging.

Improvements in cell culture and molecular markers offer new hope for cell and tissue therapy. As with gene therapy, there are as yet few examples. However, exciting developments in burn treatment, neuronal grafting, and bone marrow grafting suggest that this mode of therapy will continue to develop vigorously.

From: *Stem Cells and CNS Development*
Edited by: M. S. Rao © Humana Press Inc., Totowa, NJ

Cultured stem cells, and embryonic stem (ES) cells in particular, are among the most promising reagents for future medical innovation. They are obvious candidates for cell and organ therapy—especially in combination with genetic- and protein factor-based manipulations in culture. In addition, stem cells may serve as useful tools for discovery of new genes and protein targets. In combination with genome sequencing and analysis (genomics), biochemical pathway analysis, genetics, and high-throughput screening, ES cells may revolutionize the process of drug target identification. Powerful synergies are likely to ensue as these genomics tools collide with stem cell biology.

Here we discuss some of the possibilities afforded by the union between molecular genetics and stem cells. We argue that the prospects for radical, new, efficacious therapies are brighter now than they have been in a long time, and stem cells are a focal point for future efforts in several therapeutic areas. The range of opportunities in biology and medicine is limited mainly by imagination, as the technologies to manipulate genes and stem cells have matured to a high level.

STEM CELLS AND TISSUE THERAPY

A therapeutic approach that is perfectly suited to stem cells is the culture of cells prior to their introduction into patients for therapeutic benefit (Fig. 1). Two different approaches can be envisioned: (a) direct transfer of stem cells into specific sites in the body; and (b) culture and manipulation in vitro of stem cells or their differentiated offspring, followed by fabrication of tissues or organs for transplantation.

With direct transfer of stem cells, one hopes that the body's regulatory systems will take over and integrate the transplanted cells appropriately on site. If successful, this process would be equivalent to reforestation—supplying a fresh batch of seed that produces a full-blown forest in due course. Such transfer procedures might be used to correct tissue or organ damage, for example, to repopulate neurons lost during strokes. They might also be used to rejuvenate tissue, for instance, to supply new brain cells to aging regions of the brain. Some of the most fundamental questions relate to the capacity of existing tissue structures to serve as templates for the incoming stream of new cells. Are fresh young cells the primary deficit in severely damaged or aging organs? Or does a more global breakdown in organization preclude this therapeutic approach? Is a healthy but mature central nervous system (CNS) even capable of instructing transplanted cells with the proper developmental cues? This subject is discussed in detail in Chapter 11 of this volume.

An alternative to direct stem cell transfer involves the use of stem cells in vitro either to produce specialized cell types for reintroduction or to assemble functional organs for transplantation. This approach, at least in its

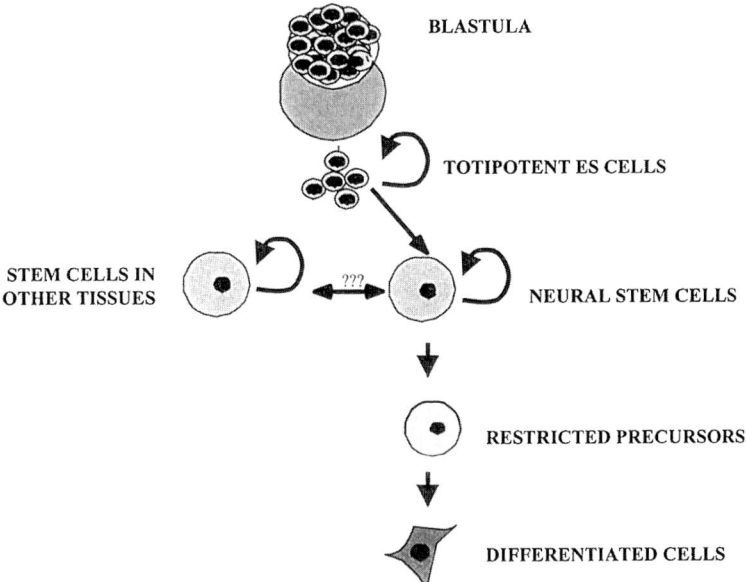

Fig. 1. Stem cell differentiation. Embryonic stem cells can be induced to differentiate in vitro, and cells arrested at different stages in development can be isolated. Arrows indicate which cell types have been isolated from mouse ES cells. Question marks indicate that whereas tissue-specific stem cells have been isolated, their ability to transdifferentiate to generate tissues derived from the precursor cells has not yet been rigorously tested.

most radical form, demands the controlled production of appropriate cell types in proper ratios. The recent finding that overexpression of telomerase, an enzyme that replicates chromosome ends, can immortalize cells suggests the possibility of using regulated telomerase expression to immortalize cells reversibly in culture *(1)*. Once sufficient numbers of cells were generated, the telomerase function could be inactivated (as it is in most normal cells), and the expanded population could be subjected to therapeutic manipulations. These expanded cell types could then be introduced into the body at the required site.

Tissue engineering enthusiasts have suggested for years that, rather than transplanting dissociated cells, it will ultimately be possible to generate entire organs—either synthetic, organic, or hybrids—that could replace damaged tissue (Fig. 2). Such extensive manipulation of human biology appears to be far off for most organs. However, some human therapies of this form already exist *(2–4)*. The best examples of tissue therapy success (using autologous tissues, not classical allogeneic transplantation) involve,

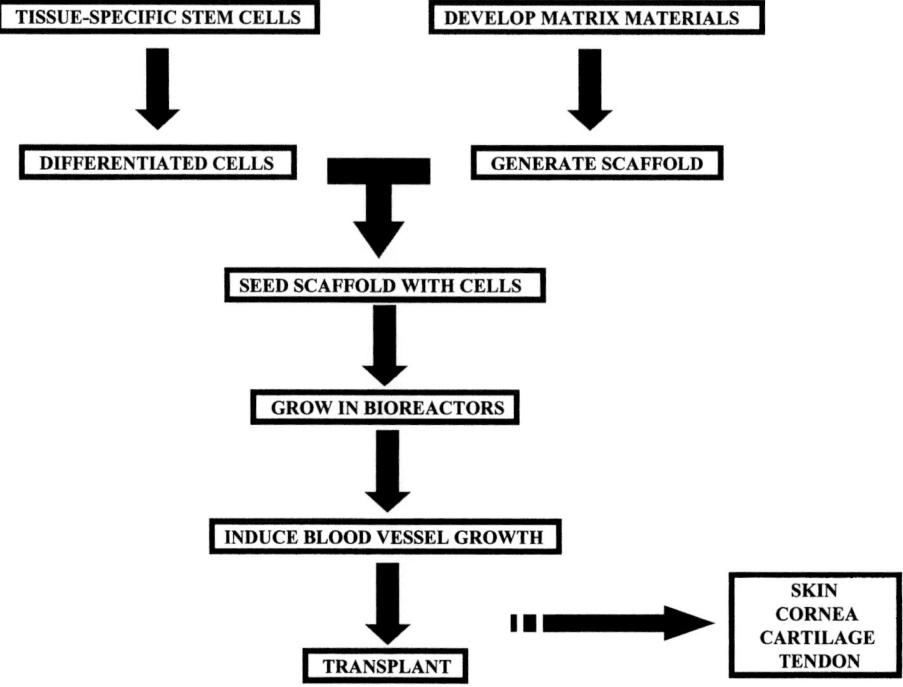

Fig. 2. Organogenesis. Reconstructing an organ requires concomitant development of appropriate biomaterials and selection of the proper numbers of cells that will reconstitute a three-dimensional organ of the appropriate size. The flow chart describes how a generic organ can be recreated.

not surprisingly, the simplest organs—blood and skin. Blood cells are not part of a rigid, complex three-dimensional structure. Skin is nearly two-dimensional and, therefore, relatively easy to culture. Thus, each is especially suited to tissue transplantation.

Advances in biomaterial engineering and stem cell technology have made the creation of biohybrid organs feasible. Indeed, biotechnology companies that market these products have billion-dollar revenues. Organogenesis and Advanced Tissues Sciences are two examples of such companies. Although a detailed discussion of recent breakthroughs is beyond the scope of this chapter, it is important to point out how some advances may expand the range of stem cell therapy. Recently biodegradable matrices have been assembled into three-dimensional structures and seeded with dividing cells to generate complex forms that closely mimic the shape of the synthetic scaffold. The scaffold is slowly digested by the growing cells, leaving a completely organic tissue replacement with the contours and functionality of the natural structure. Early applications of this methodology have been in the area of cartilaginous structures produced by mesodermal cells, for example, nasal and ear cartilage and cartilaginous implants for joint repair *(5–7)*.

More complex tissue structures have also been fabricated. Using precise ratios of different cell types, corneas have been grown in culture that are translucent and have an inflammatory response similar to that of natural corneas *(8–10)*. Heart valve production in vitro is also feasible because muscle cell precursors and endothelial cells can be harvested and grown in culture *(11)*. In certain respects, the structural complexity of these tissues is no less daunting than, for example, the spinal cord. However, the spinal cord is probably more complex in terms of cell type diversity and intercellular connections.

To help address some of these complexities, a strategy that incorporates hybrid biologic/synthetic structures has been developed *(12)*. One approach employs electrodes to bridge connections that have been lost and cannot be regenerated easily *(13–14)*. These electrode implants can lead to a partial restoration of vision or hearing by supplying sufficient bridging connections *(15–18)*. Other uses include treatment of benign essential tremor, and Parkinson's disease and bladder control *(19–21)*. In addition, various matrix materials, which include biodegradable and inert materials arranged in predetermined orientations, have been used to direct axonal bundles *(22)*.

One fanciful concept of how stem cells might be exploited entails their collection from young individuals and preservation until a time of need. Their stable karyotype, infinite life span, and syngeneic origin would ensure their use later for production of tissues and organs, as necessary. Organ replacements and tissue "refits" would be granted to individuals using their own cells at appropriate times. In this context, stem cell transfer might provide a route toward life-span extension, because aging tissues and organs might receive young replacements for senescent or functionally compromised component cells. Certainly numerous technical issues (as well as some ethical issues) must be confronted.

A potentially even more controversial therapeutic use of ES cells involves transfer of somatic nuclei—cellular cloning (Fig. 3). With the recent demonstration of whole animal cloning in sheep, this type of manipulation has incited public fears about the ethical consequences of the technology *(23)*. However, if applied on a more limited scale—for example, to the production of syngeneic organs and tissues rather than whole organisms—nuclear transfer may elicit less concern about the ultimate misuse of the technology.

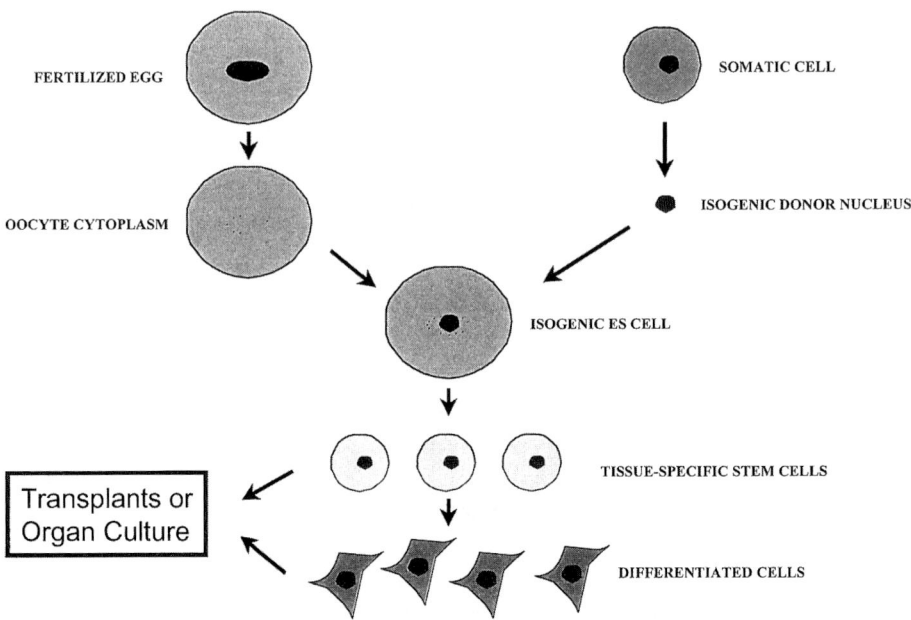

FERTILIZED EGG

SOMATIC CELL

OOCYTE CYTOPLASM

ISOGENIC DONOR NUCLEUS

ISOGENIC ES CELL

TISSUE-SPECIFIC STEM CELLS

Transplants or Organ Culture

DIFFERENTIATED CELLS

Fig. 3. Generating isogenic tissues. Nuclear transfer can be used to generate cellular clones. Isogenic ES cells can be generated, and these cells can be differentiated in culture to generate either tissue-specific stem cells or isogenic organs.

GENE THERAPY

Gene therapy comes in two varieties: in vivo and ex vivo. Each has its own special set of technical issues and advantages. In general, gene therapy must overcome a number of technical obstacles that at first glance appear rather trivial, but have in reality proved to be serious. These include design of gene transfer vehicles (e.g., viral vectors), choice of target cell type, attainment of sufficient levels of transgene expression, and appropriate tissue targeting of the transgene *(24)*.

The first clinical trials of gene therapy were initiated about 20 yr ago. These were widely criticized for their lack of rigor and proper preparation *(25)*. Subsequent developments in the field have proved the critics correct. Even though much work (and a great deal of capital) has been poured into this extremely exciting medical area, relatively little in terms of therapeutic benefit has emerged. Many companies that were founded on the promise of gene therapy have either disappeared or reinvented themselves around a different technology. Only recently is there some indication that the cusp of success may be imminent. For example, the gene for vascular endothelial growth factor (VEGF), a factor that stimulates blood vessel formation, has been successfully used in clinical trials to treat diabetes sequellae resulting from decreased blood supply *(26)*. However, a recent setback in gene therapy clinical trials is likely to dampen enthusiasm in the entire field—both clinically and financially. A few highly publicized patient deaths using adenovirus delivery vectors have prompted the Food and Drug Administration (FDA) to halt several trials *(27)*. In spite of the bumpy road and the long gestation time, some investors continue to have confidence in the overall approach. Meanwhile, other companies and academic groups have been patiently pushing ahead to develop new vectors, formulations, and protocols that are likely to help overcome the barriers that have so far frustrated gene-based therapeutic efforts.

Stem cells may be most obviously useful for ex vivo gene therapy, although in principle targeting procedures might be developed to engineer stem cells in vivo with exogenous genes. The concept for ex vivo stem cell therapy is very simple and is similar to the strategy being employed by a variety of commercial and academic groups using other cell types. For example, Transkaryotic Therapies, Inc. has developed a method for introducing expression cassettes into fibroblast explants prior to returning them to certain sites in the body as cellular factories for production of specific proteins *(28)*. The virtues of engineering stem cells are as follows: first, the transgenic cells could be expanded prior to reintroduction; second, the transplant would in theory be perpetually renewable, capable of giving rise to progeny cells that carry the transgene; third, stem cells generally possess a tremendous capacity to migrate and thus may repopulate areas distant from the site of transplantation, reducing the invasiveness and improving the outcome of single procedures; and fourth, specific tissues could be targeted via stem cells or their differentiated offspring. For example, neuronal stem cells (NEPs) might be engineered to produce acetylcholine, a neurotransmitter whose presence has been linked to improved cognitive function. These engineered NEPs could be injected into the brains of cognitively impaired individuals, creating an endless supply of cholinergic neurons and thus a source of beneficial acetylcholine. The therapeutic basis for a similar strategy has been explored *(29)*.

A different type of ex vivo gene therapy involves disruption of gene activity by homologous recombination rather than transfer of functional genes into cells. Such a strategy might be useful when dominant mutant alleles are responsible for pathology (e.g., Hirschsprung's disease; *29a*) or when diminution of gene expression might be useful for other reasons. One barrier to gene knockout therapy is the low frequency of homologous recombination in most somatic cells. However, murine ES cells have relatively high rates of homologous recombination and have served mouse geneticists for several years as vehicles for reconstitution of mutant mice with disrupted genes. If human ES cells have the same property, production of human knockout ES cells should be possible.

An alternative to homologous recombination in ES cells or gene transfer by viral vectors utilizes chimeraplasty—correction of germline defects by DNA/RNA hybrid oligonucleotides *(30)*. These molecules are designed to contain the correct version of the mutated/defective sequence as well as guide sequences that direct specific binding to the defective gene. The hybrid is recognized by the endogenous DNA repair machinery, and the incorrect nucleotide is replaced by the correct sequence. The exact mechanism of the repair process is unknown, but the process appears to work The technique has been used to correct a metabolic disorder in mice and to repair the mutation responsible for sickle cell anemia. Kimeragen, a Pennsylvania-based company, plans clinical trials for treating a rare metabolic disorder, Crigler-Najjar syndrome, which afflicts members of the Amish population *(31)*.

Combining chimeraplasty with advances in stem cell isolation from affected individuals may open new therapeutic opportunities. For example, it may be possible to correct genetic defects in vitro followed by transplantation of syngeneic cells into the donors, providing a permanent cure. This process would sidestep the problems associated with viral vector delivery and foreign gene inactivation, issues that have frustrated gene therapy.

STEM CELLS AND DRUG DISCOVERY

Drugs come in two general forms: small organic molecules and proteins (Table 1). Stem cells have a variety of applications in drug discovery for both cases, either directly as targets of such drugs or as tools for drug discovery.

The traditional avenue toward discovery of biologic activities—and hence candidate therapeutic proteins—involves bioassays, typically cell- or organism-based screens of biological extracts. Fractionation by chromatographic procedures or expression cloning forces convergence on the physiologically active agent. This general approach has led to the isolation and characterization of numerous receptors and protein factors, including the ubiquitous, medically important nerve growth factor (NGF) molecule *(32)*.

The advent of recombinant DNA technology in the late 1970s and its introduction into the commercial sector spawned a huge number of companies focused on particular therapeutic proteins. The ability to engineer cells to produce large quantities of specific proteins has generated therapeutics that are customized, cheap to produce, and human or "humanized" thereby avoiding problems associated with immune response. The blockbuster drugs of the biotechnology sector, erythropoietin, granulocyte colony stimulating factor (GCSF), human growth hormone (HGH), insulin, and tissue

Table 1
Types of Therapeutic Agents

Type	Example	Advantages	Disadvantages
Small molecule	Aspirin	Oral delivery	Restricted to "drug-able" targets
Secreted factor	NGF	Endogenous molecule	Injected delivery
Peptide	Conapeptide	Customizable Immune response	Injected delivery;
Antibody	Anti-IgE antibody	(Mostly) endogenous molecule	Injected delivery
Gene	VEGF	Treats disease cause or symptoms	Delivery; immune response; finite duration of expression

plasminogen activator (TPA), are all protein therapeutics derived from genetic engineering. From the beginning, human proteins promised specificity and natural integration within the physiologic framework of the body. After all, they are molecular components of our own tissues. Although there have been many successes, this simplistic notion of efficacy has proved inadequate, as imprecisely understood biology, toxic side effects, tissue localization, and pharmacokinetic issues have derailed a slew of once-promising protein therapeutics.

Genomics, the systematic exploration of genomes and expression products, has provided a powerful boost to protein therapeutic research. It will soon be possible to search databases that contain human sequences representing nearly the entire genome. With rapid sequence similarity search algorithms and rigorous statistical methods, homologous genes can be found with the use of a computer. Sequences characteristic of secreted proteins, growth factors, and the like are now used routinely by scientists in the pharmaceutical and biotechnology industries as new candidate proteins for specific therapeutic applications. For example, Human Genome Sciences has recently launched clinical trials based on protein therapeutics derived from their sequencing efforts *(33)*. These are probably the first potential drugs to spring directly from genomics. Expression profiling, with the capacity to examine the expression of thousands of genes in a single experiment, is clogging the downstream portion of drug target discovery with even more candidates. Whether or not the expression pattern of genes proves to be a cost-effective criterion for prioritizing target candidates remains to be seen. What is abundantly clear is that genomics has switched the rate-limiting step in drug discovery from identification of candidate proteins to physiologic validation, that is, candidates are being proposed in huge numbers, but the physiologic significance and medical utility of these proteins must still be painstakingly explored.

In the last year, therapeutic antibodies designed with the aid of molecular genetics have emerged on the scene, creating a new form of protein therapy. Monoclonal antibodies, the controlled production of which was achieved in the late 1970s, are the basis for the commercial enterprises that develop and manufacture antibody therapeutics. The utility of monoclonals as therapeutics has been aided greatly by other technologies including genetic engineering and phage display. Initial attempts to develop therapeutics were hamstrung by the use of mouse immunoglobulins that generated unacceptable side effects based mainly on their xenogeneic origin. As the field has evolved, researchers have shifted away from murine antibodies, either by creation of fully human monoclonals, or by wholesale substitution of the immunogenic mouse sequences by human sequences (hence, "humanized" monoclonals). As an example of this therapy, preliminary studies have found that anti-IgE antibodies are remarkably effective in combating asthma *(34)*.

The advantages of monoclonal antibody therapeutics are numerous. Antibodies derive from the body's own defense system and thus are expected to be well-tolerated. They are capable of exquisite specificity of binding, thus reducing the probability of side effects owing to cross-reaction. They comprise a generic scaffold for display of a wide variety of idiotypes or binding sites, thus potentially improving predictability of pharmacokinetic behaviors.

On the down side, there have been some visible failures, including a much publicized withdrawal by Biogen, Inc. of their lead monoclonal against CD40, designed to treat a variety of ailments *(35)*. In addition, monoclonals, like other protein therapeutics, are restricted to injection as a route of administration, barring some quantum advance in drug delivery technology. Once in the blood stream, they cannot penetrate cell membranes. Thus, their targets are limited to the 10% or so of the expressed proteins present on the extracellular surface of the membrane or secreted into the extracellular space.

It seems likely that stem cells will increasingly provide a crucial link in the protein therapeutic discovery process. For protein discovery, stem cells have several advantageous characteristics. First, they give rise to primary cells with normal karyotypes and with properties that are similar to those of corresponding cells in the body. A related issue is that they may be used in culture to generate homogeneous tissue samples in a controlled setting. By appropriate treatments, stem cells may be driven to differentiate down predetermined pathways, thus avoiding in theory much of the messiness of tissue explants. Stem cells thus hold great promise as a key component of bioassays. One of the major shortcomings of the current genomics-based drug discovery effort is its reliance on correlative information such as DNA sequence and expression pattern. Although such information is relatively easy to access at present, it generally lacks any causal connection to the disease of interest. Such connections must be established later using some type of bioassay. Stem cells, along with their lineages, may provide a series of cultured cell types that can be used to assay the effects of specific candidate therapeutic proteins. In association with a collection of appropriate phenotypic probes (e.g., a set of molecular markers), ES cell populations may possess sensitivity and generality as a mode of bioassay that is extremely difficult to duplicate in other formats.

Stem cells also offer great promise as cellular targets of protein therapy. Protein factors that regulate stem cell differentiation in specific ways may be used in defined therapeutic contexts to increase or decrease the proportion of a given cell type. For example, infusion of fibroblast growth factor (FGF) or epidermal growth factor (EGF) can alter the rate of stem cell division in vivo and bias the phenotype of cells produced *(36,37)*. Similarly, infusion of brain-derived neurotrophic factor (BDNF) can alter the production of neurons in the subventricular zone (SVZ) *(38)*, and Sonic Hedgehog (shh) can alter the rate of granule cell division *(39)*.

Administration of such protein factors could occur in vivo, where the physiologic setting may produce more favorable outcomes, or in vitro, where conditions can be monitored precisely prior to transplantation into the patient. In principle, a single person's stem cells could be removed, cultured, and expanded to perform the therapy. Such syngeneic cells would not provoke an immune response on reintroduction, even when driven down particular pathways of differentiation.

Small molecules are the traditional favorite of the pharmaceutical industry, and even with the emergence of protein therapy as a major commercial opportunity most large pharmaceutical companies remain convinced that small-molecule drugs are the best long-term strategy. This mindset may be partly the result of the historical development of the pharmaceutical industry, which sprang largely from chemical and dye companies. The philosophy of conservative chemists persists in the sector. However, there is undeniable supporting logic. Compounds of low molecular weight (generally below 500 kDa) have propelled the industry into one of the most lucrative businesses in the world. In addition, delivery, storage, and formulation problems inherent in protein therapy pose significant obstacles compared with chemical compounds that typically can be made cheaply and stored for long periods prior to oral administration.

Stem cells can be used in "direct screens" of chemical compounds to identify growth- or differentiation-modulating chemicals. ES cells, for example, are amenable to cell culture and high-throughput screening in a microtiter format, prerequisites for this type of screening. Cells might be engineered to contain specific reporter constructs that could be assayed rapidly and quantitatively. For example, drugs that enhance production of endogenous NGF might be identified using a cell line engineered with a fluorescent reporter downstream of the NGF promotor (i.e., DNA sequences that regulate NGF expression). The drawback of this direct approach is the potential problems associated with determining the mechanism of action of small molecules.

STEM CELLS AND GENE DISCOVERY

In the past two decades, rational drug design has been ascendant among pharmaceutical companies. In vitro screens using purified targets are nearly universal in the industry. Therefore, the large companies as well as small biotechnology firms and academic laboratories have scrambled to define good molecular targets as the prelude to drug screening. It is axiomatic that a good drug requires a good target. Thus, there is at present intense focus on discovery of novel (and hence proprietary) selective targets that can be inhibited by small molecule compounds. This goal of gene discovery may be placed in a more general context: the elucidation of physiologic functions of genes. As the lead-in to target selection, it is the first step in the drug development process.

The current fashion for discovery of genes is genomics (Fig. 4). Sequence databases replete with numerous gene family members—for instance, G-protein-coupled receptors, proteases, and ion channels—supply hundreds of potential targets to the industry. Drug company executives seem convinced that the route to new blockbuster compounds passes through homologs of proteins that are known to be good drug targets (Table 2). Comparative gene expression, which is being practiced on an increasingly grand scale, provides even more target candidates—based solely on their expression profile in tissues. Because genomics is so fecund in terms of furnishing candidate targets, the bottleneck in drug target discovery has shifted downstream to validation and testing. The critical issue is the physiologic functions of genes. Without such information, it is impossible to predict the effect of specific protein-targeted therapies.

Gene expression patterns along with sequence similarities are the preferred methods for formulating hypotheses about gene function at present (Table 3). Because this type of information is readily accessible, it has fueled a frenzy of hypothesis generation by academic and commercial scientists. There are several ways to assess gene/protein expression. Probably the most popular and technically simplest method involves microarrays of gene sequences presented on a surface at high density, either as oligonucleotides,

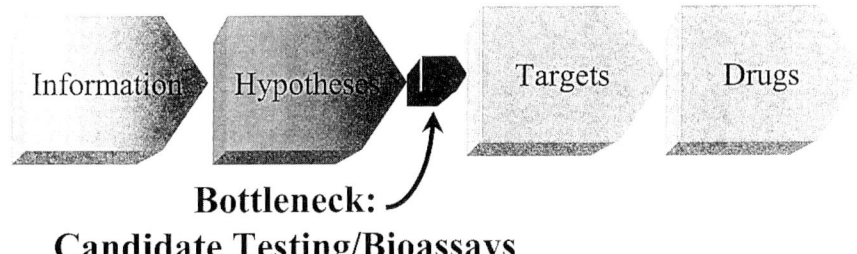

Bottleneck:
Candidate Testing/Bioassays

Fig. 4. Genomics process. Diagram illustrates the scheme for gene, target, and drug discovery based on the genomics paradigm. The upstream portions of the process have become highly efficient, thus producing a bottleneck at the stage of hypothesis testing and target validation.

Table 2
Desirable Characteristics of Drug Targets

Physiologic role in disease
Known biochemical function
Existing drugs targeting related protein(s)
Inhibitable
Selective (e.g., localized exclusively in target tissue)

plasmid clones, or polymerase chain reaction (PCR) products. Probes are generated from mRNA isolated from tissues or cell lines. These probes are then labeled and hybridized to the microarrays, which can be scanned. The scanning device measures the amounts of label in each spot on the array. The method is rapid, fairly quantitative, and generates enormous quantities of information about gene expression. Genes can be grouped by expression pattern, markers can be identified, and functions can be proposed based on correlations *(40)*. A variety of other methods to assess gene expression have been used successfully; most of these utilize some form of electrophoresis. Protein quantitation is also gaining in popularity as a form of expression analysis. This is partly because of the concern that mRNA levels may not be closely correlated with protein levels, and to recent advances in the technology to assay protein levels, including two-dimensional gel electrophoresis and mass spectrometry.

Homogeneity, infinite life span, and multipotency endow stem cells with almost unmatched quality as the source of mRNA for comparative gene expression experiments. Problems of tissue heterogeneity in surgical specimens and genomic abnormalities in somatic cell lines may be largely avoided through the use of stem cells and their progeny populations of differentiated cells. Comparative gene expression data may be used to identify candidate targets for controlling differentiation in precursor populations or for defining genes with specific patterns of regulation that may be used to create reporter constructs for bioassays. Thus, as a tool for gene discovery they have potential that may be unrivalled, at least in certain therapeutic areas such as neurologic dysfunction. As a hypothetical example, pure populations of multipotent stem cells or neuronal-restricted precursors can be isolated and their patterns of gene expression compared using one of the many available technologies (Table 3). Genes that are unique to each population can be identified and their physiologic roles determined through functional assays that give clues to their roles in differentiation.

A marriage between comparative gene expression techniques and transgenic ES cells may add robustness to approaches based on transcriptional profiling. Human precursor cells can be immortalized, and subclones that express specific receptors (or other properties) can be identified. These subclones can be used in high-throughput assays to examine the effects of compounds in a more natural—but nonetheless manipulable—environment. Clonal cell lines that differentiate solely into neurons have been developed and subclones that generate motoneurons have been described *(41)*. These lines, for example, can be used to study effects of drugs on cholinergic neurons.

The shortcoming of comparative gene expression is its reliance on correlational as opposed to causal relations. For causality—and hence true physiologic function of genes—mutational analysis is paramount. In an effort to define physiologic relevance up front, various groups have embarked on massive genetic experiments using either model systems such as the mouse or fruit fly, or human linkage analysis. These whole-organism genetic approaches have yielded information about disease pathways but have not been successful in delineating drug targets directly.

A different genetic approach involving somatic cells may hold more promise for drug target discovery. Cultured cells are beneficial when employed in a variety of screens based on growth or expression of molecular markers (e.g., differentiation markers). Such screens may be used to pick apart pathways involved in differentiation. In addition, screens of certain types can find targets that are "drug-able," that is, they can be modulated in vivo by small molecules *(42–45)*. One of the most promising methods in somatic cell genetics is expression cloning, selection of clones from expression libraries based on functional assays in mammalian cells (Fig. 5). ES cells have perfect characteristics for such screens. They can be manipulated in culture and engineered in a variety of ways. A version of expression cloning, transdominant genetics, in which peptides, protein fragments, or

Table 3
Methods of Comparative Gene Expression[a]

Name	Method	Readout	Advantages	Disadvantages
SAGE	Electrophoresis	DNA sequence not required	Gene sequence	Slow
AFLP	Electrophoresis	Fragment length/ DNA sequence	Gene sequence not required	Slow
Microarray	Hybridization	Quantitative emission label	Parallel/rapid required	Gene sequence
2D gel	Electrophoresis	Quantitative emission label	Parallel difficult to automate	Cumbersome;
Mass spectrometry	Electrospray/	Charge/mass ratio MALDI	High resolution no label required	Limited input complexity

[a]SAGE, serial analysis of gene expression; AFLP, amplified fragment-length polymorphism; MALDI, matrix-assisted laser desorption ionization mass spectrometry.

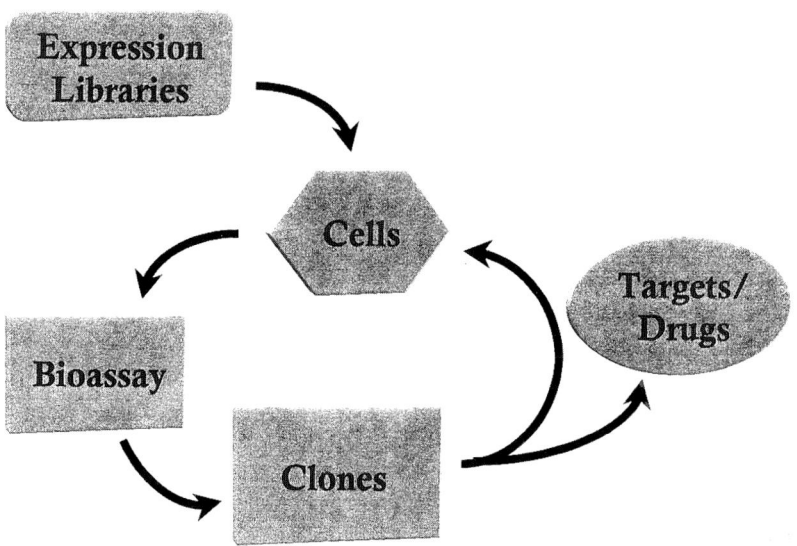

Fig. 5. Somatic cell genetics process flow. A flow chart illustrating how cultured cells can be used to identify either therapeutic targets or therapeutic agents. DNA libraries are expressed in appropriate cell lines, and subclones of cells that display a desirable trait are isolated. The process is repeated several times to ensure specificity, and the active DNA fragments are cloned, sequenced, and otherwise characterized. Proteins or peptides derived from this process can be used therapeutically. Alternatively, they can be used in high-throughput screens to identify small molecules that modulate target function.

RNAs are expressed in cells, was developed especially for somatic cell genetics in mammalian cells *(42,46)*. This approach overcomes many of the obstacles that have hamstrung somatic cell genetics, including diploidy, heterogeneity, and recovery of mutant genes.

In conjunction with neural stem cells, somatic cell genetics may reach very high levels of utility. For instance, to identify proteins (or drugs) that regulate differentiation down specific pathways, differentiation markers (or reporters linked to cell type-specific promotor elements) could be used to select agents from libraries that affect differentiation in predetermined ways.

Without question the most prolific use of stem cells in the context of gene function studies has been in the mouse. Gene knockout technology that relies on ES cells has generated a wealth of data that bear on gene functions. Hypotheses can be tested in a relatively unequivocal experimental context: comparing wild-type mice with the phenotype of mice in which gene function is completely and selectively eliminated by gene disruption. Despite the undoubted success of this knockout technology, it still remains uncertain how these experiments conducted in mice will translate into therapy in humans.

CONCLUSIONS

The unique properties of stem cells endow them with exceptional power in the context of therapy—both as tools for discovery and as therapeutic agents themselves. Indeed, the potential of stem cell biology is so revolutionary in this regard that a variety of social and ethical issues will probably need serious attention. Huge benefits will ensue, and, as with all paradigm-shifting advances in technology, care must be taken so that the dangers of irresponsible application of molecular and cellular stem cell biology will be avoided.

REFERENCES

1. Morales, C. P., Holt, S. E., Ouellette, M., Kaur, K. J., Yan, Y., et al. (1999) Absence of cancer-associated changes in human fibroblasts immortalized with telomerase. *Nature Genet.* **21,** 115–118.
2. Langer, R. and Vacanti, J. P. (1993) Tissue engineering. *Science* **260,** 920–926.
3. Langer, R. and Vacanti, J. P. (1999) Tissue engineering. *Sci. Am.* **28,** 86–89.
4. Pabst, P., Amall G., and Gregory, L. L. P. (1999) A brief review of some of the patents issuing or published in 1999. *Tissue Eng.* **5,** 597–598.
5. Angele, P., Kujat, R., Nerlich, M., Yoo, J., Goldberg, V., and Johnstone, B. (1999) Engineering of osteochondral tissue with bone marrow mesenchymal progenitor cells in a derivatized hyaluronan-gelatin composite sponge. *Tissue Eng.* **5,** 545–554.
6. Saxena, A. K., Marler, J., Benvenuto, M,, Willital, G H , and Vacanti, J. P. (1999) Skeletal muscle tissue engineering using isolated myoblasts on synthetic biodegradable polymers: preliminary studies. *Tissue Eng.* **5,** 525–532.

7. Kobashi, T. and Matsuda, T. (1999) Fabrication of branched hybrid vascular prostheses. *Tissue Eng.* **5,** 515–524.
8. Koizumi, N., Inatomi, T., Quantock, A. J., Fullwood, N. J., Dota, A., and Kinoshita, S. (2000) Amniotic membrane as a substrate for cultivating limbal corneal epithelial cells for autologous transplantation inrabbits. *Cornea* **19,** 65–71.
9. Griffith, M., Osborne, R., Munger, R., Xiong, X., Doillon, C. J., Laycock, N. L., et al. (1999) Functional human corneal equivalents constructed from cell lines. *Science* **286,** 2169–2172.
10. Ferber, D. (1999) Tissue engineering. Growing human corneas in the lab. *Science* **286,** 2051, 2053.
11. Sodian, R., Sperling, J. S., Martin, D. P., Stock, U., Mayer, J. E., Jr., and Vacanti, J. P. (1999) Tissue engineering of a trileaflet heart valve-early in vitro experiences with a combined polymer. *Tissue Eng.* **5,** 489–494.
12. Woerly, S., Petrov, P., Sykova, E., Roitbak, T., Simonova, Z., and Harvey, A. R. (1999) Neural tissue formation within porous hydrogels implanted in brain and spinal cord lesions: ultrastructural, immunohistochemical, and diffusion studies. *Tissue Eng.* **5,** 467–488.
13. Liu, X. (1999) Stability of the interface between neural tissue and chronically implanted intracortical microelectrodes. *EEE Trans. Rehabil. Eng.* **7,** 315–326.
14. McCreery, D. B., Carter, R. R., Bullara, L. A., Yuen, T. G., and Agnew, W. F. (1986) Neuronal activity evoked by chronically implanted intracortical microelectrodes. *Exp. Neurol.* **92,** 147–161.
15. Nadig, M. N. (1999) Development of a silicon retinal implant: cortical evoked potentials following focal stimulation of the rabbit retina with light and electricity. *Clin. Neurophysiol.* **110,** 1545–1553.
16. Zrenner, E., Stett, A., Weiss, S., Aramant, R. B., Guenther, E., Kohler, K., et al. (1999) Can subretinal microphotodiodes successfully replace degenerated photoreceptors? *Vision Res.* **39,** 2555–2567.
17. Zrenner, E., Miliczek, K. D., Gabel, V. P., Graf, H. G., Guenther, E., Haemmerle, H., et al. (1997) The development of subretinal microphotodiodes for replacement of degenerated photoreceptors. *Ophthalmic Res.* **29,** 269–280.
18. Humayun, M. S., de Juan, E. Jr., Dagnelie, G., Greenberg, R. J., Propst, R. H., and Phillips, D. H. (1996) Visual perception elicited by electrical stimulation of retina in blind humans. *Arch. Ophthalmol.* **114,** 40–46.
19. Pahwa, R., Lyons, K. L., Wilkinson, S. B., Carpenter, M. A., Troster, A. I., Searl, J. P., et al. (1999) Bilateral thalamic stimulation for the treatment of essential tremor. *Neurology* **53,** 1447–1450.
20. Maniglia, A. J., Abbass, H., Azar, T., Kane, M., Amantia, P., Garverick, S., et al. (1999) The middle ear bioelectronic microphone for a totally implantable cochlear hearing device for profound and total hearing loss. *Am. J. Otol.* **20,** 602–611.
21. Lanmuller, H., Sauermann, S., Unger, E., Mayr, W., and Zrunek, M. (1999) Battery-powered miniature implant for electrical nerve stimulation. *Biomed. Tech. (Berl.)* **44,** 114–119.
22. Chauhan, N. B., Figlewicz, H. M., and Khan, T. (1999) Carbon filaments direct the growth of postlesional plastic axons after spinal cord injury. *Int. J. Dev. Neurosci.* **17,** 255–264.

23. Wilmut, I., Schnieke, A. E., McWhir, J., Kind, A. J., and Campbell, K. H. (1997) Viable offspring derived from fetal and adult mammalian cells. *Nature* **385,** 810–813.

24. Verma, I. M. and Somia, N. (1997) Gene therapy — promises, problems and prospects. *Nature* **389,** 239–242.

25. Anderson, W. F. and Fletcher, J. C. (1980) Sounding boards. Gene therapy in human beings: When is it ethical to begin? *N. Engl. J. Med.* **303,** 1293–1297.

26. Rivard, A., Silver, M., Chen, D., Kearney, M., Magner, M., Annex, B., et al. (1999) Rescue of diabetes-related impairment of angiogenesis by intramuscular gene therapy with adeno-VEGF. *Am. J. Pathol.* **154,** 355–363.

27. Smaglik, P. (1999) Tighter watch urged on adenoviral vector with proposal to report all "adverse events." *Nature* **402,** 1707.

28. Heartlein, M. W., Roman, V. A., Jiang, J. L., Sellers, J. W., Zuliani, A. M., Treco, D. A., et al. (1994) Long-term production and delivery of human growth hormone in vivo. *Proc. Natl. Acad. Sci. USA* **91,** 10,967–10,971.

29. Balse, E., Lazarus, C., Kelche, C., Jeltsch, H., Jackisch, R., and Cassel, J. C. (1999) Intrahippocampal grafts containing cholinergic and serotonergic fetal neurons ameliorate spatial reference but not working memory in rats with fimbria-fornix/cingular bundle lesions. *Brain Res. Bull.* **49,** 263–272.

29a. Edery, P., Attie, T., Mulligan, L. M., Pelet, A., Eng, C., Ponder, B. A., et al. (1994) A novel polymorphism in the coding sequence of the human ret proto-oncogene. *Hum. Genet.* **94,** 579–580.

30. Stephenson, J. (1999) New method to repair faulty genes stirs interest in chimeraplasty technique. *JAMA* **281,** 119–121.

31. Kren, B. T., Parashar, B., Bandyopadhyay, P., Chowdhury, N. R., Chowdury, J. R., and Steer, C. J. (1999) Correction of the UDP-glucuronosyltransferase gene defect in the gunn rat model of Crigler-Najjar syndrome type I with a chimeric oligonucleotide. *Proc. Natl. Acad. Sci. USA* **96,** 10,349–10,354.

32. Schenkein, I., Levy, M., Bueker, E. D., and Tokarsky, E. (1968) Nerve growth factor of very high yeild and specific activity. *Science* **159,** 640–643.

33. Haseltine, W. A. (1997) Discovering genes for new medicines. *Sci. Am.* **276,** 92–97.

34. Barnes, P. J. (1999) Anti-IgE antibody therapy for asthma. *N. Engl. J. Med.* **341,** 2006–2008.

35. Dove, A. (1999) Antibody may not be immune to side effects. *Nature Biotechnol.* **17,** 1147.

36. Kuhn, H. G., Winkler, J., Kemermann, G., Thal, L. J., and Cage, F. H. (1997) Epidermal growth factor and fibroplast growth factor-2 have different effects on neural progenitors in the adult rat brain. *J. Neurosci.* **17,** 5820–5829.

37. Craig, C. G., Thropepe, V., Morsehead, C. M., Reynolds, B. A., Weiss, S., and van der Kooy, D. (1996) In vivo growth factor expansion of endogenous subependymal neural precursor cell populations in the adult mouse brain. *J. Neurosci.* **16,** 2649–2658.

38. Betarbet, R., Zicova, T., Bakay, R. A., and Luskin, M. B. (1996) Migration patterns of neonatal subventricular zone progenitor cells transplanted into the neonatal striatum. *Cell Transplant.* **5,** 165–178.

39. Scott, B. W., Wang, S., Burnham, W. M., DeBoni, U., and Wojtowicz, J. M. (1998) Kindling-induced neurogenesis in the dentate gyrus of the rat. *Neurosci. Lett.* **248,** 73–76.
40. Iyer, V. R., Eisen, M. B., Ross, D. T., Schuler, G., Moore, T., Lee, J. C. F., et al. (1999) The transcriptional program in the reponse of human fibroplasts to serum. *Science* **283,** 87–87.
41. Li, R., Thode, S., Zhou, J. Y., Richards, N., Pardinas, J., Rao, M. S., et al. (2000) Motoneuron differentiation of immortalized human spinal cord cell lines. *J. Neurosci. Res.* **59,** 342–352.
42. Caponigro, G., Abedi, M., Hurlburt, A., Maxfield, A., Judd, W., and Kamb, A. (1998) Transdominant fentetic analysis of a growth control pathway. *Proc. Natl. Acad. Sci. USA* **95,** 7508–7513.
43. Norman, T. C., Smith, D. L., Sorger, P. K., Dress, B. L., O'Rourke, S. M., Hughes, T. R., et al. (1999) Genetic selection of peptide inhibitors of biological pathways. *Science* **285,** 591–595.
44. Geyer, C. R., Colman-Lerner, A., and Brent, R. (1999) "Mutagenesis" by peptide aptamers identifies genetic network members and pathway connections. *Proc. Natl. Acad. Sci. USA* **96,** 8567–8572.
45. Hannon, G. J., Sun, P., Carnero, A., Xie, L. Y., Maestro, R., Conklin, D. S., and Beach, D. (1999) An approach to genetics in mammalian cells. *Science* **283,** 1129–1130.
46. Dunn, S. J., Park, S. W., Sharma, V., Raghu, G., Simone, J. M., Tavassoli, R., et al. (1999) Isolation of efficient antivirals: genetic suppressor elements against HIV-1. *Gene Ther.* **6,** 130–137.

Appendix A

Neural Stem Cell Companies

Advanced Cell Technology, Inc.
One Innovation Drive
Biotech Three
Worcester, MA 01605 USA
Phone: +1 508-756-1212
URL: http://www.advancedcell.com/

Advanced Tissue Sciences, Inc.
10933 North Torrey Pines Road
La Jolla, CA 92037 USA
Phone: +1 858-713-7300
URL: http://www.advancedtissue.com/

BioTransplant Incorporated
Building 75, Third Avenue
Charlestown Navy Yard
Charlestown, MA 02129 USA
Phone: +1 617-241-5200
URL: http://www.biotransplant.com/

Clonetics Corporation
PO Box 127
8830 Biggs Ford Road
Walkersville, MD 21793-0127 USA
Phone: +1 800-344-6618
URL: http://www.clonetics.com/

Clonexpress, Inc.
504 E.Diamond Avenue, Suite G
Gaithersburg, MD 20877 USA
Phone: +1 301-869-0840
URL: http://www.clonexpress.com/

From: *Stem Cells and CNS Development*
Edited by: M. S. Rao © Humana Press Inc., Totowa, NJ

Diacrin
Building 96, 13th Street,
Charlestown, MA, 02129 USA
Phone: +1 617-242-9100
URL: http://www.diacrin.com/

Geron
230 Constitution Drive
Menlo Park, CA 94025 USA
Phone: +1 650-473-7700
URL: http://www.geron.com/

Layton BioScience, Inc.
709 East Evelyn Avenue
Sunnyvale, CA 95086 USA
Phone: +1 408-732-5050
URL: http://www.laytonbio.com/

NeuralSTEM Biopharmaceuticals, Ltd.
College Park, MD USA
Phone: +1 301-405-6089
URL: http://www.neuralstem.com/

NeuroSearch A/S
93 Pederstrupvej
DK-2750 Ballerup
Denmark
Phone: +45 44-60-80-00
URL: http://www.neurosearch.com/uk/

NeuroNova AB
Floragatan 5
S-114 31 Stockholm
Sweden
Phone: +46 8-728-00-71
URL: http://www.neuronova.com/

Neuronyx, Inc.
1 Great Valley Parkway, Suite 20
Malvern, PA 19355 USA
Tel: +1 610-240-4150
URL: http://www.neuronyx.com/

Organogenesis Inc.
150 Dan Road
Canton, MA 02021 USA
Phone: +1 781-575-0775
URL: http://www.organogenesis.com/

Osiris Therapeutics, Inc.
2001 Aliceanna Street
Baltimore, MD 21231 USA
Phone: +1 410-522-5005
URL: http://www.osiristx.com/

Proneuron Biotechnology, Ltd.
P.O. Box 277
Ness-Ziona, 74101
Israel
Phone: +972 8-9409550
URL: http://www.proneuron.com/

ReNeuron
Europoint Centre
5-11 Lavington Street
London, SE1 ONZ
United Kingdom
Phone: +44 207-928-1720
URL: http://www.reneuron.com/

Signal Pharmaceuticals, Inc.
5555 Oberlin Drive
San Diego, CA 92121 USA
Phone: +1 858-558-7500
URL: http://www.signalpharm.com/

StemCells, Inc.
525 Del Rey Avenue, Suite C
Sunnyvale, CA 94086 USA
Phone: +1 408-731-8670
URL: http://www.cyto.com/

Titan Pharmaceuticals, Inc.
Post Office Plaza
50 Division Street, Suite 503
Somerville, NJ 08876 USA
URL: http://www.titanpharm.com/

WiCell Research Institute, Inc.
P.O. Box 7365
Madison, WI 53707-7365
Fax: (608) 263-1064
email: info@WiCell.org
https://www.wicell.org/index2.html

Appendix B

Stem Cells and Transplants

Transplants and the Food and Drug Administration

Food and Drug Administration
URL: http://www.fda.gov/
Summary: The Food and Drug Administration (FDA) regulates the use of stem cells for clinical therapy under the provisions of the US Public Health Service Act. Stem Cells fall under the purview of the Center for Biologics Evaluation and Research (CBER), which is the center within the FDA responsible for ensuring the safety and efficacy of blood and blood products, vaccines, allergenics, and biological therapeutics. Newer products, such as biotechnology products, somatic cell therapy and gene therapy, and banked human tissues are also regulated by the same section. However, because most biological products also meet the definition of "drugs" under the Federal Food, Drug, and Cosmetic Act (FD&C Act), they are also subject to regulation under the FD&C Act provisions.

The FDA maintains a comprehensive web-site that can be accessed for detailed information on application procedures, submission requirements and current protocols and policies.

Neural Transplantation Resources

NINDS
URL: http://www.ninds.nih.org/
Summary: The National Institute of Neurological Disorders and Stroke (NINDS), an agency of the US Federal Government and a component of the National Institutes of Health (NIH) and the US Public Health Service, is a lead agency for the Congressionally designated "Decade of the Brain," and the leading supporter of biomedical research on disorders of the brain and nervous system.

From: *Stem Cells and CNS Development*
Edited by: M. S. Rao © Humana Press Inc., Totowa, NJ

Clinical Trials Database
URL: http://clinicaltrials.gov/ct/gui
Summary: The National Library of Medicine at the National Institutes of
Health (NIH) has developed a Clinical Trials Database to provide patients,
family members, and members of the public with current information about
clinical research studies.

Rare Diseases Clinical Research Database
URL: http://rarediseases.info.nih.gov/ord/wwwprot/index.shtml
Summary: This is a searchable database that lists government-funded tri-
als on a variety of CNS and non-CNS disorders.

CenterWatch Clinical Trials Listing Service
URL: http://www.centerwatch.com/
Summary: This is an international listing of clinical research trials containing information about physicians and medical centers performing clinical research and drug therapies newly approved by the FDA.

American Society for Neurotransplantation and Repair
URL: http://www.asntr.org/
Summary: The American Society for Neurotransplantation and Repair
(ASNTR) is a society composed of basic and clinical neuroscientists who
utilize transplantation and related technologies to better understand the way
the nervous system functions and establish new procedures for its repair in
response to trauma or neurodegenerative disease.

Cell Transplant Society
URL: http://www.celltx.org/
Summary: The mission of the Cell Transplant Society is to promote
research and collaboration in cellular transplantation. The Society publishes
a cell transplantation journal that shares information on diverse research
topics of interest to transplant researchers.

The Halifax Fetal Transplantation Program
URL: http://www.mcms.dal.ca/dnts/neurotr.html
Summary: As the only program of its kind in Canada, the Halifax Fetal
Transplantation Program has been in the forefront of neural transplantation
research in this country. Clones of human brain cells are being used in laboratory experiments aimed at repairing, even re-creating, brain areas damaged by injury, disease, and birth defects.

Network of European CNS Transplantation and Restoration
URL: http://www.nectar.org/
Summary: The Network of European CNS Transplantation and Restoration
(NECTAR) is aimed at a concerted European effort to develop efficient, reliable, safe and ethically acceptable transplantation therapies for neurodegenerative diseases, in particular Parkinson's and Huntington's diseases.

MRC Cambridge Centre for Brain Repair
URL: http://www.mrc.ac.uk/
Summary: The Brain Repair Centre is an institution of the University of
Cambridge. The ultimate aim of work in the Centre is to understand, and
eventually, to alleviate and repair damage to the brain and spinal cord,
resulting from injury or neurodegenerative disease.

Other Transplant Centers and Sites

Stroke Treatment at the University of Pittsburgh
Summary: The first experimental study of human neuron implantation
for patients with paralysis after stroke was conducted here.

Neural Transplantation and Neurotrophin Mechanisms in Experimental
Epilepsy
URL: http://www.lu.se/intsek/eaeu/eaeu261.html

The University of Nebraska Medical Center and the Nebraska Health
System Transplant Programs
URL: http://www.nebraskatransplant.org/
Summary: The University of Nebraska Medical Center has an active
transplant program that includes use of mesenchymal and other stem cells.

Proneuron Biotechnology, Ltd.
URL: http://www.proneuron.com/
Summary: Proneuron has initiated a study that will be conducted in Israel
with the approval of the Israel Ministry of Health under a US Food and Drug
Administration Investigational New Drug Application. This study is designed to study autologous activated macrophage therapy in approximately
eight complete spinal cord injury patients. The investigators will follow their
post-treatment course for one year or longer.

Diacrin
URL: http://www.diacrin.com/
Summary: Diacrin has initiated phase I and phase II clinical trials using
porcine cells to treat Parkinson's and Huntington's Disease patients.

Layton BioScience, Inc.

URL: http://www.laytonbio.com/

Summary: Layton has recruited twelve stroke patients to undergo human cell transplants (HNT). Transplants will be performed by University of Pittsburgh neurosurgeon Douglas Kondziolka. In 2000 Dr. Kondziolka reported that the patients who received 6 million cells showed much more improvement than those who received 2 million. The first four received 2 million, and the remaining eight received, at random, either 2 million or 6 million cells.

Appendix C

Patents and Stem Cells

The patent situation on stem cells, progenitor cells and differentiated cells for therapy is complex and confusing and more than twenty different patents have issued in the past two years. We have listed some useful searchable sites. The reader is advised to use multiple keywords to obtain a comprehensive listing of patent filings. Given the different requirements for public release in different countries, it is often advisable to search several different databases. Some sites are listed and most other sites can be readily identified using standard search engines.

United States Patent and Trademark Office
URL: http://www.uspto.gov/
Summary: This is the only official web-site of the United States Patent and Trademark Office, a performance-based organization of the government of the United States of America.

US Patents
URL: http://patents.cos.com/
Summary: This is a fully searchable bibliographic database, accessed through the Community of Science, Inc. (COS) web-site, containing all of the approximately 1.7 million U.S. patents issued since 1975.

European Patent Office
URL: http://www.european-patent-office.org/
Summary: This is the official web-site of the European Patent Office, the executive body of the European Patent Organization.

Espacenet
URL: http://www.european-patent-office.org/espacenet/info/access.htm
Summary: Established by the European Patent Office in conjunction with the member states of the European Patent Organization and the European Commission to provide the general public with free patent information.

From: *Stem Cells and CNS Development*
Edited by: M. S. Rao © Humana Press Inc., Totowa, NJ

United Kingdom Patent Office
URL: http://www.patent.gov.uk/
Summary: The role of the UK Patent Office is to help to stimulate innovation and the international competitiveness of industry through intellectual property rights.

Canadian Patent Database
URL: http://patents1.ic.gc.ca/intro-e.html
Summary: The Canadian Patent Database lets you access over 75 years of patent descriptions and images. You can search, retrieve, and study more than 1.4 million patent documents.

Japanese Patent Office
URL: http://www.jpo-miti.go.jp/

Appendix D

Stem Cells and US Federal Guidelines

Currently, US federal law regulates the use of fetal cells and fetal tissue for research or clinical use. Use of neural stem cells, restricted precursors, and more differentiated cells derived from fetal tissue is governed by guidelines established for fetal tissue research. Detailed information on the guidelines that govern fetal tissue research is available from the National Institutes of Health (NIH) web-site.

Adult stem cells and their derivatives are not regulated by the same guidelines that govern fetal research. Individual universities have established Institutional Review Boards (IRBs) that evaluate all applications that involve patient or donor contact. Guidelines for informed consent, donor confidentiality, and donor legal rights have been established. Most universities follow similar guidelines, although minor differences may exist. Researchers can contact their IRB to obtain specific guidelines.

The rules that govern use and isolation of human pluripotent stem cells (hPSCs) or embryonic stem cells (ES cells) have been finalized. Currently, federal law prohibits the Department of Health and Human Services (DHHS) from funding research in which human embryos are created for research purposes or are destroyed, discarded or subjected to greater than minimal risk. In light of this restriction the DHHS Office of the General Counsel sought a legal opinion on whether NIH funds may be used for research utilizing human pluripotent stem cells. DHHS concluded that the Congressional prohibition does not prohibit the funding of research utilizing hPSCs because such cells are not embryos. Thus, NIH funding for research using pluripotent stem cells derived from human embryos is not legislatively prohibited. The legal opinion also clarified that hPSCs derived from fetal tissue would fall within the legal definition of human fetal tissue and are, therefore, subject to federal restrictions on the use of such tissue. NIH funding for research to derive or utilize hPSCs from fetal tissue is permissible, subject to applicable law and regulation.

From: *Stem Cells and CNS Development*
Edited by: M. S. Rao © Humana Press Inc., Totowa, NJ

Final guidelines have been published in the Federal register. A copy of the guidelines obtained from http://www.nih.gov/news/stemcell/stemcellguidelines.htm is appended below. More information on stem cells is available at http://www.nih.gov/news/stemcell/index.htm. Readers are advised to check for updates prior to initiating their experiments. Investigators should also note that different guidelines will apply to deriving new cell lines, use of existing cell lines and use of primordial germ cells.

NATIONAL INSTITUTES OF HEALTH GUIDELINES FOR RESEARCH USING HUMAN PLURIPOTENT STEM CELLS

Summary

The National Institutes of Health (NIH) is hereby publishing final *National Institutes of Health Guidelines for Research Using Human Pluripotent Stem Cells*. The *Guidelines* establish procedures to help ensure that NIH-funded research in this area is conducted in an ethical and legal manner.

Effective Date

These *Guidelines* are effective on August 25, 2000. The moratorium on research using human pluripotent stem cells derived from human embryos and fetal tissue put in place by the Director, NIH, in January 1999, will be lifted on August 25, 2000.

Summary of Public Comments on Draft Guidelines

On December 2, 1999, the NIH published *Draft Guidelines* for research involving human pluripotent stem cells (hPSCs) in the *Federal Register* for public comment. The comment period ended on February 22, 2000.

The NIH received approximately 50,000 comments from members of Congress, patient advocacy groups, scientific societies, religious organizations, and private citizens. This Notice presents the final *Guidelines* together with NIH's response to the substantive public comments that addressed provisions of the *Guidelines*.

Scope of Guidelines and General Issues

Respondents asked for clarification of terminology used in the "Guidelines" and some commented that the language was not appropriate or was too technical, particularly the informed consent sections. The NIH agrees that these *Guidelines* should be clear and understandable. Changes, including some reorganization of the sections, were made to this end. The *Guidelines* are written primarily for the purpose of informing investigators of the conditions that must be met in order to receive NIH funding for research using hPSCs and, therefore, some technical language is required. The *Guide-*

lines do not define the precise language that should appear in informed consent documents because these should be developed by the investigator/clinician specifically for a particular study protocol or procedure for which the consent is being sought. Existing regulatory provisions require (45 CFR 46.116) that the language in informed consent documents be understandable to prospective participants in the study.

Respondents suggested that NIH funding for research using hPSCs would be in violation of the DHHS appropriations law and that derivation of hPSCs cannot be distinguished from their use. For this reason, a number of respondents asked that the NIH withdraw the "draft Guidelines." The NIH sought the opinion of the Department of Health and Human Services (DHHS) General Counsel, who determined that "federally funded research that utilizes hPSCs would not be prohibited by the HHS appropriations law prohibiting human embryo research, because such cells are not human embryos." Comments questioning this conclusion did not present information or arguments that justify reconsideration of the conclusion.

Respondents commented that the "Guidelines" are too restrictive or that there is no need for Federal Guidelines for this arena of research. Comments asserted that federally funded research using hPSCs should go forward without formal requirements, in the same manner as in the private sector. In order to help ensure that the NIH-funded research using hPSCs is conducted in an ethical and legal manner, the NIH felt it was advisable to develop and implement guidelines. To this end, the NIH Director convened a Working Group of the Advisory Committee to the Director, NIH (ACD), to advise the ACD on the development of guidelines and an oversight process for research involving hPSCs. The NIH Director charged the Working Group with developing appropriate guidelines to govern research involving the derivation and use of hPSCs from fetal tissue and research involving the use of hPSCs derived from human embryos that are in excess of clinical need.

Respondents commented regarding the sources of stem cells. Some respondents stated that research on hPSCs was unnecessary because stem cells from adults, umbilical cords, and placentas could be used instead. Other respondents asked the NIH to restrict Federal funding for hPSC research to those cells derived from fetal and adult tissue but not embryos. Other respondents asked that the Guidelines encompass research using stem cells from adult tissues. As stated under Section I. Scope of *Guidelines*, the *Guidelines* apply to the use of NIH funds for research using hPSCs derived from human embryos or human fetal tissue. The *Guidelines* do not impose requirements on Federal funding of research involving stem cells from human adults, umbilical cords, or placentas.

Given the enormous potential of stem cells to the development of new therapies for the most devastating diseases, it is important to simultaneously pursue all lines of promising research. It is possible that no single source of stem cells is best or even suitable/usable for all therapies. Different types or sources of stem cells may be optimal for treatment of specific conditions. In order to determine the very best source of many of the specialized cells and tissues of the body for new treatments and even cures, it is vitally important to study the potential of adult stem cells for comparison to that of hPSCs derived from embryos and fetuses. Unless all stem cell types are studied, the differences between adult stem cells and embryo and fetal-derived hPSCs will not be known.

Moreover, there is evidence that adult stem cells may have more limited potential than hPSCs. First, stem cells for all cell and tissue types have not yet been found in the adult human. Significantly, cardiac stem cells or pancreatic islet stem cells have not been identified in adult humans. Second, stem cells in adults are often present in only minute quantities, are difficult to isolate and purify, and their numbers may decrease with age. For example, brain cells from adults that may be neural stem cells have been obtained only by removing a portion of the brain of an adult with epilepsy, a complex and invasive procedure that carries the added risk of further neurological damage. Any attempt to use stem cells from a patient's own body for treatment would require that stem cells would first have to be isolated from the patient and then grown in culture in sufficient numbers to obtain adequate quantities for treatment. This would mean that for some rapidly progressing disorders, there may not be sufficient time to grow enough cells to use for treatment. Third, in disorders that are caused by a genetic defect, the genetic error likely would be present in the patient's stem cells, making cells from such a patient inappropriate for transplantation. In addition, adult stem cells may contain more DNA abnormalities caused by exposure to daily living, including sunlight, toxins, and errors made during DNA replication than will be found in fetal or embryonic hPSCs. Fourth, there is evidence that stem cells from adults may not have the same capacity to multiply as do younger cells. These potential weaknesses may limit the usefulness of adult stem cells.

Respondents were concerned that these are guidelines and not requirements or regulations. Although these are guidelines and not regulations, they prescribe the documentation and assurances that must accompany requests for NIH funding for research utilizing hPSCs. If the funding requests do not contain the prescribed information, funding for hPSC research will not be provided. Compliance with the *Guidelines* will be imposed as a condition of grant award.

Respondents commented that there had not been enough widespread public disclosure/discussion of this research or the "Guidelines". Prior to the development of draft *Guidelines*, there were two Congressional hearings on hPSCs. In a further effort to ensure substantial discussion and comment, the NIH convened a Working Group of the Advisory Committee to the Director, NIH (ACD), to advise the ACD on the development of these *Guidelines*. The Working Group was composed of scientists, patients and patient advocates, ethicists, clinicians, and lawyers. The Working Group met in public session on April 8, 1999, and heard from members of the public, as well as professional associations and Congress. In developing the draft *Guidelines*, the NIH also considered advice from the National Bioethics Advisory Commission (NBAC). *Draft Guidelines* were published for public comment in the *Federal Register* on December 2, 1999, for 60 days, and, in response to public interest, the comment period was extended an additional 28 days. Approximately 50,000 comments were received. NIH issued a national press release announcing the *Federal Register* notice and many of the Nation's newspapers carried articles on this area of research and on the *Guidelines*. Patient groups, scientific societies, and religious organizations convened meetings and discussion groups and disseminated materials about this area of research and about the *Guidelines*.

Comment was received about whether the "Guidelines" apply to hPSC lines developed outside of the United States. The *Guidelines* make no distinction based upon the country in which an hPSC line is developed. All lines to be used in hPSC cell research funded by NIH must meet the same requirements.

Derivation and Use of hPSCs From Fetal Tissue

Respondents made the point that the NIH has specified certain requirements for the use of human fetal tissue to derive hPSCs in addition to those imposed on other areas of human fetal tissue research. These respondents suggested that the section of the "Guidelines" pertaining to fetal tissue sources be omitted. In order to ensure uniformity in NIH's oversight of research using hPSCs, the *Guidelines* were extended to govern hPSCs derived from both human embryos and fetal tissue.

Use of hPSCs Derived From Human Embryos

Respondents suggested that the "Guidelines" refer to "fertility treatment" rather than to "infertility treatment" in order to clarify that they allow the use of human embryos from treatments that employ assisted reproductive technologies to facilitate reproduction in fertile, as well as in infertile, individuals. The *Guidelines* have been changed accordingly.

Respondents suggested dropping the word "early" throughout the document or more clearly defining "early." The word "early" in reference to human embryos has been deleted; the *Guidelines* make it clear that NIH funding of research using hPSCs derived in the private sector from human embryos can involve only embryos that have not reached the stage at which the mesoderm is formed.

Some respondents were concerned that embryos might be created for research purposes. Other respondents stated there should be no distinction between embryos created for research purposes and those created for fertility treatment. Investigators seeking NIH funds for research using hPSCs are required to provide documentation, prior to the award of any NIH funds, that embryos were created for the purposes of fertility treatment. President Clinton, many members of Congress, the NIH Human Embryo Research Panel, and the NBAC have all embraced the distinction between embryos created for research purposes and those created for reproductive purposes.

Respondents were concerned about the creation of a "black market" for human embryos, and expressed concerns that individuals will be coerced into donating embryos. The *Guidelines* state that there can be no incentives for donation and that a decision to donate must be made free of coercion. In addition, the *Guidelines* set forth conditions that will help ensure all donations are voluntary. For example, with regard to hPSCs derived from embryos, research using Federal funds may only be conducted if the cells were derived from frozen embryos that were created for the purpose of fertility treatment and that were in excess of clinical need.

Respondents commented on the requirement that human embryos be frozen in order to qualify for derivation of hPSCs to be used in NIH-funded research. Respondents suggested that the freezing requirement would preclude the use of hPSCs derived from embryos that are genetically and chromosomally abnormal, since such embryos are usually not frozen for reproductive purposes. While the NIH acknowledges that research on hPSCs derived from such embryos could yield important scientific information, limiting research to hPSCs derived from frozen human embryos will help ensure that the decision to donate the embryo for hPSC research is distinct and separate from the fertility treatment.

Financial Issues

Respondents expressed concern regarding the sale of fetal tissue for profit and whether hPSC research would encourage such activity. Respondents also were concerned about whether clinics or doctors would profit from the derivation of hPSCs and/or their sale. Section 498B of the Public Health Service Act prohibits any individual from knowingly acquiring or selling human fetal tissue for "valuable consideration." In addition, the *Guidelines* prohibit any inducement for the donation of human embryos for research purposes. The *Guidelines* also call for an assurance that the hPSCs to be used in NIH-funded research were obtained through a donation or through a payment that does not exceed the reasonable costs associated with the transportation, processing, preservation, quality control and storage of the hPSCs. All grantees must sign an assurance that they are in compliance with all applicable Federal, State, and local laws. Each funded research institution is responsible for monitoring compliance by individual investigators with any such applicable laws.

Respondents questioned the prohibition against embryo donors benefitting financially from their donation. This clause was retained in the final *Guidelines* to help ensure that the donating individuals are offered no inducements to donate and that all donations are voluntary.

Respondents suggested that the "Guidelines" be strengthened to include a waiver of intellectual property rights. This proposed change would be inconsistent with 45 CFR 46.116 of the regulation for the protection of human subjects of research, which provides that no informed consent may include language through which the subject waives or appears to waive any of the subject's legal rights.

Respondents questioned the reference in the requirements for informed consent related to the commercial potential of donated material. The paragraphs providing for disclosure in the informed consent of the possibility that the donated material could have commercial potential were modified. The reference in these paragraphs to "donated material" did not accurately reflect the intent of the provision. The *Guidelines* now make clear that the "results of research on the human pluripotent stem cells may have commercial potential."

Ineligible Research

Respondents objected to the areas of research that the NIH has deemed ineligible, particularly research that is not restricted by statute or regulation, such as research utilizing hPSCs that were derived using somatic cell nuclear transfer, i.e., the transfer of a human somatic cell nucleus into a human egg. The NIH determined that, at this time, research using hPSCs derived from such sources has not received adequate discussion and consideration by the public and is, therefore, ineligible for NIH funding.

Separation of Fertility Treatment and Abortion From Research

Respondents were concerned that hPSC research would encourage abortion. The law and the *Guidelines* guard against encouraging abortion by requiring that the decision to have an abortion be made apart from and prior to the decision to donate tissue.

Respondents objected to the condition in the "Guidelines" that the fertility physician could not be the same person as the researcher deriving stem cells. Some respondents stated that the Institutional Review Board (IRB) or an independent physician would be able to guard against this conflict of interest. The restriction was designed so that the person treating the individuals seeking fertility treatment, who is involved in decisions such as how many embryos to produce, is not the person seeking to derive hPSCs. This separation will help ensure that embryos will not be created in numbers greater than necessary for fertility treatment.

Respondents suggested that the clauses regarding donation of fetal tissue or human embryos for derivation of stem cells for eventual use in transplantation be changed explicitly to prevent directed donation. This change has been made.

Identifiers

Respondents were concerned about removing identifiers. There was concern that the investigator would not be able to document compliance with the "Guidelines" requirements without identifiers, or that the removal of identifiers would make it impossible to conduct certain genetic studies or develop therapeutic materials. The *Guidelines* have been modified to clarify that the term "identifier" refers to any information from which the donor(s) can be identified, directly or through identifiers linked to the donors. However, since information identifying the donor(s) may be necessary if the tissue or cells are to be used in transplantation, the *Guidelines* have also been modified to state that the informed consent should notify donor(s) whether or not identifiers will be retained.

Respondents commented that DNA is an identifier and that all donors of human embryos or fetal tissue should be told that identifiers such as DNA will be retained with the samples. Although DNA can be used to determine the individual from whom a tissue sample was taken, this can be done only when one has a sample from both the tissue in question and the putative donor; it cannot be used to identify an individual out of a population. Moreover, it is difficult to identify a donor using tissue derived from a fetus or embryo, since the tissue is not genetically identical to the donor.

Informed Consent and IRB Review

Respondents asked why investigators were expected to provide documentation of IRB review of derivation from human embryos, but not for derivation from fetal tissue. Respondents suggested that the requirements be changed so that protocols for both sources of hPSCs must be approved by an IRB. The *Guidelines* have been changed to make clear that the IRB review requirements regarding the derivation of cells from fetal tissue and human embryos are the same.

Comment was received expressing concern that the informed consent explicitly state that the donor will have no dispositional authority over derived pluripotent stem cells. The *Guidelines* state that donation of human embryos should have been made without any restriction regarding the individual(s) who may be the recipient of the cells derived from the hPSCs for transplantation. Such a statement is consistent with the statutory provision applicable to the donor informed consent for the use of fetal tissue for transplantation. The *Guidelines* now provide for the inclusion of a statement to this effect in the informed consent.

Respondents urged that the "Guidelines" be revised to remove the prohibition on potential donors receiving information regarding subsequent testing of donated tissue in the situation when physicians deem disclosure to be in the donors' best interest. This change has been made.

Respondents requested clarification regarding the persons from whom consent for donation of embryos for research must be obtained. The *Guidelines* call for informed consent from individual(s) who have sought fertility treatment. Only the individual(s) who were part of the decision to create the embryo for reproductive purposes should have been part of the decision to donate for the derivation of hPSCs.

Respondents urged that fertility clinics should be able to discuss with patients the option of donating embryos for research at the beginning of the IVF process. The *Guidelines* do not delineate the timeframe during which the general option of donating embryos for research can be discussed. However, according to the *Guidelines*, obtaining consent for donation of embryos for the purpose of deriving hPSCs should not occur until after the embryos are determined to be in "excess of clinical need."

Oversight

Respondents stated that the NIH's oversight in this area of research was very important to the legal and ethical conduct of this research, and asked for more information regarding the oversight process. Information about the operations of the Human Pluripotent Stem Cell Review Group (HPSCRG) can be found in the final *Guidelines* and on the NIH Web page.

Respondents were concerned about whether and how NIH would monitor research after a researcher receives NIH funds. Compliance with the *Guidelines* will be largely determined prior to the award of funds. Follow-up to ensure continued compliance with the *Guidelines* will be conducted in the same manner as for all other conditions of all other NIH grant awards. It is the responsibility of the investigator to file progress reports, and it is the responsibility of the funded institution to ensure compliance with the NIH *Guidelines*. NIH staff will also monitor the progress of these investigators as part of their regular duties.

Respondents asked about penalties for not following the "Guidelines." The following actions may be taken by the NIH when there is a failure to comply with the terms and conditions of any award: 1) Under 45 CFR 74.14, the NIH can impose special conditions on an award, including increased oversight/monitoring/reporting requirements for an institution, project or investigator; 2) Under 45 CFR 74.62, if a grantee materially fails to comply with the terms and conditions of the award, the NIH may withhold funds pending correction of the problem or, pending more severe enforcement action, disallow all or part of the costs of the activity that was not in compliance, withhold further awards for the project, or suspend or terminate all or part of the funding for the project. Individuals and institutions may be debarred from eligibility for all Federal financial assistance and contracts under 45 CFR Part 76 and 48 CFR Subpart 9.4, respectively. Because these sanctions pertain to all conditions of grant award, the NIH did not reiterate them in the *Guidelines*.

Respondents suggested that the HPSCRG hold periodic Stem Cell Policy Conferences (similar to the Gene Therapy Policy Conferences conducted by the Recombinant DNA Advisory Committee ("RAC")) in order to solicit and consider public comment from interested parties on the scientific, medical, legal, and ethical issues arising from stem cell research. Members of the HPSCRG will serve as a resource for recommending to the NIH any need for Human Pluripotent Stem Cell Policy Conferences.

Other Changes

Because compliance materials may be made public prior to funding decisions, we have added a sentence requiring the principal investigator's written consent to the disclosure of such material necessary to carry out public review and other oversight procedures.

The draft *Guidelines* required HPSCRG review of proposals from investigators planning to derive hPSCs from fetal tissue. Because the *Guidelines* address proposals for NIH funding for the use of hPSCs, this requirement has been removed from the *Guidelines*.

The text of the final *Guidelines* follows.

NATIONAL INSTITUTES OF HEALTH GUIDELINES FOR RESEARCH USING HUMAN PLURIPOTENT STEM CELLS

I. Scope of Guidelines

These *Guidelines* apply to the expenditure of National Institutes of Health (NIH) funds for research using human pluripotent stem cells derived from human embryos (technically known as human embryonic stem cells) or human fetal tissue (technically known as human embryonic germ cells). For purposes of these *Guidelines*, "human pluripotent stem cells" are cells that are self-replicating, are derived from human embryos or human fetal tissue, and are known to develop into cells and tissues of the three primary germ layers. Although human pluripotent stem cells may be derived from embryos or fetal tissue, such stem cells are not themselves embryos. NIH research funded under these *Guidelines* will involve human pluripotent stem cells derived 1) from human fetal tissue; or 2) from human embryos that are the result of *in vitro* fertilization, are in excess of clinical need, and have not reached the stage at which the mesoderm is formed.

In accordance with 42 Code of Federal Regulations (CFR) § 52.4, these *Guidelines* prescribe the documentation and assurances that must accompany requests for NIH funding for research using human pluripotent stem cells from: (1) awardees who want to use existing funds; (2) awardees requesting an administrative or competing supplement; and 3) applicants or intramural researchers submitting applications or proposals. NIH funds may be used to derive human pluripotent stem cells from fetal tissue. NIH funds may not be used to derive human pluripotent stem cells from human embryos. These *Guidelines* also designate certain areas of human pluripotent stem cell research as ineligible for NIH funding.

II. Guidelines for Research Using Human Pluripotent Stem Cells that is Eligible for NIH Funding

A. Utilization of Human Pluripotent Stem Cells Derived from Human Embryos

1. Submission to NIH

Intramural or extramural investigators who are intending to use existing funds, are requesting an administrative supplement, or are applying for new NIH funding for research using human pluripotent stem cells derived from human embryos must submit to NIH the following:

a. An assurance signed by the responsible institutional official that the pluripotent stem cells were derived from human embryos in accordance with the conditions set forth in Section II.A.2 of these *Guidelines* and that the institution will maintain documentation in support of the assurance;

b. A sample informed consent document (with patient identifier information removed) and a description of the informed consent process that meet the criteria for informed consent set forth in Section II.A.2.e of these *Guidelines*;

c. An abstract of the scientific protocol used to derive human pluripotent stem cells from an embryo;

d. Documentation of Institutional Review Board (IRB) approval of the derivation protocol;

e. An assurance that the stem cells to be used in the research were or will be obtained through a donation or through a payment that does not exceed the reasonable costs associated with the transportation, processing, preservation, quality control and storage of the stem cells;

f. The title of the research proposal or specific subproject that proposes the use of human pluripotent stem cells;

g. An assurance that the proposed research using human pluripotent stem cells is not a class of research that is ineligible for NIH funding as set forth in Section III of these *Guidelines*; and

h. The Principal Investigator's written consent to the disclosure of all material submitted under Paragraph A.1 of this Section, as necessary to carry out the public review and other oversight procedures set forth in Section IV of these *Guidelines*.

2. Conditions for the Utilization of Human Pluripotent Stem Cells Derived From Human Embryos

Studies utilizing pluripotent stem cells derived from human embryos may be conducted using NIH funds only if the cells were derived (without Federal funds) from human embryos that were created for the purposes of fertility treatment and were in excess of the clinical need of the individuals seeking such treatment.

a. To ensure that the donation of human embryos in excess of the clinical need is voluntary, no inducements, monetary or otherwise, should have been offered for the donation of human embryos for research purposes. Fertility clinics and/or their affiliated laboratories should have implemented specific written policies and practices to ensure that no such inducements are made available.

b. There should have been a clear separation between the decision to create embryos for fertility treatment and the decision to donate human embryos in excess of clinical need for research purposes to derive pluripotent stem cells. Decisions related to the creation of embryos for fertility treatment should have been made free from the influence of researchers or investigators proposing to derive or utilize human pluripotent stem cells in research. To this end, the attending physician responsible for the fertility treatment and the researcher or investigator deriving and/or proposing to utilize human pluripotent stem cells should not have been one and the same person.

c. To ensure that human embryos donated for research were in excess of the clinical need of the individuals seeking fertility treatment and to allow potential donors time between the creation of the embryos for fertility treatment and the decision to donate for research purposes, only frozen human embryos should have been used to derive human pluripotent stem cells. In addition, individuals undergoing fertility treatment should have been approached about consent for donation of human embryos to derive pluripotent stem cells only at the time of deciding the disposition of embryos in excess of the clinical need.

d. Donation of human embryos should have been made without any restriction or direction regarding the individual(s) who may be the recipients of transplantation of the cells derived from the human pluripotent stem cells.

e. Informed Consent

Informed consent should have been obtained from individuals who have sought fertility treatment and who elect to donate human embryos in excess of clinical need for human pluripotent stem cell research purposes. The informed consent process should have included discussion of the following information with potential donors, pertinent to making the decision whether or not to donate their embryos for research purposes.

Informed consent should have included:

(i) A statement that the embryos will be used to derive human pluripotent stem cells for research that may include human transplantation research;

(ii) A statement that the donation is made without any restriction or direction regarding the individual(s) who may be the recipient(s) of transplantation of the cells derived from the embryo;

(iii) A statement as to whether or not information that could identify the donors of the embryos, directly or through identifiers linked to the donors, will be removed prior to the derivation or the use of human pluripotent stem cells;

(iv) A statement that derived cells and/or cell lines may be kept for many years;

(v) Disclosure of the possibility that the results of research on the human pluripotent stem cells may have commercial potential, and a statement that the donor will not receive financial or any other benefits from any such future commercial development;

(vi) A statement that the research is not intended to provide direct medical benefit to the donor; and

(vii) A statement that embryos donated will not be transferred to a woman's uterus and will not survive the human pluripotent stem cell derivation process.

f. Derivation protocols should have been approved by an IRB established in accord with 45 CFR §46.107 and §46.108 or FDA regulations at 21 CFR §56.107 and §56.108.

B. Utilization of Human Pluripotent Stem Cells Derived From Human Fetal Tissue

1. *Submission to NIH*

Intramural or extramural investigators who are intending to use existing funds, are requesting an administrative supplement, or are applying for new NIH funding for research using human pluripotent stem cells derived from fetal tissue must submit to NIH the following:

a. An assurance signed by the responsible institutional official that the pluripotent stem cells were derived from human fetal tissue in accordance with the conditions set forth in Section II.A.2 of these *Guidelines* and that the institution will maintain documentation in support of the assurance;

b. A sample informed consent document (with patient identifier information removed) and a description of the informed consent process that meet the criteria for informed consent set forth in Section II.B.2.b of these *Guidelines*;

c. An abstract of the scientific protocol used to derive human pluripotent stem cells from fetal tissue;

d. Documentation of IRB approval of the derivation protocol;

e. An assurance that the stem cells to be used in the research were or will be obtained through a donation or through a payment that does not exceed the reasonable costs associated with the transportation, processing, preservation, quality control and storage of the stem cells;

f. The title of the research proposal or specific subproject that proposes the use of human pluripotent stem cells;

g. An assurance that the proposed research using human pluripotent stem cells is not a class of research that is ineligible for NIH funding as set forth in Section III of these *Guidelines*; and

h. The Principal Investigator's written consent to the disclosure of all material submitted under Paragraph B.1 of this Section, as necessary to carry out the public review and other oversight procedures set forth in Section IV of these *Guidelines*.

2. Conditions for the Utilization of Human Pluripotent Stem Cells Derived From Fetal Tissue.

a. Unlike pluripotent stem cells derived from human embryos, DHHS funds may be used to support research to derive pluripotent stem cells from fetal tissue, as well as for research utilizing such cells. Such research is governed by Federal statutory restrictions regarding fetal tissue research at 42 U.S.C. 289g-2(a) and the Federal regulations at 45 CFR § 46.210. In addition, because cells derived from fetal tissue at the early stages of investigation may, at a later date, be used in human fetal tissue transplantation research, it is the policy of NIH to require that all NIH-funded research involving the derivation or utilization of pluripotent stem cells from human fetal tissue also comply with the fetal tissue transplantation research statute at 42 U.S.C. 289g-1.

b. Informed Consent

As a policy matter, NIH-funded research deriving or utilizing human pluripotent stem cells from fetal tissue should comply with the informed consent law applicable to fetal tissue transplantation research (42 U.S.C. 289g-1) and the following conditions. The informed consent process should have included discussion of the following information with potential donors, pertinent to making the decision whether to donate fetal tissue for research purposes.

Informed consent should have included:

(i) A statement that fetal tissue will be used to derive human pluripotent stem cells for research that may include human transplantation research;

(ii) A statement that the donation is made without any restriction or direction regarding the individual(s) who may be the recipient(s) of transplantation of the cells derived from the fetal tissue;

(iii) A statement as to whether or not information that could identify the donors of the fetal tissue, directly or through identifiers linked to the donors, will be removed prior to the derivation or the use of human pluripotent stem cells;

(iv) A statement that derived cells and/or cell lines may be kept for many years;

(v) Disclosure of the possibility that the results of research on the human pluripotent stem cells may have commercial potential, and a statement that the donor will not receive financial or any other benefits from any such future commercial development; and

(vi) A statement that the research is not intended to provide direct medical benefit to the donor.

c. Derivation protocols should have been approved by an IRB established in accord with 45 CFR §46.107 and §46.108 or FDA regulations at 21 CFR §56.107 and §56.108.

III. Areas of Research Involving Human Pluripotent Stem Cells that are Ineligible for NIH Funding

Areas of research ineligible for NIH funding include:

A. The derivation of pluripotent stem cells from human embryos;

B. Research in which human pluripotent stem cells are utilized to create or contribute to a human embryo;

C. Research utilizing pluripotent stem cells that were derived from human embryos created for research purposes, rather than for fertility treatment;

D. Research in which human pluripotent stem cells are derived using somatic cell nuclear transfer, i.e., the transfer of a human somatic cell nucleus into a human or animal egg;

E. Research utilizing human pluripotent stem cells that were derived using somatic cell nuclear transfer, i.e., the transfer of a human somatic cell nucleus into a human or animal egg;

F. Research in which human pluripotent stem cells are combined with an animal embryo; and

G. Research in which human pluripotent stem cells are used in combination with somatic cell nuclear transfer for the purposes of reproductive cloning of a human.

IV. Oversight

A. The NIH Human Pluripotent Stem Cell Review Group (HPSCRG) will review documentation of compliance with the *Guidelines* for funding requests that propose the use of human pluripotent stem cells. This working group will hold public meetings when a funding request proposes the use of a line of human pluripotent stem cells that has not been previously reviewed and approved by the HPSCRG.

B. In the case of new or competing continuation (renewal) or competing supplement applications, all applications shall be reviewed by HPSCRG and for scientific merit by a Scientific Review Group. In the case of requests to use existing funds or applications for an administrative supplement or in the case of intramural proposals, Institute or Center staff should forward material to the HPSCRG for review and determination of compliance with the *Guidelines* prior to allowing the research to proceed.

C. The NIH will compile a yearly report that will include the number of applications and proposals reviewed and the titles of all awarded applications, supplements or administrative approvals for the use of existing funds, and intramural projects.

D. Members of the HPSCRG will also serve as a resource for recommendations to the NIH with regard to any revisions to the *NIH Guidelines for Research Using Human Pluripotent Stem Cells* and any need for human pluripotent stem cell policy conferences.

About the Authors

Defining Neural Stem Cells and Their Role in Normal Development of the Nervous System

Sally Temple

Sally Temple received her BA from Cambridge, UK. As a PhD student with Martin Raff at University College London, she began clonal analysis of brain progenitor cells, focusing on oligodendrocyte lineages. She studied spinal cord development as a postdoctoral fellow with Tom Jessell at Columbia, and completed her training with John Barrett in Miami who introduced her to neuronal cell culture. In Miami she extended her clonal studies of brain progenitor cells, leading to the description of stem-like cells in the embryonic CNS. In 1990 she joined the faculty of Albany Medical College and is now an Associate Professor in the Center for Neuroscience and Neuropharmacology. She is continuing her studies aimed at discovering the role of stem cells and other progenitor cells in forebrain development.

Multipotent Stem Cells in the Embryonic Nervous System

John A. Kessler, Mark F. Mehler, and Peter C. Mabie

John A. Kessler is a graduate of Princeton University and Cornell University Medical College. After completing clinical training in internal medicine and neurology, Dr. Kessler joined the faculty at Cornell Medical College where he began his studies of neuronal development. He then moved to Albert Einstein College of Medicine where he became the Alpern Professor of Neurology and Neuroscience and Director of the Rose Kennedy Research Center. Recently he has moved to Northwestern University Medical School as the Boshes Professor and Chair of the Department of Neurology. Dr. Kessler is the author of more than 175 research articles, an editor of two Journals, and the recipient of numerous research awards.

From: *Stem Cells and CNS Development*
Edited by: M. S. Rao © Humana Press Inc., Totowa, NJ

Multipotent Stem Cells in the Adult Nervous System

Luca Bonfanti, Angela Gritti, Rossella Galli, and Angelo L. Vescovi

Angelo L. Vescovi received his PhD from the University of Milan. He spent two years as a post-doctoral fellow at the University of Calgary before returning to Italy in 1992 to head the laboratory of Cellular Neuropharmacology at the National Neurological Institute C. Besta in Milan. From 1995 to 1998 he was the head of the "Tissue Culture Laboratory" at Neurospheres Ltd., in Calgary, and Adjunct Professor of the Department of Physiology at the University of Milan. He has supervised projects on stem cell research for the Italian Ministry of Health. He is currently Director of the Stem Cell Research Institute in the Department of Biotechnology at the Hospital San Raffaele in Milan. Dr. Vescovi's main research interests concern the study of basic stem cell physiology, particularly regarding the epigenetic regulation of the fate of the neural stem cell. More recently, he has been studying stem cell plasticity and their trans-differentiation into cells of a different germ layer origin. He is also conducting experimental studies on neural transplantation with human stem cells.

Glial Characteristics of Adult Subventricular Zone Stem Cells

Daniel A. Lim and Arturo Alvarez-Buylla

Arturo Alvarez-Buylla received his PhD from Rockefeller University in New York where he studied the mechanism of migration of primordial germ cells. As a graduate student at Rockefeller University, Dr. Alvarez-Buylla demonstrated the migrating of young neurons in adult canary brain and suggested that radial glia are neuronal precursors. From 1989 until 2000, he was both an Associate and Assistant Professor at Rockefeller University. His laboratory research there has focussed on characterizing a germinal zone, demonstrating long term migration and differentiation of new neurons in the adult mammalian brain, and on the discovery of a novel form of neuronal migration (chain migration,) characterized by an extensive network of pathways for chain migration and the identity of the neural stem cells in the adult mammalian brain. Dr. Alvarez-Buylla joined the faculty of the University of California, San Francisco in September 2000, where he is a Professor.

Neuron Restricted Precursors

Giri Venkataraman and Marla B. Luskin

Marla B. Luskin, a Professor in the Department of Cell Biology at Emory University School of Medicine in Atlanta, has had a long standing interest in the development of the mammalian forebrain. Dr. Luskin's studies have focussed on (1) the progenitor cells, including their lineage relationships, which generate the diverse cell types in the cerebral cortex, as well as (2) the specialized neuronal progenitor cells of the postnatal anterior portion of the proliferative subventricular zone (SVZa) lining the lateral ventricles. Recent research from her lab has revealed that the cells of the SVZa are distinct among neural progenitor cells in they are phenotypically neurons that have the capacity to undergo division. Ongoing studies are also investigating how the differentiation of the unique SVZa neuronal progenitor cells and their progeny are controlled, and the possible mechanisms underlying their ability to divide the neonatal and adult brain. Dr. Luskin and her colleagues are also exploiting the potential of using the SVZa neuronal progenitor cells for transplantation to therapeutically treat the injured and diseased central nervous system.

Glial Restricted Precursors

Mark Noble and Margot Mayer-Pröschel

Mark Noble is a Professor of Genetics at the Center for Cancer Biology of the University of Rochester Medical Center. He participated in the discovery of the first precursor cell to be identified in the CNS, the oligodendrocyte-type-2 astrocyte progenitor cell, and has since been a pioneering contributor to the field of stem cell and precursor cell biology. His research contributions also include the discovery of adult-specific precursor cells, studies on cell-intrinsic and cell-extrinsic regulation of developmental timing, the role of redox regulation in modulating the balance between cell division and differentiation, the use of 1H-NMR spectroscopy to study normal cells and cancer cells, the development of the H-2KbtsA58 transgenic mouse as a source of conditionally immortal cell lines from multiple tissues, and the use of cell transplantation for CNS repair.

PNS Precursor Cells

Tanya A. Moreno and Marianne Bronner-Fraser

Marianne Bronner-Fraser received her ScB from Brown University in 1975 and her PhD from Johns Hopkins University in 1979. Her doctoral work examined the lineage and migration of neural crest cells, an important cell type that gives rise to most of the peripheral nervous system. This topic has remained a central research direction throughout her career. Dr. Bronner-Fraser assumed a faculty position at the University of California, Irvine, in 1980, where she remained until 1996, at which time she joined the faculty in the Division of Biology at the California Institute of Technology. Her current research efforts focus on elucidating the molecular mechanisms underlying the formation and evolution of neural crest and placode cells.

Neural Progenitor Cells of the Adult Human Brain

Steven A. Goldman

Steven A. Goldman received his BA, PhD and MD degrees from the University of Pennsylvania, the Rockefeller University and Cornell University Medical College, respectively. At Rockefeller, his thesis work with Fernando Nottebohm included the discovery of neurogenesis in the adult songbird brain, a system which he has continued to study, and which has strongly influenced his lab's studies of human neural progenitor cells. Dr. Goldman then trained in Medicine and Neurology at the New York Hospital-Cornell Medical Center, and joined the faculty of Neurology there in 1988. In 1997 he became Professor of Neurology and Neuroscience at Cornell and New York Presbyterian Hospital, where he is also an Attending Neurologist.

ES Cells and Neurogenesis

John W. McDonald

John W. McDonald, III, MD, PhD is Assistant Professor of Neurology and Neurological Surgery at Washington University School of Medicine and Director of the Barnes-Jewish Hospital Spinal Cord Injury Program. His basic science research, clinical research, and clinical practice focus are on spinal cord injury. The emphasis of his research is on regeneration utilizing embryonic stem cells as a strategy for neural transplantation. His group is developing remyelination as an approachable strategy for restoration of function in the injured nervous system.

Mobilizing Endogenous Stem Cells

Theo D. Palmer, Sophia Colamarino, and Fred H. Gage

Fred H. Gage is a Professor in the Laboratory of Genetics at the Salk Institute for Biological Studies. He concentrates his research on studying the cellular, molecular, and environmental influences that regulate neurogenesis in the adult brain and spinal cord. Dr. Gage has made several seminal contributions to our understanding of neural stem cells. Dr Gage and his colleagues have provided compelling evidence for ongoing neurogenesis in the adult nervous system and his laboratory continues to analyze the factors that regulate adult neurogenesis. Dr. Gage has also developed techniques to harvest clonal populations of stem cells, modify them genetically, and follow their behavior after transplantation. Dr. Gage maintains an active and diverse research program and is widely regarded as one of the pioneers in the field of stem cell biology.

Transplant Therapy

Barbara A. Tate, Kate A. Bower, and Evan Y. Snyder

Evan Y. Snyder is both a neurobiologist and a clinician. He is an Assistant Professor of Neurology at Harvard Medical School and an Assistant in Neurology (Neuroscience), Medicine (Neonatology), and Neurosurgery (Neuroscience Research) at Children's Hospital. He received his MD and PhD at the University of Pennsylvania in a combined degree program. His thesis suggested a pivotal role for insulin in neural development. He pursued clinical training in pediatrics, child neurology, and newborn intensive care at Children's Hospital, Boston, the Harvard Longwood Neurological Training Program, and the Harvard Joint Program in Neonatology, respectively concurrently with post-doctoral research training in the Dept. of Genetics, Harvard Medical School in the use of retroviruses for studying neural development. His lab in the Division of Neuroscience, Dept. of Neurology, Children's Hospital and Harvard Medical School studies neural stem cell biology from both a developmental and therapeutic perspective, an area of investigation in which he has been active for almost 15 years. He also maintains a clinical practice in newborn medicine and neurology at that institution.

Drug Discovery and Gene Discovery

Alexander Kamb and Mahendra S. Rao

Alexander Kamb received his BA from Harvard University in 1982 and his PhD from the California Institute of Technology in 1988. His postdoctoral work in protein crystallography was carried out at the University of California, San Francisco. In 1992 Dr. Kamb joined Myriad Genetics, Inc., a fledgling genomics company in Salt Lake City, Utah. Until 1996 he served as Myriad's Director of Research and was involved in directing groups that identified genes responsible for familial melanoma and breast cancer. In 1996 Dr. Kamb founded Ventana Genetics, Inc., also in Salt Lake City. As President, Chief Executive Officer, and Chief Scientific Officer, he currently directs research and business efforts focused on discovery of novel pharmaceutical targets.

About the Editor

Mahendra S. Rao is an Associate Professor in the Department of Neurobiology and Anatomy at the University of Utah. His laboratory is interested in defining the molecular and cellular interactions that instruct a pluripotent cell to differentiate into cells restricted to a particular phenotype. He has published extensively on classes of precursor cells present in the nervous system and the factors that regulate their differentiation. Dr. Rao was recently awarded the C. J. Herrick Young Investigator award for his work on neural stem cells.

Index